ecoDESIGN

éco DESIGN

REVISED EDITION

the sourcebook

Alastair Fuad-Luke

CHRONICLE BOOKS
SAN FRANCISCO

Acknowledgments

Special thanks go to all the designers, designer-makers, manufacturers and other organizations who supplied information and photographic or illustrative materials, without whose kindness this second edition would not have been possible. Certain individuals provided invaluable support in sourcing material, including: Tom Johnson, co-founder of the Design Resource Institute that operates the International Design Resource Awards (IDRA), Seattle, USA; Silke Becke, iF Press Office, Neumann + Luz, Köln, Germany; Bas Berck, Coordinator Cultural & Commercial Affairs, Design Academy Eindhoven, Eindhoven, the Netherlands; Jackie Dehn, Johnathan Chapman and Anne Chick, Rematerialize project, Kingston University, UK. Thanks are also extended to all those who responded to my continuous stream of queries and requests for information. I would also like to acknowledge the support of Lucas Dietrich and Alice Park, Thames & Hudson, London, UK. Finally, a big thanks and hug to my wife, Dina, for her encouragement and support.

I dedicate this book to all those who have contributed by their designs and products herein to inspire us towards a more sustainable future.

First published in the United States in 2006 by Chronicle Books LLC.

ISBN-10: 0-8118-5532-5
ISBN-13: 978-0-8118-5532-7

The Library of Congress has catalogued the previous edition under the following ISBN: 0-8118-3548-0

Manufactured in China.

Distributed in Canada by Raincoast Books
9050 Shaughnessy Street
Vancouver, British Columbia V6P 6E5

10 9 8 7 6 5 4 3 2 1

Chronicle Books LLC
85 Second Street
San Francisco, California 94105

www.chroniclebooks.com

This book was printed on 130 gsm Munken Pure paper, produced by the Swedish mill Munkendals, which has a long tradition of environmental responsibility. Munkendals' papers are created to minimize the impact on the environment. For example, they have not been subject to chlorine bleaching; pulp is taken from sustainable forests; and production processes have been developed to reduce harmful effects and effluents.

PO39850   10/19/06

Preface

The new edition of *ecoDesign: The Sourcebook* continues to take an eco-pluralistic approach to design, celebrating diverse thinking by designers and manufacturers from over thirty countries who are challenging the direction of contemporary design, rejecting the dictatorial 'one-model-fits-all' philosophy of many twentieth-century design movements. Eco-pluralist designs range from those that embrace minor modifications of existing products (such as the use of recycled rather than virgin materials) to radical new concepts and the 'dematerialization' of existing products into services.

During the past two years the eco (r)evolution has undoubtedly progressed. An increasing number of international companies are placing sustainability at the heart of their vision. Companies are embracing young concepts such as corporate social responsibility (CSR) and integrating environmental issues into day-to-day business.

There are signs that the work of designers is likely to change over the next decade. Designers are being asked to meet consumers' needs in different ways, for example by creating product-service-systems (PSS) – products or systems of products, services, supporting infrastructures and networks (often ICT-based) that have less environmental impact than individually acquired consumer products. Designers are also re-engaging with lapsed concepts such as universal or inclusive design (design accessible to all). New centres of debate are emerging about the relationship between design and emotion, designing experiences (rather than products) and exploration of new paradigms such as 'slow design' (www.slowdesign.org), and stimulation of new design approaches is being encouraged by systems thinking and scenario planning. One of the most enlightening projects was 'Sustainable Everyday: Scenarios of Urban Living', at the Milan Triennale, December 2003.

A common thread runs through the selection of products, buildings and materials in this book. Each is an attempt to improve on the status quo, in small or large increments, and to increase our well-being while reducing the inherent environmental impact of these 'designed' things. The book provides you with the options but it leaves you with the decision of using or not using them. Before you rush to purchase that eco-efficient car you've read about in the book, ask yourself, 'Do I really need it? Or shall I use the car I already have for only half the time?' Designing more sustainable ways of production and consumption requires a concomitant change in consumer behaviour and ways of living. Here are some objects that can help you on your way.

How to use this book

Each product or material is accompanied by a caption and a line box with all or some of the following icons:

✎ The name and nationality of the designer/designer-maker (see pages 304–10), or manufacturer and country if the products/materials are designed in-house (see pages 311–23).

⚙ The name and country of the manufacturer (see pages 311–23) or designer-maker (see pages 304–10).

🖥 The main materials and/or components (see pages 280–301 for examples of materials with reduced environmental impacts, i.e., ecomaterials).

♺ The main eco-design strategies applied to the design of the product (see pages 324–28).

⚑ Important design awards recognizing eco-design (see pages 330–31).

The page numbers in the line boxes permit rapid cross-referencing and enable the reader quickly to find designers/designer-makers, manufacturers and eco-design strategies.

Contents

2.0 Objects for Working   192

3.0 Materials   280

4.0 Resources   302

Design for a Sustainable Future

More than forty years ago Rachel Carson, in her seminal book *Silent Spring*, documented the devastating effects of pesticide use on mammals and birds in the USA. Today traces of organo-phosphorus pesticides are found in organisms throughout the globe, including in human beings. At the 1967 UNESCO Intergovernmental Conference for Rational Use and Conservation of the Biosphere, the concept of ecologically sustainable development was first mooted. Paul Ehrlich's 1968 book, *The Population Bomb*, linked human population growth, resource degradation and the environment and pondered the carrying capacity of the planet. By 1973 the Club of Rome, in its controversial report, *Limits to Growth*, was predicting dire consequences for the world if economic growth was not slowed down. This report accurately predicted that the world population would reach six billion by the year 2000, although its more frightening predictions of the exhaustion of resources such as fossil fuels were less accurate. Such warnings were, by and large, ignored, with the result that during the last thirty years people have continued to poison the planet with pesticides and other toxic chemicals, which has led to the destruction of ecosystems and the extinction of many species. More recently people have realized that they too are now threatened by human actions. Unfettered use of the internal combustion engine and the burning of fossil fuels to generate electrical power have catalyzed action on climate change. Significant minorities in different places around the globe face the real risk that the land on which they depend will be inundated by rising sea-levels.

In 1950 the world car fleet numbered fifty million vehicles and global fossil fuel use was 1,715 millions of tonnes of oil equivalent. Today there are over five hundred million vehicles and consumption of fossil fuels exceeds 8,000 million tonnes of oil equivalent. For all the individual freedom it confers, the car is making a huge collective negative impact on the environment, specifically the balance of gases, particulate matter and carcinogens in the atmosphere. For every one of the millions of products we use to 'improve' the quality of our lives there are associated environmental impacts. While some products have a small environmental impact, others consume finite resources in vast quantities.

The ultimate design challenge of the twenty-first century is to avoid or minimize the adverse impacts of all products on the environment. Like all challenges, this constitutes both a demand and an opportunity – to steer the debate on more sustainable patterns of production and consumption. Designers need to be an integral part of the debate rather than remain on the fringe or be subject to the whim of the political and commercial forces of the day.

A brief history of green design

Green design has a long pedigree and before the Industrial Revolution it was the norm for many cultures. Goods such as furniture and utility items tended to be made locally by craftsmen such as blacksmiths, wheelwrights and woodland workers, from readily available local resources. Innovation in farming machinery in Europe, particularly Britain, destabilized the natural employment structure of rural areas and in the first half of the nineteenth century almost half of the rural population in Britain migrated to towns to work in factories. Throughout the twentieth century this pattern was repeated around the world as countries industrialized and created new urban centres.

The founders of the British Arts and Crafts movement (1850–1914) were quick to note the environmental degradation associated with the new industries. Their concerns about the poor quality of many mass-manufactured goods and the associated environmental damage prompted them to examine new methods combining inherently lower impact with increased production. For various social and technical reasons, only a small section of society reaped the benefits of the Arts and Crafts movement but the seeds were sown for development of the early modernist movements in Europe,

Plaky table designed by Christopher Connell

notably in Germany (the Deutsche Werkbund and later the Bauhaus), Austria (the Secession and the Wiener Werkstätte) and the Netherlands (De Stijl). The modernists insisted that the form of an object had to suit its function and that standardized simple forms facilitated the mass-production of good-quality, durable goods at an affordable price, thus contributing to social reform.

Economy of material and energy use went hand in hand with functionalism and modernism. Marcel Breuer, an eminent student at the Bauhaus between 1920 and 1924, applied new lightweight steel tubing to furniture design, arriving at his celebrated Wassily armchair and B-32 cantilever chair. Breuer's 1927 essay, 'Metal Furniture', conveys his enthusiasm for the materials and reveals his green credentials. He saw the opportunity to rationalize and standardize components, allowing the production of 'flat-pack' chairs that could be reassembled (and so save on transport energy) and were durable and inexpensive (and so help improve the lives of the masses).

The early proponents of organic design promoted a holistic approach, borrowing from nature's own model of components within systems. In the USA the architect Frank Lloyd Wright was the first to blend the functionality of buildings, interiors and furniture into one concept. In the 1930s the Finnish architect and designer Alvar Aalto also achieved a synergy between the built environment and his curvilinear bent plywood furniture that evoked natural rhythms. At a landmark competition and exhibition in 1942, entitled Organic Design for Home Furnishings, organized by the Museum of Modern Art, New York, the winners, Charles Eames and Eero Saarinen, firmly established their biomorphic plywood furniture as a means of satisfying the ergonomic and emotional needs of the user. These designs often incorporated laminated wood or plywood to obtain more structural strength with greater economy. With the rapid evolution of new materials such as plastics in the 1960s and 1970s more ambitious expressions of biomorphism were achieved.

Ironically, one of the early advocates of a more sustainable design philosophy, Richard Buckminster Fuller, originated from the USA, a country renowned for both prolific production and consumption. One of Buckminster Fuller's early ventures, the Stockade Building System, established a method of wall construction using cement with waste wood shavings. Building inspectors of the day did not approve of this innovation and the venture faded. Not easily to be deterred, he soon set up a new design company, 4-D, whose name makes reference to the consequence (to humanity) of 3-D objects over time. 'Dymaxion' was the term he coined for products that gave maximum human benefit from minimal use of materials and energy. His 1929

Dymaxion house, later developed as a commercial product in the metal prefabricated Wichita house (1945), and 1933 teardrop-shaped Dymaxion car were both radical designs. The car had a capacity of up to a dozen adults, fuel consumption of 10.7 km/litre (30 mpg) and the ability to turn within its own length thanks to the arrangement of the three wheels. Remarkable as it was, the car was plagued with serious design faults and never became a commercial reality. The Wichita house could have been a runaway commercial success as nearly forty thousand orders poured in but delays in refining the design led to the collapse of the company. Buckminster Fuller persevered and in 1949 developed a new method of construction based on lightweight polygons. The geodesic dome was suitable for domestic dwellings or multipurpose use and its components were readily transported, easily erected and reusable. His legacy inspired new endeavours such as the Eden Project, near St Austell in Cornwall, UK (2001), in which the world's largest biomes house eighty thousand plant species from tropical to temperate climates.

From 1945 to the mid-1950s most of Europe suffered from shortages of materials and energy supplies. This austerity encouraged a rationalization of design summed up in the axiom 'less is more'. The 1951 Festival of Britain breathed optimism into a depressed society and produced some celebrated designs including Ernest Race's Antelope chair, which used the minimum amount of steel rod in a lightweight curvilinear frame.

During the 1950s European manufacturers such as Fiat, Citroën and British Leyland extolled the virtues of the small car. Economical to build, fuel-efficient (by standards of the day) and accessible to huge mass markets, these cars transformed the lives of almost nine million owners. By contrast, the gas-guzzling, heavyweight, shortlived Buicks, Cadillacs and Chevrolets of America may have celebrated American optimism but were the very antithesis of green design.

The hippie movement of the 1960s questioned consumerism and drew on various back-to-nature themes, taking inspiration from the dwellings and lives of nomadic peoples. Do-it-yourself design books sat alongside publications such as The Whole Earth Catalog, a source book of self-sufficiency advice and tools that is still produced annually. Out of this era emerged the 'alternative technologists' who encouraged the application of appropriate levels of technology to the provision of basic needs such as fresh water, sanitation, energy and food for populations in developing countries. And within Europe young designers experimented with new forms using recycled materials and examined alternative systems of design, production and sales.

In 1971 the rumblings of the first energy crisis were felt and by 1974, when the price of a barrel of oil hit an all-time high, the technologists began designing products that consumed less energy and so decreased reliance on fossil fuels. This crisis had a silver lining in the form of the first rational attempts to examine the life of a product and its consequent energy requirements. Lifecycle analysis (LCA), as it became known, has since been developed further into a means of examining the 'cradle to grave' life of products to determine not only energy and material inputs but also associated environmental impacts.

In his 1971 book, *Design for the Real World*, Victor Papanek confronted the design profession head on, demanding that they face their social responsibilities instead of selling out to commercial interests. Although he was pilloried by most design establishments of the day, his book was translated into twenty-one languages and remains one of the most widely read books on design. Papanek believed that designers could provide everything from simple, 'appropriate technology' solutions to objects and systems for community or society use.

By the 1980s three factors, improved environmental legislation, greater public awareness of environmental issues and private-sector competition, ensured that 'green consumers' became a visible force. In the UK in 1988 John Elkington and Julia Hailes wrote *The Green Consumer Guide*, which was purchased by millions of people keen to understand the issues and exercise their 'consumer power'. Designers and manufacturers applied themselves to the task of making their products 'environmentally friendly', not always with genuine zeal or success. Unsubstantiated claims on product labels soon disillusioned an already sceptical public and green design got buried in an avalanche of market-driven, environmentally unfriendly products from the emerging capitalist-driven 'global economy'. Then the pendulum swung back, resulting in more stringent environmental legislation, greater regulation and more uptake of eco-labelling, energy labels and environmental management standards.

Against the grain of the high-tech, matt-black 1980s, a few notable designer-makers blended post-modernism with low environmental-impact materials and recycled or salvaged components. In London Ron Arad produced eclectic works ranging from armchairs made from old car seats to stereo casings of reinforced cast concrete; while Tom Dixon created organic chair forms using welded steel rod covered with natural-rush seating, a design that is still manufactured by Cappellini SpA, Italy, today.

The green design debate gathered momentum following the publication of the Brundtland Report, *Our Common Future*, prepared by the World Commission on Environment and Development in 1987, which first defined 'sustainable development', and also as a result of collaborative work between governments, industry and academia. Dorothy McKenzie's 1991 book, *Green Design*, reported initiatives by individual designers and the corporate world to tackle the real impact of products on the environment.

In the early 1990s in the Netherlands, Philips Electronics, the Dutch government and the University of TU Delft collaborated to develop lifecycle analysis that could be widely used by all designers, especially those in the industrial sector. Their *IDEMAT* LCA software provided single eco-indicators to 'measure' the overall impact of a product. *IDEMAT* was rapidly followed by three commercial options, *EcoScan*, *Eco-It* and a higher-grade package, *SimaPro*. Today there are tens of different LCA and lifecycle inventory (LCI) packages, which can help designers minimize the impact of their designs from cradle to grave.

Over the past ten years academic communities around the world have evolved new terminology to describe particular types of 'green' design, such as Design for environment (DfE), DfX – where X can be assembly, disassembly, reuse and so on – eco-efficiency, eco-design and EcoReDesign. (Refer to the Glossary for full definitions of these terms.)

Along with the sustainable-development debate has come the concept of sustainable product design (SPD). Most definitions of SPD embrace the need for designers to recognize not only the environmental impact of their designs over time but their social and ethical impacts too. Buckminster Fuller and Papanek would recognize the issues but perhaps wonder why it took so long for the design community at large to take them up.

Our imperilled planet

Twenty-five per cent of the world's population of six billion people account for eighty per cent of global energy use, ninety per cent of car use and eighty-five per cent of chemical use. By 2050 there may be up to twenty billion people on the planet, ten times more than at the beginning of the twentieth century. Scientists estimate that human activities to date have been responsible for increases in atmospheric temperature of between 1.5 and 6 Celsius degrees (2.9–10.8 Fahrenheit degrees). Global warming on an unprecedented scale has melted ice caps and permafrost, with consequent rises in sea-level by up to 60 centimetres (2 ft).

It is not an equable world. A typical consumer from the developed 'North' consumes between ten and twenty times more resources than a typical consumer from the developing 'South'. Both types of consumer can sustain their lives but the quality of those lives is substantially different. Almost one billion people suffer from poverty, hunger or water shortages. At present rates of production and consumption the earth can sustain two billion people at 'Northern' standards of living. Could it support twenty billion people at 'Southern' standards of living? Or is there an urgent need to address the way 'Northern' populations consume and examine the true impact of each product's life?

The impact of global production and consumption

Between 1950 and 1997 the production of world grain tripled, world fertilizer use increased nearly tenfold, the annual global catch of fish increased by a factor of five and global water use nearly tripled. Fossil-fuel usage quadrupled and the world car fleet increased by a factor of ten. During the same period destruction of the environment progressed on a massive scale. There was a reduction in biodiversity. For example, the world elephant population decreased from six million to just 600,000 and total tropical rainforest cover decreased by twenty-five per cent. Average global temperature rose from 14.86°C to 15.32°C (58.75–59.58°F), largely owing to an increase in carbon dioxide emissions from 1.6 billion tonnes per annum in 1950 to 7 billion tonnes in 1997. CFC (chlorofluorocarbon) concentrations rose from zero to three parts per billion, causing holes in the protective ozone layer at the North and South poles.

In the North ownership of such products as refrigerators and televisions has reached almost all households. More than two in three households own a washing machine and a car. The North is indeed a material world. It also generates huge quantities of waste. According to *The Green Consumer Guide*, even back in 1988 an average British person generated two dustbins of waste each week, used two trees a year in the form of paper and board and disposed of 90 drinks cans, 70 food cans, 35 petfood cans, 107 bottles and jars and 45 kg (99 lb) of plastics. By 2000 local authorities in Britain were recycling on average only twenty-five per cent of domestic waste and such valuable resources as glass, metal and plastics were shamefully neglected by disposal in landfill sites or incineration. Furthermore, landfill sites generate methane and contribute to the accumulation of greenhouse gases and rising global temperatures.

The big environmental issues

In 1995 the European Environment Agency defined the key environmental issues of the day as: climate change, ozone depletion, acidification of soils and surface water, air pollution and quality, waste management, urban issues, inland water resources, coastal zones and marine waters, risk management (of man-made and natural disasters), soil quality and biodiversity. Recognition that the planet was

fast reaching a perilous state galvanized 172 governments to gather in Rio de Janeiro, Brazil, in 1992 for the United Nations Conference on Environment and Development. The achievements of the 'Earth Summit' were considerable. The Rio Declaration on Environment and Development set forth a series of principles defining the rights and responsibilities of states, a comprehensive blueprint for global action called Agenda 21 was published, guidelines for the management of sustainable forests (Forest Principles) were set and the UN Convention on Biodiversity and the UN Framework on Climate Change (UNFCC) were ratified. The conference set the foundations for establishing the UN Commission for Sustainable Development (UNCSD), which produces annual progress reports, and adopted the Precautionary Principle, which states that 'lack of full scientific certainty shall not be used as a reason for postponing cost-effective measures to prevent environmental degradation'.

Europe's cutting-edge environmental legislation

In 1972 the then members of the European Economic Community (now the European Union), recognizing that environmental damage transgresses national boundaries, agreed that a common transnational policy was required in Europe. Since then the European output of legislation and regulatory measures to combat environmental degradation has been prolific.

Regulations passed by the European Council become effective law for all member states immediately, whereas directives, which are also legally binding, do not come into force in the member states until carried into national law by individual governments. Important legislative advances include the Directive on Conservation of Wild Birds 1979, the Directive on the Assessment of the Effects of Certain Public and Private Projects on the Environment 1985, the Directive on the Conservation of Natural Habitats and Wild Flora and Fauna 1992 and the Directive on Integrated Pollution Prevention and Control (IPPC) 1996. A range of other directives is of great relevance to manufacturers and designers, including on vehicles, electronic equipment, toxic and dangerous waste and packaging and packaging waste. The effect of these regulations is felt well beyond Europe, as transglobal companies manufacturing cars, electronic goods, packaging and chemical products have to meet these stringent standards.

Europe's collaborative efforts to introduce environmental legislation and regulation provide a model to other regions of the world for international cooperation, for example, North America and the 'Tiger' economies of South-east Asia (ASEAN).

The real lives of products

Freedom and death

The car is the ultimate symbol of personal freedom for the twentieth century. It confers unending choices for the user but condemns many to death, directly as accident victims and indirectly as the recipients of pollutants causing asthma (from particulate matter), brain damage (from lead) and cancer (from carcinogens). It also contributes towards climate change via emissions of carbon dioxide, marine pollution in the event of oil tanker spillage or accidents, and noise pollution. Most societies feel that the personal freedom outweighs the collective price but recently several European cities such as Paris and Milan have banned cars on selected days.

One-way trip

Some products lead short, miserable lives, destined for a one-way trip between the retail shelf and burial in a landfill site. Packaging products are the prime example of one-trip products but there are many others – kitchen appliances, furniture, garden accessories and all the paraphernalia of the modern world.

Everyday products quietly killing

Quietly humming away in the corner of millions of kitchens worldwide is the humble refrigerator. It protects by keeping food fresh, but it is a killer too. Coolants using CFCs (chlorofluorocarbons) or HCFCs (hydrochlorofluorocarbons) are the main culprits in precipitating rapid degradation of the layer of ozone gas, which keeps out harmful radiation from space. Not only are there substantial seasonal holes in the ozone layer at the North and South poles but the layer has thinned considerably in other parts of the world. Thus inhabitants receive higher doses of radiation with an increased risk of contracting skin afflictions and cancer.

Everyday inefficient products

The efficiency of products that have become a way of life needs to be challenged continually. The European eco-label for washing machines lays down threshold values for energy consumption of 0.24 kWh per kilogram and water consumption of 15 litres per kilogram of clothes (1.5 gals/lb). Yet only a few companies apply for this eco-label and many European retailers sell machines that do not meet the standards, even though they obviously have the technological means to do so. Failure to apply the best technology available means unnecessary daily consumption of massive quantities of electricity and water.

Occasional use

The developed world's preoccupation with DIY home improvements means that each household owns specialist tools, such as electric drills and screwdrivers, which are rarely used.

Novelties and gimmicks

Many of the products available through mail order catalogues are in fact gimmicks that will do no more than provide temporary amusement.

Small but dangerous

Many small electronic devices, such as personal stereos and mobile phones, have a voracious appetite for batteries. While more devices are offered these days with rechargeable batteries, the older models still consign millions of batteries to landfill sites, where cadmium, mercury and other toxic substances accumulate. In the European Union the disposal of certain battery types is illegal but in many parts of the world it continues unabated.

Industry visions and reality

Although the wastage of resources associated with the planned obsolescence in the US car industry in the 1950s is no longer tolerated, the lifetime of the average family vehicle remains less than ten years. Furthermore, the global car industry is geared up to keep adding to the existing five hundred million cars worldwide at the same level of production. More fuel-efficient cars that can be disassembled at the end of their lives have been produced and some are already on the market, but many manufacturers will not roll out this technology into new models until they have extracted the returns on their capital investment in current models. Moreover, most are concerned to maintain their market share by providing customers with choice, often in the form of fuel-inefficient, prestige or luxury cars.

Both hardware and software companies are obsessed with doubling the speed of personal computers every eighteen months as chip technology continues its meteoric development. Users are seduced into buying faster machines even though they use only a small fraction of the computing power available. Basic functionality, such as being able to adjust the height of a monitor or arrange a keyboard to suit individual needs, remains inadequate. Yet the computer industry conjures up a vision of a future in which we can programme our house to cook the dinner before we arrive back from work, of a wired-up 'information age' in which everyone has access to the Internet. The reality is that ninety-four per cent of the world's population does not have access to the Internet. The building of ever bigger and faster networks and workstations involves considerable consumption of finite resources and the use of toxic substances during manufacture and disposal.

The brand thing

Companies with internationally recognized brands aspire to increase their market share in individual nations in order to claim world dominance. Expectation, in the form of the brand promise, often delivers a transient moment of satisfaction for the purchaser. Whatever happened to products that were guaranteed to 'last a lifetime'? Where is the long view in the companies that sell these brands? The big brands have the potential to reduce the environmental impact of their activities, but not if they persist in encouraging their customers to consume more, not less.

Moving commerce toward sustainability

Evolving environment management systems (EMS)

The flagship international standard that encourages organizations to examine their overall environmental impact arising from production (but not the impact of their products during usage) is ISO14001 compiled by the International Standards Organization in Geneva, Switzerland. Companies that achieve this independently certified EMS have integrated management systems into their business to reduce environmental impacts directly and have agreed to publication of an annual environmental report from an audited baseline, so reductions in impact can be measured. Other independently certified standards exist, such as the Eco-Management and Audit Scheme (EMAS) for companies in EU member states.

Sustainable production and consumption

In 1995 the World Business Council for Sustainable Development (WBCSD), a coalition of 120 international companies committed to the principles of economic growth and sustainable development, published a report entitled *Sustainable Production and Consumption: A Business Perspective*. It defined sustainable production and consumption as 'involving business, government, communities and households contributing to environmental quality through the efficient production and use of natural resources, the minimization of wastes and the optimization of products and services'. The United Nations Commission on Sustainable Development (UNCSD), formed at Rio in 1992, sees the role of business as crucial since it requires the integration of environmental criteria into purchasing policies (green procurement), the design of more efficient products and services, including a longer lifespan for

durable goods, better after-sales service, increased reuse and recycling and the promotion of more sustainable consumption by improved product information and by the positive use of advertising and marketing. This represents an important change in the way businesses operate.

Model solutions

WBCSD members are encouraged to adopt measures to improve their eco-efficiency, that is, greater resource productivity, by maximizing the (financial) value added per unit of resource input. This means providing more consumer performance and value from fewer resources and producing less waste. Amory Lovins *et al* of the Rocky Mountain Institute in the USA proposed the concept of 'Factor 4' – a doubling of production using half the existing resources, with a consequent doubling of the quality of life. Researchers at the Wuppertal Institute in Germany find Factor 4 inadequate to deal with the expected doubling or trebling of world population by 2050 and so propose 'Factor 10' as a more appropriate model for the developed North to achieve equable use of resources for populations in the North and developing South.

Another model that is finding favour with business is called 'The Natural Step' (TNS). It sets out four basic 'system conditions' for businesses to adopt. First, substances from the earth's crust, the lithosphere, must not be extracted at a greater rate than they can reaccumulate – thus there must be less reliance on 'virgin' raw materials. Second, man-made substances must not systematically increase but should be biodegradable and recyclable. Third, the physical basis for the productivity and diversity of nature must not be systematically diminished – renewable resources must be maintained and ecosystems kept healthy. Fourth, we must be fair and efficient in meeting basic human needs – resources should be shared in a more equable manner. Companies as diverse as carpet manufacturers, water suppliers and house builders have taken up TNS.

Early adopters and new business models

International companies from Europe, the USA and Japan are exploring new business models that take a long view enmeshed with the concept of sustainable development. For example, Mitsubishi considered the ecology of the tropical rainforest system, which is highly productive in terms of biomass on a fixed amount of nutritional resources. Waste becomes other organisms' food in the rainforest. Mitsubishi mimic this ecology by ensuring their industrial system meets eco-efficient parameters. Where possible waste should be consumed within the company. This model could be extended to ensure that materials are returned to the manufacturer at the end of their lives, keeping the materials in a closed loop and ensuring that the manufacturer retains control of these resources. At the same time, consumers should be discouraged from buying products and instead encouraged to lease product services.

Triple bottom line

Companies engaging in sustainability have a healthy 'triple bottom line', as improved environmental and social performance mirrors itself in increased profits. It is not surprising that there is an increase in the number of companies seeking membership of the Dow Jones Sustainability Indexes (www.sustainability-indexes.com), where financial performance often outstrips the more conventional market indexes. Sustainability is, at last, being seen as an opportunity to do business differently. Design, the faithful translator of human, financial, natural and social capital into goods since the start of the Industrial Revolution, is a key catalyst. Design is critical to social acceptance and market longevity.

Designers save the earth

Designers actually have more potential to slow environmental degradation than economists, politicians, businesses and even environmentalists. The power of designers is catalytic. Once a new, more environmentally benign design penetrates markets its beneficial effects multiply. Businesses spend less on raw materials and production and so realize better profits, users enjoy more efficient, better-value products, governments reduce spending on regulatory enforcement and the net gain is an improved environment and quality of life. The vivid examples in this book demonstrate the capability of design, and hence designers, to shape the future and save the earth.

A robust tool kit

Today's designer has a powerful array of tools to assist him or her to meet the challenge of reducing environmental impacts at the design stage, such as simple checklists, impact matrices, lifecycle matrices, eco-wheels, Lifecycle Inventory (LCI) and Lifecycle Analysis (LCA) software. Checklists can be found in the publications included in Further Reading (p. 340) and a full list of organizations and agencies offering information and software to assist designers is given in the Green Organizations section (p. 329).

A manifesto for eco-pluralistic design … designs that tread lightly on the planet

The thoughtful designer of the twenty-first century will design with integrity, sensitivity and compassion. He or she will design products/materials/service products that are sustainable, i.e. they serve human needs without depleting natural and man-made resources, without damage to the carrying capacity of ecosystems and without restricting the options available for present and future generations. An eco-pluralistic designer will:

1. Design to *satisfy real needs* rather than transient, fashionable or market-driven needs.
2. Design to *minimize the ecological footprint* of the product/material/service product, i.e., reduce resource consumption, including energy and water.
3. Design to *harness solar income* (sun, wind, water or sea power) rather than use non-renewable natural capital such as fossil fuels.
4. Design to *enable separation of components* of the product/material/service product at the end of life in order to encourage recycling or reuse of materials and/or components.
5. Design to *exclude the use of substances toxic or hazardous* to human and other forms of life at all stages of the product/material/service product's lifecycle.
6. Design to *engender maximum benefits to the intended audience* and to educate the client and the user and thereby create a more equable future.
7. Design to *use locally available materials and resources* wherever possible (thinking globally but acting locally).
8. Design to *exclude innovation lethargy* by re-examining original assumptions behind existing concepts and products/materials/service products.
9. Design to *dematerialize products into services* wherever feasible.
10. Design to *maximize a product/material/service product's benefits to communities*.
11. Design to *encourage modularity in design* to permit sequential purchases, as needs require and funds permit, to facilitate repair/reuse and to improve functionality.
12. Design to *foster debate and challenge the status quo* surrounding existing products/materials/service products.
13. *Publish eco-pluralistic designs in the public domain* for everyone's benefit, especially those designs that commerce will not manufacture.
14. Design to *create more sustainable products/materials/ service products for a more sustainable future*.

1.0 Objects for Living

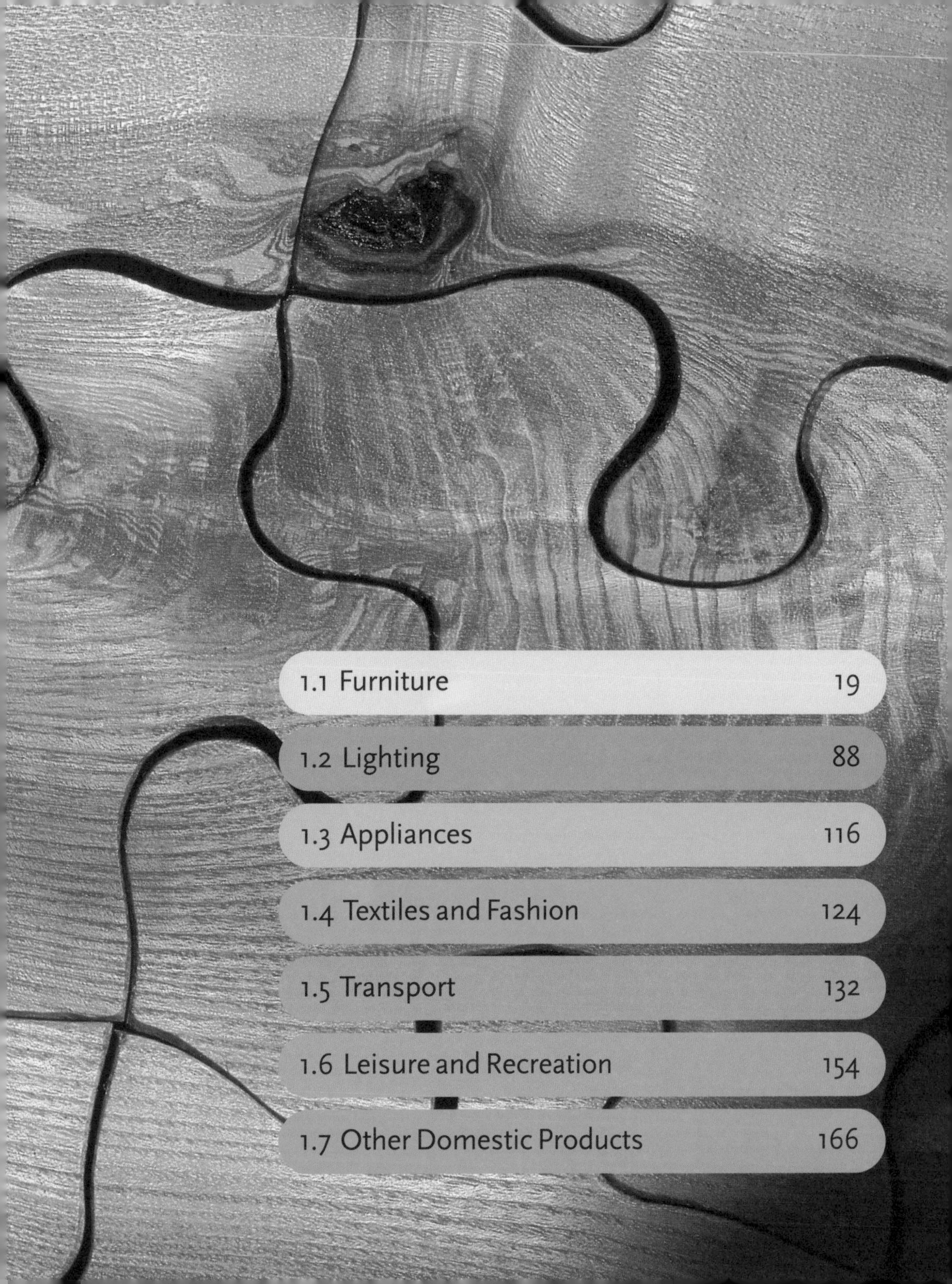

Living or Lifestyles

In a media-driven world, where brands promise a lifestyle guaranteed to satisfy your desires, it is difficult to step back and honestly appraise your real needs for living. The word 'lifestyle' implies not just a way of life but also choice. For many people around the globe lifestyle choices are simply not available, as the basic needs of life – clean water, clean air, sufficient food, shelter and medical care – are absent. In today's global economy international brands, such as Coca-Cola soft drinks and Nike trainers, exist cheek by jowl with locally or nationally made products. Designers need to reappraise their role in the production of fashionable lifestyle products or at least strive to minimize the impact of these ephemeral goods, by concentrating on durable, multi-user, multi-purpose designs.

Essential products
The car has become the ultimate symbol of our freedom to move around, yet this 'impact-use' product, which only twenty per cent of the world's population own, impinges on the collective freedom of all people to enjoy clean air and unpolluted water. Over the last twenty-five years the fuel efficiency of the average car has improved only eighteen per cent. The car is a classic example of design innovation lethargy. Only a paradigm shift in the design of cars will remove the environmental burden of this product. But this must also be accompanied by more innovation in alternative modes of mobility. Improvements in personal modes of transport – the push scooter, bicycle and motorbike or scooter – must be accompanied by radical improvements in systems of public transport, such as the provision of flexible mobility paths for individual and group users.

With increasing reliance on electronic equipment and networks to deliver information, to control third-party equipment remotely and to entertain, it is possible to reduce mobility needs. Yet again a small proportion only of the world's population is wired in to the information networks such as the Internet or cable TV. Furthermore, building and maintaining the infrastructure of the information superhighway requires vast physical resources, including metals, chemicals and electricity. Virtual moments may provide some of the needs for some of the world's population for some of the time but the real cost to the environment and societies still needs computing.

Each individual requires different products to sustain life. Aside from the essential physical resources, humans need 'comfort' products to achieve a level of emotional, spiritual and social well-being. These products may permit or provide improved mobility, specialist recreational activities, communal meeting-places or spiritual contemplation. Since comfort products tend to be used over a longish time, rather than being ephemeral, the design parameters can embrace durability and therefore judicious use of resources.

**Living lightly –
a sustainable day**
As the products in this section illustrate, it is possible to tread more lightly on the planet, to consume and waste less, yet to maintain or even improve the quality of life. A double responsibility falls on the developed countries of the North. The North must rapidly evolve more sustainable patterns of consumption and production. Further, the North must offer the South the assistance and the means to avoid bad practice and reap the benefits of a more sustainable way of life, sooner rather than later.

A sustainable day in 2025 might involve the following products...

Bluebelle

Designing a chair remains the quintessential test of any furniture designer. The imaginative form of the Bluebelle chair carefully models itself like a prosthesis, supporting and caressing the seated body. Rigid seat and arms are made from one type of polypropylene to which is clipped a more flexible polypropylene forming the comfortable backrest.

Easily disassembled into component parts, the PP and metal frames can be recycled.

✏	Ross Lovegrove, UK	308
⚙	Driade SpA, Italy	313
▣	Polypropylene, metal	
🎧	• Improved ergonomics • Design for disassembly • Recyclable	327 325 324

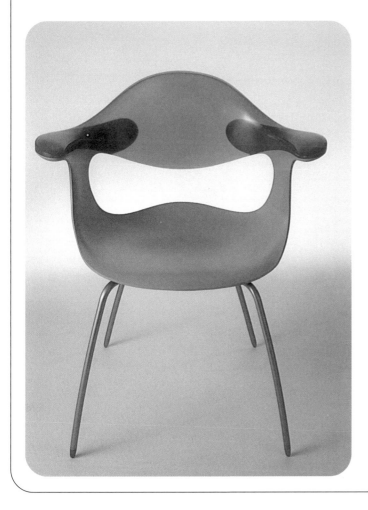

Agatha Dreams

Pillet's elegant chaise longue combines the eclecticism of craft with the technical skills of the workers at Ceccotti's factory, a labour force with a long history of 'craft technology'. Renewable materials are bought to a state of refinement that will encourage owners to cherish this design and confer a degree of longevity. High-quality manufacturing using nature's materials will always be a sustainable business model, as long as raw materials are procured from managed forests.

✏	Christophe Pillet, France	308
⚙	Ceccotti Collezioni, Italy	313
▣	Layered timber, solid cherry wood	
🎧	• Renewable materials	325

Favela

Each Favela is unique, hand-made from hundreds of pieces of recycled wood. With typical exuberance the Campana brothers focus the viewer on the nature and origins of the object while suggesting a whole world of questions beyond it: it is a commentary, a work of art, a crafted one-off and a manufactured product. It challenges our perceptions of manufacturing, renews our faith in the hand-made and poses a final question: should we make our own chairs?

✏	Fernando and Humberto Campana	305
⚙	edra, Italy	314
📰	Recycled wood, nails, glue	
♻	• Recycled materials • Hand-made, low-energy production	324 325

Slick Slick

Stackable, injection-moulded, polypropylene chairs are produced by numerous manufacturers for the contract furniture market. Unfortunately ugliness is often a common design denominator of this genre. Starck rescues the concept with this elegant design requiring minimal materials, creating a chair suitable for conference or office seating and domestic use.

✏	Philippe Starck, France	309
⚙	XO, France	323
📰	Polypropylene	
♻	• Single recyclable material • Multifunctional	324/ 325 327

Sushi sofa and chair

Design is the transformation of financial, natural, social and human capital into saleable products, producing economic gain, providing employment, and, occasionally, meeting our real, human, needs. Today, there is an expectation that design will, unfailingly, deliver the (consumer) goods, and so the excitement from design can be short-lived. Campana Brothers' designs deliver much more. Their use and understanding of a wide palette of materials is uplifting. The Sushi sofa and Sushi chair celebrate everyday materials by repositioning them centre stage. Offcuts of carpet, felt, rubber, cotton and plastic find a new life. There's a sterility with certain types of mass production that leaves a gap in our souls; Sushi and other unique low-volume manufacturing fill that void.

✏	Fernando and Humberto Campana, , Brazil	305
⚙	edra, Italy (Sushi chair), Studio Campana (Sushi sofa)	314 305
📁	Plastics, various textiles, rubber, felt, EVA, carpet, re-used offcuts	
♺	• Recycled materials • Low-energy producion	324 325

Storvik

From the cone-like 'capsule' chairs of the 1950s to the present day, the love affair with rattan continues. IKEA's interpretation Storvik poses somewhere between a floor cushion and an armchair, hinting at a hippy chillout revival but with urbanite overtones. The beauty of rattan is its consistent diameter that facilitates an even weave and an 'engineered' finish.

✏	Carl Öjerstam, IKEA, Sweden	308
⚙	IKEA, Sweden	316
📜	Rattan	
♻	• Renewable materials	325

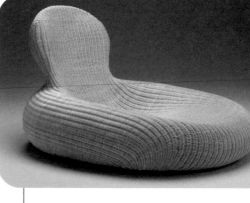

Airbag

Since the 1960s inflatable chairs have come and gone but Suppanen and Kolhonen have added an extra comfort dimension by placing balls of EPS inside the nylon outer cover, at the same time as allowing the chair to be deflated when not in use. Nylon is tough and resists puncturing better than other polymers.

✏	Ilkka Suppanen and Pasi Kolhonen, Finland	309
⚙	Ilkka Suppanen and Pasi Kolhonen, Finland	309
📜	Expanded polystyrene, nylon	
♻	• Lightweight materials • Reduced energy use during transport	324 326

Bastian

Brown wrapping paper and softwood, both inexpensive renewable materials sourced locally, are hand-crafted into a lightweight chair with matching footstool. Clean lines reinforce the simplicity of the construction technique and materials, borrowing from the long tradition of Far Eastern wood and paper manufacturing, but in harmony with a Western design ethos.

✏	Robert A. Wettstein, Switzerland	310
⚙	One-off/small batch production, Robert A. Wettstein	310
📜	Paper, wood	
♻	• Renewable • Low-embodied-energy materials • Low-energy construction	325 324 325

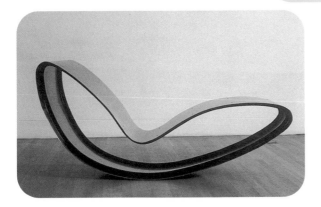

Blotter

'Keeping it simple' is the message delivered by this bent-steel chaise longue with its soft, cushioning skin of rubber to keep out the cold.

✏️	Marre Moerel (Netherlands), USA	308
⚙️	One-off	
📜	Stainless and mild steel, rubber	
🎧	• Recyclable materials • Reduction in materials used	324 325

Body Raft

Local wych elm is bent with steam to create a curved frame to which further curved lathes are attached. This organic shape is visually appealing. Hand-crafted furniture of this kind can contribute to sustaining local economies.

✏️	David Trubridge, New Zealand	310
⚙️	Cappellini SpA, Italy	312
📜	Wych elm	
🎧	• Renewable materials	325

bolla 10, 11, 15, 16

Rattan harvested from forests in southeast Asia is a wonderfully flexible and expressive material. More uniform than European willow, woven rattan produces smooth surfaces, making it a perfect material for sculptural and modernist gestures in furniture and lighting. Bolla 10 and 11 are combined side-tables cum ottomans made entirely of rattan, while bolla 15 and 16 offer an additional function with a storage volume underneath the MDF white laminate lid. Companies like Gervasoni offer a valuable outlet for the consummate skills of the rattan weavers and in doing so participate in the on-going evolution of the craft.

✏️	Michael Sodeau, UK	309
⚙️	Gervasoni SpA, Italy	315
📜	Rattan, MDF	
🎧	• Renewable materials • Retention of craft skills	325 324

Waka, Ruth 1 and 2, Glide

David Trubridge continues his experimentation with new forms using native New Zealand or Australian timber from sustainably managed plantations. Waka is a bench seat made from plantation-grown Australian hoop pine plywood and is finished in a mixture of natural oil and earth pigment. The rocking chairs, Ruth 1 and 2, are made from the same plywood in combination with plantation-grown ash or oak. Glide is made from a mono-material, plantation grown-ash. As a designer Trubridge is concerned with the footprint he leaves and has embarked on a new project called 'Structures for Survival' that addresses the issue of design ethics and an artefact's life cycle.

✏	David Trubridge, New Zealand	310
⚙	One-offs and small batch production	
▭	Various timbers	
↻	• Renewable materials from managed sources • Natural finishes	325

Garden bench

Bey brings nature indoors by taking plant waste from the garden and using high-pressure extrusion containers to generate benches of dried grass, leaves and woody prunings. Durability and longevity of the seating depend on the extent of use and the inherent strength of the compressed raw materials. At the end of its natural lifespan the furniture can be broken up and left to rot on the

garden compost heap. Bey's designs perhaps represent the current best practice in biodegradable furniture.

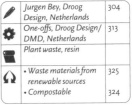

	Jurgen Bey, Droog Design, Netherlands	304
	One-offs, Droog Design/ DMD, Netherlands	313
	Plant waste, resin	
	• Waste materials from renewable sources	325
	• Compostable	324

Furnature™ range

The entire output of the Furnature™ range adheres to strict green supply chain management of materials by selecting hand-picked and organically grown cotton, organic canvas for hard-wearing arms, hard rock kiln-dried maple for frames and 97% pure natural rubber for cushions and padding. Natural rubber contains its own dust-mite repellent and anti-bacterial substances; the only additives included are natural zinc and soda ash to extend durability. Exposed frames are stained using low VOC content AfM Safecoat and wax-based finishes. All slip-covers are removable for cleaning. There is a choice of traditional or more contemporary sofas and loveseats, armchairs, recliners, ottomans and dining chairs.

	Furnature™, USA	315
	Furnature™, USA	315
	Maple wood, natural rubber, organic cotton and natural textiles, AfM Safecoat stains	
	• Renewable and recycled materials	324/ 325
	• Avoidance of toxic substances	325
	• Improved health	327
	• Durable	327

Model 290F

For over 150 years the manufacturer Gebrüder Thonet has mass-produced elegant bentwood chairs, combining good design with economical use of local (European) materials to produce modular 'flat-pack' designs facilitating distribution with basic, yet customizable options. In 1849 at Michael Thonet's factory in Vienna 'Chair No. 1', the Schwarzenberg chair, made of four prefabricated components that could be reassembled in different configurations, was the precursor of a design ideally suited to industrial production. Thonet chairs graced many a café and restaurant from Paris to Berlin and London in the late nineteenth and early twentieth centuries and created the definitive archetype for the café chair. 'Chair No. 14', later known as the 'Vienna coffee-house chair', was one of the most successful products of the nineteenth century and probably remains the world's best-selling chair, with over fifty million sold in 1930 alone. The roll-call of iconic designers, such as Mies van der Rohe, Mart Stam, Marcel Breuer and Verner Panton, ensured that Thonet always explored designs driven by new movements and schools of thought. Yet Thonet remain aware of their traditions and currently produce modern variants using well-tested principles and materials,

such as steamed and bent solid beech wood. Model 290F epitomizes the Thonet philosophy: the designers, Wulf Schneider and Partners, use three pieces to create a robust, durable and repairable chair. A single piece of solid bent beech forms the front legs and back stay, a cut-and-drilled, moulded, laminated beech section forms the back legs and back rest, and both pieces are fixed to the laminated seat with cast-aluminium angled brackets and screws. Nineteenth-century examples of Thonet chairs turn up in the prestigious sale rooms of Sotheby's, Christie's and Bonhams, attesting to their durability. It is quite likely that Model 290F will become a sought-after antique, validating it as a good and green design.

✏️	Prof. Wulf Schneider and Partners, Germany	309
⚙️	Gebrüder Thonet GmbH, Germany	315
📜	Beech wood, aluminium	
🎧	• Renewable materials	325
	• Reduced-energy transport and assembly	326
🔍	iF Ecology Design Award, Germany, 1999	330

Stoop™ Social Seating

Sitting on a door stoop (door step in the UK) in Philadelphia's row homes inspired Jamie Salm to examine the social role of these stoops in city life. Stoop™ takes the opportunity to 'stoop' to a multitude of indoor and outdoor environments. Stackable, portable and lightweight, this seating is suitable for small studios,

apartments, balconies or patio rooftop gardens where it is easily stowed away when not needed. A rubber grommet lines the finger hole for easy transport while the brightly coloured powder-coated spun aluminium ensures durability and easy cleaning. Although designed to encourage better body posture, it could be argued that the main benefit is to foster an informal body language that encourages social discourse.

Yogi Family low chair, bench and low table

Young plays with his audience, mixing the aesthetics of children's play-things with the dimensions of usable adult furniture. This blurring of ergonomic boundaries creates a fresh appeal to this family of indoor/outdoor moulded plastic furniture and challenges the need to produce separate furniture ranges for children and adults. The solid mono-material is not easily damaged, which ensures

longevity and also means that the plastic can be recycled when the object no longer can fulfil its promised function.

✏	Michael Young, UK	310
⚙	Magis SpA, Italy	318
🗋	Plastic	
♺	• Single material • Recyclable • Durable	325 324 327

✏	Jamie Salm, MIO, USA	309
⚙	MIO, USA	318
🗋	Spun aluminium, rubber	
♺	• Multifunctional • Improved social well-being	327 327

✎	Olav Eldøy, Stokke Gruppen, Norway	306
⚙	Stokke Gruppen, Norway	321
▤	Aluminium profiles, polymers, polyester fabric	
↻	• Multifunctional	327
	• Improved ergonomics	327

Stokke™ Peel™ and Peel II™

Recliner, chaise longue, armchair and footstool, Peel, and its mirror-version Peel II, is designed to cocoon its occupant, provide free rotation and three angled positions using a controlled double spring system. This armchair embraces like a supporting prosthetic but relieves pressure on the lower spine and ensures breathing and circulation are not impeded. A minor shift in body weight allows seamless movement between the pre-selected positions but the sculptural configuration of the chair permits endless variations in posture. A laminated beech base provides stability and the polyurethane fabric-covered form cushions the body.

Faiver Low Chair

This multipurpose chair, positioned close to the ground, respects Japanese traditional sentiments while creating a new visual style. Its economy of line is reflected in the simple geometry of the MDF and post-consumer recycled board. Purposeful restraint seems to have driven the vision for this chair, which is a salutary lesson to designers who regularly indulge in extravagant use of resources.

✏️	Tsujimura Hisanoubu Design Office, Japan	310
⚙️	One-off	
📄	MDF, Recycled paper board	
🎧	• Renewable and recycled materials	324/325
⚡	IDRA award, 2000–2001	330

Stokke™ Tripp Trapp®

✏️	Peter Opsvik, Stokke Gruppen, Norway	308
⚙️	Stokke Gruppen, Norway	321
📄	Beech wood, steel	
🎧	• Multifunctional	327
	• Universal design	327
	• Durable	327

This modern classic emerged from the design studio some 30 years ago, but still fulfils its original purpose to provide a chair that grows with the child. Made from solid beech wood from cultivated forests, the Tripp Trapp can accommodate from six-month-old babies up to teenagers or even adults by moving the relative height of the seat and footrest components in the frame. The chair encourages good posture by allowing the sitting position to be adjusted to the table height. This arrangement also permits freedom of movement. Wooden components are guaranteed for seven years and a range of safety rails and cushions enable further customization. Here's one chair you can raise your entire family in.

Maggi

Hinting at an imminent (environmental) meltdown, Bär + Knell's chair is formed of plastic packaging waste, dyed black then surface finished with Maggi plastic carrier bags. The branding lives on but not as the marketeers intended. This chair was made in 1995 but the design partnership experimented with many variants from 1993 to 1997, including sofas and furniture for children. Their playful yet deliberate exposure of the raw-waste medium contrasts with the highly controlled, stylized injection- or blow-moulded designs of the 1960s and 1970s. This reincarnation poses the question: 'how can we throw away such a useful resource?'

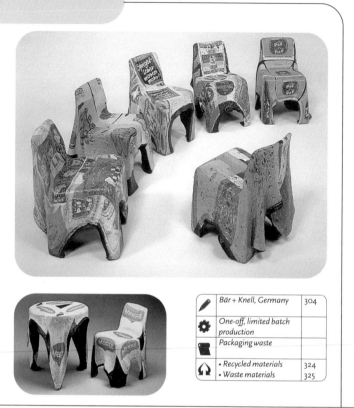

✏️	Bär + Knell, Germany	304
⚙️	One-off, limited batch production	
📜	Packaging waste	
↻	• Recycled materials	324
	• Waste materials	325

Gallery

This moulded plywood stool is a module that functions in its own right or can be joined to others to form a continuous bench or rows of seats. Efficient use of materials is achieved through simplicity and strength of form.

✏️	Hans Sandgren Jakobsen, Denmark	307
⚙️	Fredericia Furniture A/S, Denmark	315
📜	Plywood	
↻	• Reduction in materials used	325
	• Renewable materials	325
	• Dual-function seating	327
🏆	iF Design Award, 2000	330

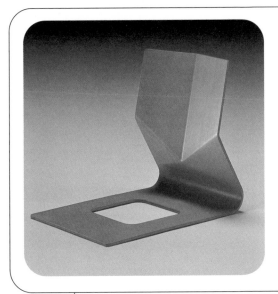

Origami Zaisu

A single sheet of plywood is bent and cut to form a simple floor seat. In Japanese culture sitting on the floor is the norm but perhaps the practice should be adopted more widely, since the omission of legs that form a conventional chair saves materials and energy.

✏️	Mitsumasa Sugasawa, Japan	309
⚙️	Tendo Co. Ltd, Japan	322
📇	Plywood	
♺	• Reduction in materials used	325
	• Renewable materials	325

Mirandolina

Reviving a technique first used by the designer Hans Coray for his pressed-aluminium 'Landi' chair designed in 1938, Pietro Arosio has produced an economical yet elegant stacking chair from a single sheet of aluminium. Cut and pressed into its final form, the Mirandolina shouts efficiency. The use of one material, aluminium, facilitates recycling waste offcuts and ensures it is easy to recycle or repair.

✏️	Pietro Arosio, Italy	304
⚙️	Zanotta SpA, Italy	323
📇	Aluminium	
♺	• Recyclable single material	324/325
	• Efficient use of raw materials	325

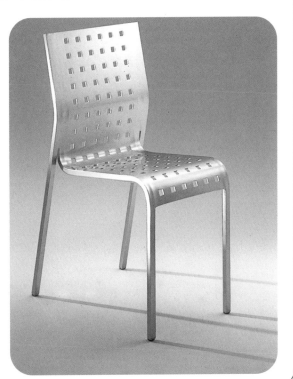

Honey-pop chair

First presented at the 2002 Milano Salone, the Honey-pop delights and surprises. Made entirely of paper, folded and joined, the flat-pack, planar chair unfolds like a magic lantern. The complex structure responds to support each individual. Yoshioka reveals a knowledge and respect for his chosen material and imbues the finished product with lightness, fun and a sculptural aesthetic. Driade invested in the prototype to create a unique manufactured, lightweight paper armchair delivering new experiences to the concept of seating.

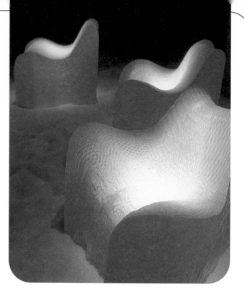

✏	Tokujin Yoshioka, Japan	310
⚙	Driade SpA, Italy	313
📜	Paper	
🎧	• Lightweight	324
	• Low-energy materials	324
	• Low transportation energy	326
	• Recyclable	324

C1 Recliner and Footstool

Utilizing steam-bent English ash wood allows economical use of materials without sacrificing strength and ensures low-embodied-energy manufacturing. The chair frame is adjustable to three positions and is cushioned with padded linen. An alternative version, using rattan, is available. Colwell merges the traditions and durability of bent ash with a fresh aesthetic and a green policy for the procurement of raw materials.

✏	David Colwell and Roy Tam, UK	305
⚙	David Colwell Design, UK	305
📜	Solid ash wood, linen or rattan	
🎧	• Renewable materials with stewardship sourcing	325
	• Low-energy construction techniques	325

Tula

Made predominantly from ash thinnings harvested from small local woodlands in Somerset, Tula is part of the evolving story of 'stick and nail' furniture. Ash slats and boards are steam-bent, scorched black then wire-brushed and waxed. Smaller diameter ash is worked for the legs and two year old cultivated willow withies are used for the hoop and struts. Economical, elegant and eclectic, this type of small-scale production is inherently sustainable by encouraging management of local resources and generating designs with regional identity.

Cardboard Chair

Letting the materials deliver the (environmental) message is a theme common to furniture designers using recycled or recyclable materials. So Jane Atfield (UK) speaks with plastic, Frank O Gehry (USA) with cardboard and Lièvore (Spain) with maderon. The Campana brothers combine a robust, solid, iron-rod frame with a laminated cardboard seat and back to create a dining chair that demonstrates how unpretentious materials can encourage a healthy hybrid of modernism and craft.

✏	Fernando and Humberto Campana, Brazil	305
⚙	Limited batch production	
🎞	Iron, cardboard	
🎧	• Renewable, recyclable materials	324/ 325

✏	Guy Martin Furniture, UK	306
⚙	One-off and small batch production	
🎞	Ash wood, willow withies, copper nails	
🎧	• Renewable and locally sourced materials	325/ 324

Tokyo-Pop sofa and table-stools

As plastics continue to form an integral part of modern living it is important that we elevate the cultural value of plastic artefacts, so we cherish them rather than carelessly dispose of them in landfill sites. Over recent years the Italian manufacturers Driade have encouraged responsible, yet innovative, designs of one-piece furniture made from single synthetic polymers. The respect for material and obvious technical expertise in rotational- and heat-moulding shown by the designer and manufacturer somehow elevate the status of the polyethylene to a valued material once it emerges as a sculptural sofa

and as multifunctional table-stools. The Tokyo-Pop range is versatile indoor or outdoor furniture. An indoor version of the sofa, Tokyo-Soft, is available with Trevira or wool fabric removable and washable covers. These objects are inherently durable; they represent a temporary materialization of a unique man-made resource that is destined to run out one day. Will furniture manufacturers operate a take-back policy when Tokyo-Pop and similar items end their useful life? It makes environmental sense to and, one day, it may be a legal requirement to 'close the loop' and reclaim plastic for future manufacturing.

✏	Tokujin Yoshioka, Japan	310
⚙	Driade SpA, Italy	313
▤	Polyethylene, Trevira or wool fabric	
♻	• Single material	325
	• Multifunctional	327
	• Recyclable	324

OTO

Cut and bent from a single sheet of laminated beech, Karpf's graceful chair avoids the need for any other components, keeping the production process efficient and reducing waste.

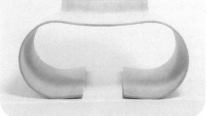

✏	Peter Karpf, Denmark	307
⚙	Iform, Sweden	316
▤	Laminated beechwood	
♻	• Single, renewable material	325
	• Low-energy manufacturing	325

Stokke™ Gravity II™

Half rocking chair, half armchair, the original Gravity by Stokke challenges our perceptions. Constructed of laminated wooden runners, this unique form enables the user to move seamlessly from an upright position, with the knees tucked behind the kneepads, through two intermediate positions to the fully inclined position where the blood circulation is encouraged in the legs and the back is completely supported. Ideal for those who spend long hours at a PC, the Gravity II offers instant opportunities for reducing stress and improving overall well-being. The ideal all-in-one solution to instant cat naps!

✏	Olav Eldøy, Stokke Gruppen, Norway	306
⚙	Stokke Gruppen, Norway	321
▤	Aluminium profiles, polymers, polyester fabric	
♻	• Multifunctional	327
	• Improved ergonomics	327

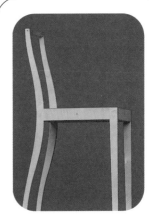

Ply Chair

Avoiding excessive usage or wastage of materials should be a guiding principle of any design in the twenty-first century. The Ply Chair is the latest answer, demonstrating restraint, grace, economy, strength and character.

✏	Jasper Morrison, UK	308
⚙	Vitra AG, Germany	322
▤	Aeronautical-quality plywood	
♻	• Reduction in materials used	325

Vuw

Part of the Iform 'Voxia', the 'voice of nature' collection (see Eco and OTO; pp. 40 and 34), Vuw is a tour de force in 3D-curved wood surfaces, made from one continuous piece carefully bent and formed. Beech wood originates from a local, sustainably managed forest and is rotary-cut to minimize waste from the harvested trees. Cut-away waste is also minimized by planning the veneer pattern in the laminated wood. The result is a thin, sensuously curved surface efficient in materials consumption and energy of manufacture.

	Peter Karpf, Sweden	307
	Iform, Sweden	316
	Beech veneer laminate	
	• Renewable and sustainably managed material	325
	• Reduction in materials and energy used	325

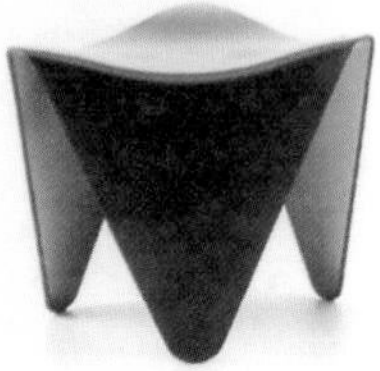

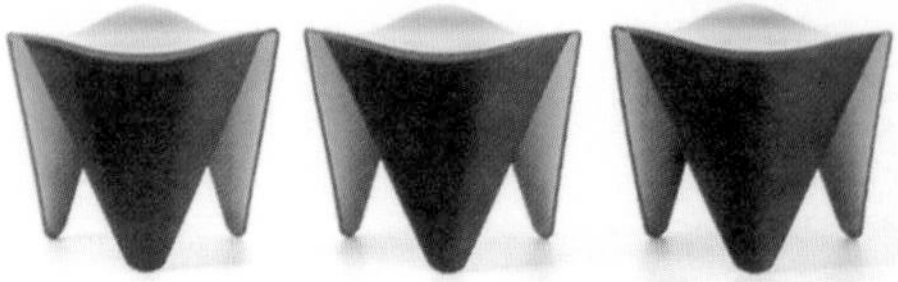

Daybed

This company has been manufacturing furniture, using special techniques for weaving twisted paper, since the beginning of the twentieth century. This daybed combines a contemporary shape with a traditional material by making the most of the manufacturer's extensive experience with this medium.

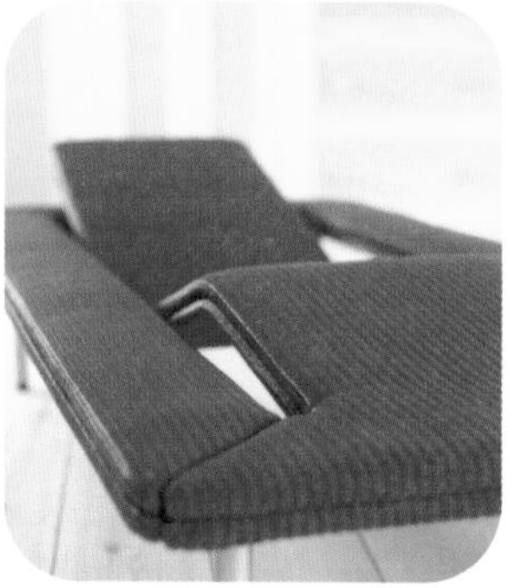

	Nigel Coates, UK	305
	Limited batch production by Lloyd Loom of Spalding, UK	317
	Twisted paper and steel wire	
	• Recyclable and renewable materials	324/325

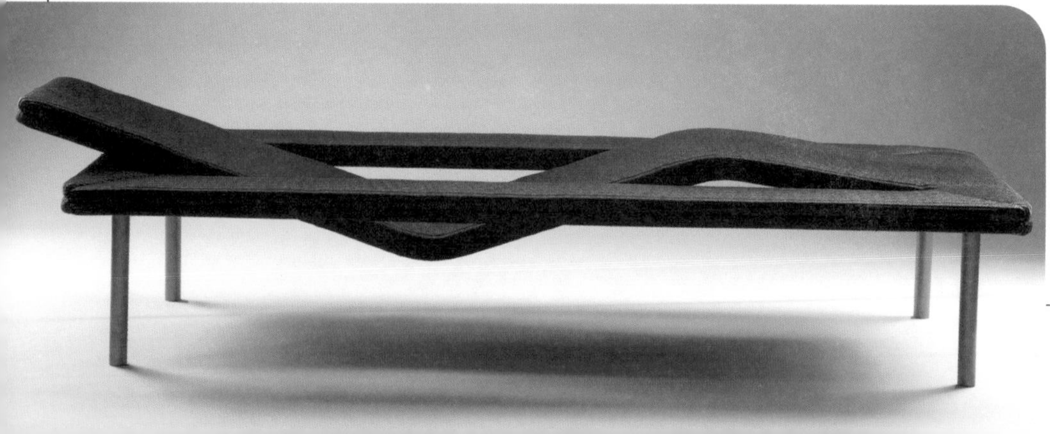

An affair with a chair

There is a rich, untapped seam in the design arena, which could be called 'innate design', that is, design that awakens individuals' ability to design and innovate. Schaap recognizes that everyday designs are modified by the consumer, subverting the well-defined functionality of the designed object to make something with extra functionality or personality. These 'standard' wooden chair frames with drilled holes invite the consumer to experiment and find novel ways of realizing the full potential of the original design. This represents the true potential of customization, not the watered-down customization of industrial design where you can choose the colour and optional extras for your car. Improved aesthetic and emotional bonds guarantee a long life for these chairs, a life beyond fashion.

✎	Natalie Schaap, Design Academy Eindhoven, The Netherlands	305
⚙	Small batch production	
▤	Wood, various materials	
🎧	• Customizable, personalization	327
	• Renewable materials	325

Yolanda Collection

Creating a welcome relief from steam-bent and moulded temperate plywoods, Minakawa's design celebrates the aesthetic of bamboo in an unexpected form. Here laminated bamboo sheets create a striking sculptural form so different from traditional round, split and woven bamboo chairs. Finished surfaces are protected using non-biocide, plant-based oils and sealers.

✎	Gerard Minakawa, USA	308
⚙	Small batch production	
▤	Bamboo laminate	
🎧	• Renewable material	325
🔍	IDRA award, 2002–2003	330

BUNSON chair and NYA armchair

BUNSON and NYA are by Lino Codato, whose motivation and ethos is to investigate new applications of natural materials for contemporary furniture. Water hyacinth and Yan lipao are abundant plants in tropical climates, the former being a native of South America that has become a weed and has invaded watercourses in the southern United States and other warm climes. Harvested, prepared and dried, it transforms into an excellent weaving material. Yan lipao, a fern vine, is harvested from the jungles of Thailand, its outer pith peeled off then polished and smoothed before weaving. As a traditional material for basketry it ensures that a skilled labour pool is available for creating other artefacts. This furniture represents that wonderful boundary between mass-production and one-off craft, supporting use of local materials and skills.

✏	Lino Codato Collection, Italy	308
⚙	Lino Codato Collection, Italy	308
🗒	Water hyacinth, Yan lipao, oak	
🎧	• Renewable materials • Conservation work by weed removal	325 328

Aspetto chair and lounge chair

Minimal manufacturing is a key objective for Lorenz*Kaz. Both aspetto chairs are produced from bending and welding 5 mm (¹⁄₅ in) aluminium sheets. Surfaces are powder-coated or the sheet is encased in leather. This minimalist approach produces a high-embodied energy product but it is very durable, easy to clean and, in the case of the lounge chair, multifunctional furniture. Green supply management, using recycled aluminium and leather from suppliers who raise organic cattle and use traditional oak-tanning processes, could further reduce environmental impacts.

Lorenz*Kaz, Italy	307
Lorenz*Kaz, Italy	307
Aluminium, powder coatings, steel	
• Reduced-process manufacturing	
• Durable	327
• Multifunctional	327

chair (Lusty's Lloyd Loom)

Lusty's Lloyd Loom Company engaged Schwendtner, a former student of the Royal College of Art, London, to create these occasional/dining chairs. Her designs build on traditional designs, so strongly associated with tearooms and Empire, and open more contemporary modern markets to the manufacturer. Economical use of materials combines with durability to produce a serviceable dining/occasional chair.

Gitta Schwendtner	309
Lusty's Lloyd Loom, UK	318
Lloyd Loom woven paper & wire, steel	
• Reduction in materials used	325
• Renewable materials	325

Eco

These stackable chairs are cut from a single piece of veneer-faced ply and follow in the Scandinavian tradition of working with bent ply, such as the designs of Gerald Summers for the firm Makers of Simple Furniture based in London in the late 1930s. Simplicity, economy and functionality meet in this award-winning design.

✏️	Peter Karpf, Denmark	307
⚙️	Iform, Sweden	316
📓	Plywood	
🎧	• Renewable materials	325
	• Reduction in materials used	325
	• Low-energy production	325
💬	Winner of the iF Ecology Design Award, 2000	330

LoveNet

During the 1950s Europe and North America witnessed a boom in woven 'capsule' chairs on spindly metal frames, appearing on the glossy pages of interior design and architecture magazines. This legacy lives on in Lovegrove's homage to economical usage of technosphere materials, the LoveNet. Opting for lightweight woven polyethylene (PE), rather than a natural material, such as rattan, increases the embodied energy of the materials for the seating but if the strength of the PE permits greater longevity then the embodied energy per lifespan year of both options may not be so different. At the end of its life the galvanized and powder-coated steel under the webbing and stainless steel frame can be separated for recycling, although questions remain regarding the feasibility of recycling the PE.

✏	Ross Lovegrove, UK	308
⚙	Moooi, the Netherlands	318
🗒	PE, steel and stainless steel	
♻	• Reduction in materials used	325
	• Durable	327

Naseweis

This visual pun surprises and pleases: upon close examination you can see that the 'skin' of this rocking chair is brown paper stretched over a wooden frame. Contrary to expectations, this combination of materials, one processed cellulose fibre, the other a cellulose-based composite, delivers strength. Presumably maintenance is easily affected if the paper is damaged.

✏	Robert A. Wettstein, Switzerland	310
⚙	Small batch production	
🗒	Wood, paper	
♻	• Renewable materials	325
	• Low-embodied-energy materials	324

B. M. Horse Chair

B. M. refers to the Bell Metal Project, 1998–2000, for which Pakhalé applied and refined the process of the *cire perdue* or lost-wax casting technique popular in central India for hundreds of years. Building on his experiences creating an eclectic range of tableware (see B. M. vase, fruitbowl and spoons; p. 166), this chair is laden with primal and symbolic forms bridging the past and present. The alchemy extends beyond the mixed recycled brass, bronze and tin to the combination of hand, computer and technical skills used to make this piece. Sand blasting the surface is the last technique applied to realize the final form. A blurring of high-craft with high industrial design shows fresh opportunities for the future.

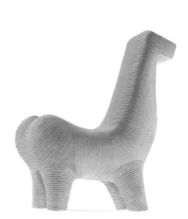

✏	Satyendra Pakhalé, the Netherlands and India	308
⚙	Atelier Satyendra Pakhalé, the Netherlands	308
📜	Brass, bronze, copper scrap	
↻	• Recycled metals	324
	• Low-energy manufacturing	325
	• Innovation of traditional technology	325

Blähtonhocker

Clay is an elemental material indelibly bound with mankind's history and creativity. Robert Wettstein's expanded clay stool is proof that the relationship continues to evolve. Working in that energy charged zone between art, craft and design, this object challenges our perceptions and stimulates our senses.

✏	Robet A. Wettstein, Switzerland	310
⚙	One-off and small batch production	
📜	Clay	
↻	• Abundant geosphere materials	324

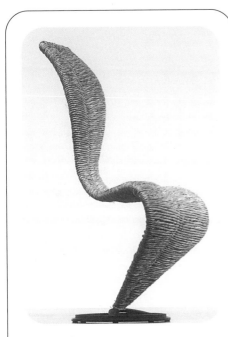

S chair

Following his experimentation in the 1980s with one-offs using salvaged materials, Tom Dixon designed this elegant cantilever chair. A steel frame is wrapped with woven rushes, creating a sculptural form. At the end of the chair's life materials are easily separated for recycling (steel) or composting (rushes).

✏	Tom Dixon, UK	305
⚙	Cappellini SpA, Italy	312
▤	Steel, rushes	
↻	• Reduction in materials used	325
	• Renewable and recyclable materials	324/325

SE 18

It is difficult to pinpoint exactly how or when a product becomes a 'classic' design. Classic implies that the design is respected, valued and maintains its socio-cultural relevance. Such a design is also likely to be well made, durable and repairable. Originally designed in 1952 by Professor Egon Eiermann, the SE 18 received the 'Good Design' Award 1953 and was immediately acquired for the Museum of Modern Art's collection in New York. Fifty-two years after the first chairs left the factory, the efficient production system, with its emphasis on hand-finishing for quality, has hardly changed. Here is a philosophy of manufacturing that is inherently more sustainable

than the fashion output of numerous furniture manufacturers. There is no need to keep spending on a continuous cycle of re-tooling and marketing. This chair virtually markets itself by appealing to consumers' design instinct for knowing a classic when they see one.

✏	Prof. Egon Eiermann, Germany	305
⚙	Wilde & Spieth GmbH, Germany	322
▤	Beech wood, metal	
↻	• Low-energy manufacturing	325
	• Classic design	324

Trinidad No.3298

'Industrial craft' production will undoubtedly prosper in the twenty-first century if the workmanship and graphical form of this ash chair are a measure of the output of today's furniture manufacturers.

✏	Nanna Ditzel, Denmark	305
⚙	Fredericia Furniture A/S, Denmark	315
📱	Ash wood, metal	
↻	• Renewable material	325

Corks

Tactile, warm, giving and durable all characterize cork, from *Quercus suber*, cork oak. Morrison approaches this material with his eye for economy and subtlety to produce a versatile object that encourages a little leisure – for sitting, leaning or for resting your coffee cups. In a gesture the embattled cork industry of the Iberian peninsula might appreciate, here's a great use for cork to offset recent gains made by plastic corks in the global wine industry.

Bubo

There is an increasingly large mountain of tough, impact-resistant ABS plastic being generated as a by-product of our avaricious consumption of computers, white goods and household appliances. Some recycled ABS ends up in the crash fenders and bumpers of automobiles but little is used in the furniture industry. Bubo eloquently demonstrates the potential of this recyclate to create durable, easily cleaned, domestic seating.

✏	David Burke, Australia	305
⚙	One-off and batch production	
📱	Recycled ABS, metal	
↻	• Recycled materials	324
◉	IDRA award, 2000–2001	330

✏	Jasper Morrison, Germany and Britain	308
⚙	Moooi, the Netherlands	318
📱	Cork	
↻	• Renewable material	325
	• Low-energy material	324
	• Reuseable or recyclable	328

Ghost

Purity of form and function can often be achieved by focusing on the properties of one particular material. Cini Boeri and Tomu Katayanagi have taken a single piece of 12 mm- (½ in-) thick toughened glass and cut and moulded it into an extraordinary object. They juxtapose the contradictory characteristics of the material – its fragility and toughness – and create a durable, rather timeless design. Ghost provides food for thought on how other familiar materials can be modified or mutated to fit new forms and functions. Being composed of a single material facilitates recycling at the end of the product's life and encourages closed-loop recycling, where the manufacturer uses its own recycled materials to produce new goods.

✏	Cini Boeri and Tomu Katayanagi, Italy and Japan	304
⚙	Fiam Italia SpA, Italy	314
▣	Glass	
☊	• Recyclable single material	324/325
	• Durable	327

Chair and ottoman

Slabs of heavy-grade industrial felt, typically used for noise insulation in military vehicles, are bolted together to create an archetypal armchair. An ottoman emerges from the offcuts. The original felt slabs are transformed from the utilitarian to the purposeful, yet retain their honesty of origin. Like the Danish designer Niels Hvass, who has made a similar chair from used newspapers, Atfield reminds us to keep it simple and create zero waste.

✏	Jane Atfield, UK	304
⚙	One-offs	
▣	Felt, steel	
☊	• Reduction in materials used	325
	• Zero waste production	325

IKEA a.i.r./MUJIAIR/ SoftAir sofas

IKEA are peering into the future and testing the way forward for sustainable product design. This example by Jan Dranger sets the stage for manufacturing furniture using recyclable plastics and interchangeable covers, as wear and tear or fashion decrees. As resource scarcity bites in the twenty-first century, manufacturers will not only have to use recyclable materials but also to develop business models that ensure that product take-back keeps materials in a closed recycling loop.

✎	Jan Dranger, Dranger Design AB, Sweden	305
⚙	News Design DfE AB for IKEA	318 316
🗋	Plastics, nylon or polyester, cotton	
♻	• Recyclable materials	324
	• Reduction in materials used	325
	• Low-energy transport and assembly	326

chair (Lloyd Loom)

Responding to market needs is always challenging for furniture manufacturers using traditional materials. This easy chair reflects the renewed interest in organic and biomorphic forms while showing off the raw materials – Lloyd Loom fabric of wire and twisted paper – to the maximum. The durability of these materials is well documented – pre-World War II chairs are still going strong. These new designs allow younger generations to acquire seating that will survive to the next generation and beyond. This slowing down of the flow of resources, avoiding the short-termism of fashion cycles, is to be welcomed.

✎	Jane Dillon and Tom Grieves	305
⚙	Lloyd Loom of Spalding, UK	317
🗋	Lloyd Loom woven paper & wire, steel	
♻	• Reduction in materials used	325
	• Renewable materials	325

Vagö

First-time home-makers are loyal consumers with IKEA, one of the world's largest retail furniture companies. Vagö is an ideal purchase for those seeking robust, versatile furniture suitable for indoor or outdoor use. Its low-slung line complements modern perceptions of urban living while also being playful – design should be accessible and fun. Its materials and construction potentially confer a long life but, like many a 1960s plastic chair, its contemporary nature might mean a relatively short life before it ends up back in the recycling loop.

✏	Thomas Sandell, IKEA, Sweden	309
⚙	IKEA, Sweden	316
🗋	Polypropylene	
♺	• Recyclable mono-material	324/ 325

Raita Bench

This eclectic bench is made from DURAT®, a polyester-based plastic that contains 50% recycled material and is itself fully recyclable. Warm and smooth to the touch, the colourful stripes indicate the range of standard DURAT® colours. Two versions are available, 120 cm or 160 cm (47 in or 63 in) long, both standing 45 cm (17¾ in) high and 40 cm (15¾ in) deep. These dimensions ensure it can double up as a coffee table too. Durable and easy to clean, this bench should give many years of service.

✏	Tonester, Finland	322
⚙	Tonester, Finland	322
🗋	Recycled and virgin polyester-based plastics	
♺	• Recycled content • Recyclable	324 324

chair (Natanel Gluska)

For Gluska freedom of expression is a huge tree trunk and a chainsaw. His highly eclectic, one-piece, solid wood chairs exhibit geometric and curvilinear lines, idiosyncratic whorls and decorative texture. Harking back to an age when the sound of the carpenter's adze could be heard in every village, these designs remind us of human endeavour and the creative talent latent in all of us, and of the importance of time for nature. In what might be an ironic twist,

these chairs could outlive many of the forests ravaged by man's insatiable appetite for timber. Perhaps, for every chair born we should plant a tree.

✏	Natanel Gluska, Switzerland	306
⚙	One-offs	
🗒	Wood	
⏎	• Renewable materials	325
	• Low-energy manufacturing	325

Pancras

Without sacrificing the comfort levels associated with a traditional armchair, the Pancras collection is a slimmed-down version to reduce material consumption. Oak and beech veneer for the laminate is rotary cut from Swedish trees to minimize waste and transport energy. The laminate is compression-moulded to form the support and attachment points for the supporting polished, rather than chromed, stainless steel frame. A removable Polytex or leather cushion completes the construction. An ottoman is also available with a detachable cushion, enabling it to be converted into an occasional table. Swedish traditions, contemporary detail and care for the environment meet in these designs.

✏	Tore Borgersen and Espen Voll, Sweden	305
⚙	Iform, Sweden	316
🗒	Oak or beech veneer laminate, Polytex or leather	
⏎	• Renewable and sustainably managed material	325
	• Reduction in materials and energy used	325

Krogh chair

The single-piece back support and arms are made of pre-compressed wood. Many types of wood will respond to pre-compression treatment. The process involves mollifying the fibres with steam and then compressing them, which causes the fibres to adopt an accordion-like zig-zag structure. This results in an overall shortening of the wood by 20%, which reduces to 5% when the compressing force is removed. The wood is malleable at this stage and can be bent before being allowed to dry. Once dry it maintains its original strength and flexibility.

✏	Erik Krogh Design, Denmark	306
⚙	One-off	
🗒	Pre-compressed wood	
⟲	• Renewable material with improved characteristics	325

Opuna

Taut leather stretched over a Douglas fir frame gives an inviting tension to this cross between deck chair and chaise longue. Put your feet up on the multifunctional ottoman/pouffe/table/stool. The surfaces of these well-chosen materials will patinate with time, especially the leather as it is moisturized and buffed with saddle soap.

✏	Startup Design, UK	309
⚙	One-off and small batch manufacturing	
🗒	Wood, leather, horse and hog hair, rubber birch plywood	
⟲	• Renewable materials • Multifunctional	325 327

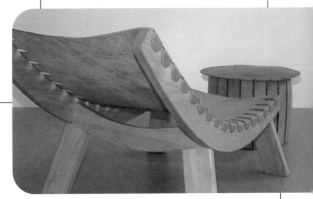

Come a little bit closer

From the 'Me, Myself and You' collection by Droog Design, aimed at creating objects that encourage or discourage interaction, 'Come a little bit closer' by Nina Farkache actively engages participants in her moving bench. The act of sitting suddenly takes on a fresh dynamic and opens new possibilities for play and discourse as the users glide along on MDF seat pads. Marbles, a familiar childhood toy, are presented en masse and in doing so are transformed into a unique tactile and audio experience. How much are these joys of discovery worth? How do these designs contribute to our well-being? Whatever the answers, design reveals itself to be most powerful when it is not restricted by commercial imperatives.

✏	Nina Farkache, Droog Design	313
⚙	One-off	
🗒	Steel, glass, MDF	
⟲	• Interactive design • Multifunctional	327 327

Amoeba

Currently residing in the permanent collection of the Brooklyn Museum of Art, New York, Amoeba finds new expression for post-consumer HDPE sheeting. Supported by a ply form, the recycled plastic is elevated to higher cultural status, dispelling its image as a cheap, substitute material. This rocking lounger received an IDRA Award in 1996 but still sets the standard for other designers striving to find the right vocabulary for recycled materials.

✏️	Isabelle Moore, USA	308
⚙️	One-offs	
📷	Recycled HDPE, plywood	
♻️	• Recycled and renewable materials	324/325
🏆	IDRA award, 1996	330

Stühl, SE 68, 1998

An economical 1950s design, which originally used plywood seating and back rest, has been reproduced using multicoloured plastic sheeting originating from waste packaging. Where renewable materials are in short supply or costly, recycled plastic offers a viable alternative.

✏️	Professor E. Eiermann, Germany	305
⚙️	Wilde & Spieth GmbH, Germany	322
📷	HDPE sheeting, steel	
♻️	• Recycled and recyclable materials	324

Chiaro di luna (moonlight)

Meda worked intuitively with polypropylene and uses Serralunga's rotational moulding technology for the construction of stiffened double-wall plastic structures to create a lightweight small bench with a free-floating visual signature. It can be used indoors or outdoors and can be easily moved between these environments. This is eco-efficient production and should last for many years before returning to the recycling loop.

✏️	Alberto Meda, Italy	308
⚙️	Serralunga, Italy	321
📷	Polyethylene	
♻️	• Reduction in materials used	325
	• Single material	325
	• Recyclable	324

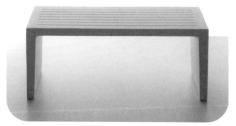

injection moulding, with its hollow arms and 'angle-iron' legs. Starck brings his usual wit and economy of line to this chair, which is equally happy in a garden, an urban loft or a café. As it is fabricated entirely from PP with a small, easily removable stainless steel plug around the drainage hole in the seat, it is easy to recycle the materials at the end of the item's life. By 2030 manufacturers may even be requesting that their products be returned by the current custodian for dismantling and recycling of components and materials. The material content of the Toy Chair will then be valued as much as the comfort and pleasure given throughout its lifetime.

✏	Philippe Starck, France	309
⚙	Driade SpA, Italy	313
▤	Polypropylene	
♺	• Recyclable single material	324/ 325
	• Reduction in materials used	325

Toy Chair

Tough, durable and colourful, polypropylene has been a favoured material with designers for nearly half a century. Toy Chair is a wonderful celebration of technological progress in single-piece

Thinking Man's Chair

Tubular and flat steel are combined in a deliberately 'engineered' look, further enhanced by the red oxide-type finish complete with written dimensions. A durable design for indoor or outdoor use, which, being made from a single material, is easily recycled.

✏	Jasper Morrison, UK	308
⚙	Cappellini SpA, Italy	312
▤	Steel	
♺	• Recyclable single material	324/ 325

Tron plaz'owy (beach throne)

Poland has a long tradition of wickerwork, from baskets to furniture, and still retains an active institute where enthusiastic apprentices can learn from master craftsmen and -women. Grunert's work searches for contemporary uses of willow. This fast-growing biomass woody plant requires little fertilizer, delivers a cash crop and locks up lots of carbon dioxide. So it makes sense to explore how new designs can stimulate production. This armchair plays with the metaphor of the drying willow bundles,

a feature of all willow growing areas in the early winter months, and delights with its 'secret' seat. Like the straw Orkney chairs that kept the cold North Sea winds from their occupants, this chair cocoons the sitter, ready to send him or her to dreamland.

✏	Pawel Grunert, Poland	306
⚙	One-offs	
▦	Willow rods	
♻	• Renewable material	325

Przecias (the draught)

Wicker is imaginatively combined with steel for this exciting range of chairs. Traditional weaving is abandoned in favour of methods in which the withies are held tightly together with steel or inserted into solid beech. The results bring nature, with all its innate variability, into the living space.

✏	Pawel Grunert, Poland	306
⚙	Limited batch production	
▦	Steel, wicker	
♻	• Renewable and compostable materials	324/325
	• Recyclable materials	324

XL1 kit chair

This lightweight chair, weighing just 2.2 kg (4.7 lb), won the 1999 Jerwood Applied Arts Prize coordinated by the Crafts Council, UK. Marriott combines materials and ready-made components, which are readily available from local builders' merchants and DIY stockists, into an honest, economical, multipurpose chair. In the UK, DIY, interior decoration, building and gardening are obsessions, so it is refreshing to see a designer encouraging such enthusiasts to apply themselves to designing furniture.

✏	Michael Marriott, UK	308
⚙	Self-assembly chair design	
▤	Beech, birch plywood, zinc-plated mild steel	
♻	• Renewable and recyclable materials	324/ 325
	• Reduction in materials used	325
	• Use of ready-made components	326

Wiggle series

Originally designed in 1972 as economical furniture and manufactured by Jack Brocan in the USA, the Wiggle side chair has been reproduced by Vitra since 1992. Each layer of corrugated cardboard is placed at an angle to the next layer to provide significantly increased durability compared with the folding cardboard chairs by the likes of Peter Raacke and Peter Murdoch in the 1960s.

✏	Frank O Gehry, USA	306
⚙	Vitra AG, Switzerland	322
▤	Cardboard, glues	
♻	• Renewable materials	325
	• Low-energy manufacturing	325

Love Seat

Dolphin-Wilding preserves the quirks of nature's patterns in her unique wooden chairs, letting the natural forms dictate the structure of her hand-crafted reincarnations. In doing so she takes us back to days before 'craft' work became a highly skilled profession or before 'industrial design' produced technologically refined furniture.

✏	Julienne Dolphin-Wilding, UK	305
⚙	One-off/small batch production	
▤	Timber	
♻	• Renewable materials	325
	• Low-energy fabrication	325

Ceramic chair project,
2001–2002

It is rare that traditional materials, ubiquitous today and throughout history, cause as much excitement or attract the design profession in quite the same way as seductive techno-materials such as high-strength composites, shape-memory metals and synthetic plastics. Yet Pakhalé suffuses new life and brings out exciting qualities from an abundant, but highly variable geosphere material, clay. 'Playing with clay' was a collaboration with EKWC (the European Ceramics Centre) in the Netherlands, drawing on the skills of craftsmen and -women and technical knowledge about the properties of different clays, especially in relation to consistent crack-free firing. Hand-made prototypes paved the way to developing pressure-cast moulds for industrial production. Where uprights and legs required joining to other parts, a special glued joint was designed using computer modelling. Technical achievements aside, each chair (Ceramic Pottery Chair, 2001; Flower Offering Ceramic Chair, 2001; Roll-Ceramic Chair, 2000–2001) possesses its own spirit and presence, reflecting new possibilities where craft is in symbiosis with industry, perhaps encouraging others to re-examine the potential in this symbiotic relationship. Conversations between highly skilled computer technicians, artisans and designers seem to offer fresh hope in revitalizing socio-cultural value and meaning in our industrial production. The Ceramic chair project is proof of a subtle re-balancing and suggests myriad opportunities.

✏	Satyendra Pakhalé, the Netherlands and India	308
⚙	Atelier Satyendra Pakhalé, the Netherlands	308
▤	Clay, PU glue	
🎧	• Abudant geosphere materials	324
	• Low-embodied-energy materials	324
	• Low energy manufacturing	325
	• Innovation of traditional technologies	325

Hudson 3, Navy 1006 2 and Emeco stools

Founded after World War II, Emeco harnessed the skills of immigrant German craftsmen to sculpt aluminium into the 1006 chair. This quickly became a classic with the US government and military and exists today as the Navy 1006 2. In 2000 the French designer Philippe Starck was involved in a commission for the new Hudson Hotel in New York and began collaborating with Emeco to evolve the 1006 into a pure, reductive form. Reducing weight but not sacrificing strength, and using 60–70% post-consumer aluminium, the latest incarnation, the

Hudson 3, celebrates the raw material, the combination of production and craft skills while exuding durability. Emeco stools reflect Starck's economical eye but his latest addition, the Kong bar stool, reverts to more traditional post-modern whims and fancies. Is this luxury design or a 'yours eternally' approach? Or is it simply providing a temporary resting place for high-embodied-energy materials patiently waiting until the next reincarnation.

✏	Philippe Starck, France, and Emeco, USA	309
⚙	Emeco, USA	314
▰	Recycled and virgin aluminium	
↻	• Single material	325
	• Recycled material	324
	• Recyclable	324
	• Reduced materials	325

Pm

This solid natural beech, cherry or maple tabletop is extendible from 180 cm to 230 cm (71 to 90½ in) by the insertion of a lightweight core tamburato mat extension that sits underneath the original top. A detachable square section steel frame finished in 'silver' or 'titanium' epoxy powder completes the minimalist purity of this design.

✒	Pallucco Italia, Italy	319
⚙	Pallucco Italia, Italy	319
▤	Solid wood, steel, tamburato mat	
⌂	• Multifunctional	327
	• Durable	327

✒	Setsu Ito, Italy	307
⚙	Front Corporation, Japan	315
▤	Steel	
⌂	• Single material	325
	• Durable	327

Saita

Almost 90% of the steel in circulation has been recycled at some time, so it is refreshing to see steel being used with great sculptural panache in this table design. Long the preserve of architectural and structural engineers, steel offers fresh perspectives for furniture designers.

Schraag

This table-cum-desk minimizes materials used by using three legs, not the traditional four, for each lightweight aluminium trestle. A range of standard 2 x 0.9 m (6 ft 7 in x 3 ft) tops can be chosen from laminated bamboo, glass or red multiplex. Simplicity and ease of assembly bring the old-fashioned trestle table into the modern world.

✏	Maarten van Severen, Belgium	310
⚙	Bulo Office Furniture, Belgium	312
🗋	Aluminium with bamboo, glass or multiplex	
🎧	• Renewable material option	325
	• Recyclable materials	324
	• Design for disassembly	325

Handy

Portability and indoor/outdoor use were the priorities for these strong, lightweight polypropylene tables. The double slots accommodate the underside supports for a trays or cushions, depending on the user's needs. Flexible, unobtrusive and durable, these tables provide functional service yet permit customization.

✏	Luisa Bocchietto , Italy	304
⚙	Serralunga, Italy	321
🗋	Polyethylene	
🎧	• Multifunctional	327
	• Single material	325
	• Recyclable	324

Jig

This table simply slots together; the plywood pieces are cut from one sheet with minimal waste offcuts. This is an excellent exercise in the rigour of keeping things simple by design. Perhaps this type of design, which rarely attracts the big furniture manufacturers, could be offered as a cutting pattern to households around the world via the Internet. In the mid-1970s Victor

✏	Leo Scarff, Ireland	309
⚙	One-off and small batch production	
🗋	Plywood	
🎧	• Single material	325
	• Reduction in materials	325
	• Renewable materials	325
	• Low-energy construction	325

Papanek suggested that 10% of designers' work should be in the non-commercial domain, design work for a larger social and environmental good. This 'design democracy' could realize people's latent ability to make domestic objects for themselves, guided by the experience and catalyzing effect of professional designers.

table

Hertz reveals the workability of Syndecrete® in his individualistic, sculptural tables.

Side table

Syndecrete® is a lightweight composite concrete utilizing mixed industrial and post-consumer waste, pulverized fly ash and PP fibre waste. The material is well suited to detailed moulding and can be polished to reveal a terrazzo-like surface.

	David Hertz, USA	307
⚙	Syndesis, Inc., USA	321
▤	Syndecrete®, glass	
↻	• Recycled materials	324

	David Hertz, USA	307
⚙	Syndesis, Inc., USA	321
▤	Syndecrete®	
↻	• Recycled materials	324

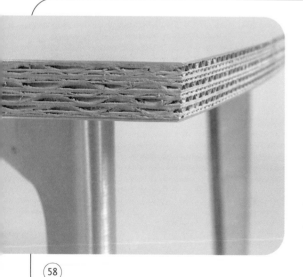

Cardboard as a medium

Cardboard is a ubiquitous packaging material that does its job very effectively, but it is only when we examine it more closely that its fascinating properties can really be appreciated. Shigeru Ban, Frank O Gehry and the Campana Brothers all fell in love with cardboard. Looyen creates lightness with durability by sandwiching his cardboard table-top in sheet aluminium and providing an anchor point for the tubular aluminium legs. This entire table can be made of recycled materials and is recyclable when its useful life ends.

	Rick Looyen, graduate student 2003, Design Academy Eindhoven, the Netherlands	305
⚙	One-off	
▤	Cardboard, aluminium	
↻	• Lightweight strong materials	324
	• Recyclable materials	324

Tischbockisch

Minimal machining is required to create the solid, untreated ash wood horizontal ribs and legs, fitted with non-slip rubber stops, for each trestle frame. Pairs of ribs are available in different lengths so users can define and make their own table tops, although the manufacturer will supply laminated MDF tops to order if required. This pared-down design enables customers to source appropriate local materials, helping to reduce the transport energy per table compared with centralized manufacturing and distribution. This is a rich seam of design opportunities that helps users to become part of the design process to realize the full function of the manufactured components.

✏	Jakob Timpe, Germany	310
⚙	Nils Holger Moormann GmbH, Germany	318
📜	Untreated ash wood, rubber	
♻	• Low-energy manufacturing	325
	• Reduced transportation energy	326
	• User involvement	327

tables (Houshmand)

Juxtaposing massive cuts of reclaimed or virgin timber with thick sheets of plate glass, Houshmand makes minimal interventions to his materials to gain maximum impact. These tables are strong on formal geometry, nature's quirky patterning and simple jointing. They are built to last and surprising in their contrasts. Unusual textures and patterning are a feature of the old growth heart pine, black walnut, elm and spalted maple that is the raw material of each unique piece.

✏	John Houshmand, USA	307
⚙	One-offs	
📜	Reclaimed wood, American hardwoods, glass	
♻	• Renewable materials	325
	• Low-embodied-energy materials	324
	• Cold construction	325

Take away

Laptop computers expanded the concept of work to the commuter train, aeroplane or café. This idea has not been wholly successful, with its blurring of boundaries between different activities. Nor does a laptop constitute a desk – where do you put the coffee? Enter the portable one-person office, a lightweight plywood shell and desktop sit on foldable aluminium legs, with neat stowaway compartments in the hinge-up cover. Take away is not the exclusive domain of the laptop, rather it feels receptive to recalcitrant pencil-and-paper creatives, or to covert watercolour painters. It is ideal for setting up and folding away in small urban flats or homes and yet small enough to take with you on a working holiday.

✏️	Beat Karrer, Switzerland	307
⚙️	Prototype	
📜	Plywood, aluminium	
♻️	• Multifunctional	327
	• Lightweight	326

Ash round table

Combining excellent rigidity and ample leg-room, the simplicity of this design relies on the strength of the solid ash, which comes from local English woodlands. A range of table sizes to seat three to ten people is manufactured to the same basic design. David Colwell successfully blends traditional furniture-making techniques with a modern aesthetic to produce durable, quality seating, tables and shelving.

✏️	David Colwell and Roy Tam, UK	305
⚙️	David Colwell Design, UK	305
📜	Solid ash wood	
♻️	• Renewable materials with stewardship sourcing	325
	• Low-energy construction techniques	325

Kite and Ghost

Kite was made for the Radical Plastics exhibition at the Geffrye Museum, UK. It is typical of one-off tables designed and made by Malcolm Baker, with conscious attention to detail. Kite uses a particularly striking co-mingled mix of plastic waste streams sourced from Smile Plastics, UK (pp. 297 and 298). Opportunities abound for designers to work with plastic recycling companies to develop bespoke and signature solids, boards and laminates to create unique design artefacts. Ghost is another example of Baker's low-environmental-impact tables designed for a long life. He uses locally grown oak or certified European or North American hardwoods, and stainless steel legs with easy separation of components for end-of-life reuse or recycling. Surface finishes include Bonakemi Mega lacquer (water-based, 85% rapeseed oil) and Osmo natural oil or wax finishes. Timber offcuts go for kindling.

✎	Malcolm Baker Furniture Design, UK	308
✿	One-off, small batch	
▬	Recycled plastics, certified timber	
⌂	• Recycled materials	324
	• Low-embodied-energy materials	324
	• Design for disassembly	325

Stol z wikliny (table of wicker)

A desk, a table, a sculpture, or all three? Grunert confronts us with the beauty in ordinary materials by juxtaposing them in unexpected ways. Like a wheat field before harvesting, the willow withies stand tall and strong while the glass sits tentatively, fragile, on its natural bed. Sitting at this object it is easy to imagine that creative writing would flow, food would taste better and the world would seem alive. In our love affair with expensive technological objects we have forgotten the pleasures of everyday materials. This design is an ambassador for re-awakening the senses by letting humble materials sing.

✏	Pawel Grunert, Poland	306
⚙	One-offs	
📖	Willow rods, glass	
♻	• Renewable material	325

Baley

It is poignant to reflect on the future of the traditional hand bale, as the British countryside becomes dominated by big business and the sight of one-tonne bales of straw at harvest-time. In a gesture that simultaneously brings the countryside to urban living while reminding us of our separation from the land, the Baley weaves a complex set of emotional responses. Straw bales are sized and treated with fire-retardant then encased in a protective transparent cover. As the straw degrades through usage it can be replaced with a new bale. Future designs might explore use of translucent biodegradable polymers for the cover.

✏	Neil Barron, Gusto Design, UK	304
⚙	Small batch manufacture, Gusto Design, UK	316
📖	Straw, fire-retardant polymer	
♻	• Renewable and compostable materials • Low-energy manufacturing	324/325 325

newangLe and crop circle tables

An angled steel frame is brought to life by the warm, tactile, visually variable surface of dense straw panels forming the table top for newangLe. Edged in oak, these panels are available in standard wheat and oat straw with an 'art infill' panel – a bespoke panel that can include plant parts (leaves, flowers, stalks) or even bits of fabric, metal and plastic – as an option. These additions to the straw are bonded into the surface or body of the straw during the making of the panels. The crop circle table is available as a straw top edged in oak, or as a walnut top with circular straw board inserts, attached to an oak frame. Contrasting textures in the walnut top suggest ever-ready coasters for wine glasses or coffee cups. These tables reveal new opportunities for exploring a humble and abundantly available material: straw.

✏	Nigel Pearce, Shortstraw, UK	308
⚙	Shortstraw, UK	321
🗒	Steel, oak, walnut, straw and binders	
🎧	• Renewable materials	325
	• Waste agricultural material	325

cardboard shelving

Easy to transport, assemble and disassemble, this shelving unit of concentric cardboard tubes held together with strong rubber bands is ideal for metropolitan wanderers. It speaks volumes about keeping things simple, functional and down to earth. At the same time it generates its own unique aesthetic.

✏	Pil Bredahl, Denmark	305
⚙	One-off and small batch production	
📔	Cardboard, rubber band	
🎧	• Low-embodied-energy materials	324
	• Recyclable materials	324
	• Re-configurable system	

Bookcase

This concertina-like bookcase has an interesting juxtaposition of natural materials. The craft aesthetic has always embraced experimentation with nature's primary materials but further possibilities are emerging to create a new 'industrial craft' production.

✏	Jan Konings and Jurgen Bey, the Netherlands	307
⚙	DMD, the Netherlands	313
📔	Maplewood, paper, linen	
🎧	• Innovative use of natural, renewable materials	325

zehn hoch

Experimenting with Finnish plywood led to the creation of the 'zehn hoch' drawer modules. Each module is finished in melamine laminate, which ensures toughness and creates an easy-glide surface for each drawer. Modules can be arranged as steps, a chest of drawers or a sideboard, according to one's needs.

✏	Haberli/Huwiler/ Marchand, Switzerland	306
⚙	Röthlisberger, Switzerland	320
📔	Plywood, melamine	
🎧	• Multifunctional	327

I Just Moved In

The layers of meaning are as numerous as the shelves in this intriguing design, which promotes individualism and juxtaposes the old, the new and the banal. The design raises a question about the attitudes of designers: should the construction details of Wiesendanger's bookshelf be published so that anyone can assemble a bookshelf on similar principles but using locally found materials? Or is the limited batch production an exclusive process that guarantees status to the purchasers?

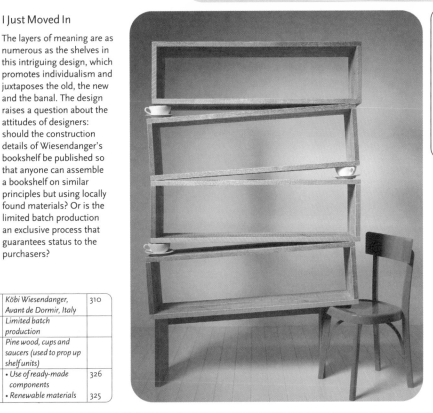

	Köbi Wiesendanger, Avant de Dormir, Italy	310
	Limited batch production	
	Pine wood, cups and saucers (used to prop up shelf units)	
	• Use of ready-made components	326
	• Renewable materials	325

Brosse

Our curiosity is aroused by Sempé's intriguing storage unit based on a metal frame whose contents are hidden by a curtain of industrial brushes. It is the unexpected application of these ready-made brushes that challenges our reactions. While the use of ready-mades does not confer instant low environmental impact it is an interesting route for designers to explore. Existing manufacturing capacity can be more efficiently deployed if its output has suitable applications in other sectors, although this concept needs careful application.

	Inga Sempé, France	309
	edra SpA, Italy	314
	Lacquered aluminium, propylene industrial brushes, metal	
	• Use of existing manufacturing capacity	324

Plus Unit

Flexibility and easy assembly and disassembly are provided by the simple, extruded polished aluminium 'X' connectors that lock each ABS injection-moulded drawer or carcass unit. Heavyweight castors attached to the base create a mobile storage unit. Users or owners can upgrade and expand their system as they can afford, or as new needs arise. The solid, well-made and elegant design encourages consumers to take a longer view. This is design as an investment, design for a lifetime.

✏	Werner Aisslinger, Germany	304
⚙	Magis SpA, Italy	318
📃	ABS, Aluminium	
☊	• Recyclable, mono-material components	324/325
	• Modular	327
	• Customizable	327
	• Durable	327

Aluregal

This lightweight high-strength modular stacking shelving system is 3D-formed from 2 mm anodized aluminium sheet, each shelving unit being tensioned with galvanized steel cross-struts. Each module is 1.6 m (5 ft 3 in) long, 33 cm (13 in) deep and 36 cm (14 in) tall, with a maximum stacking height of five modules. Components promise long life but are easily separated for removal, maintenance or recycling at the end of their life.

✏	Atelier Alinea, Switzerland	311
⚙	Atelier Alinea, Switzerland	311
📃	Aluminium, steel	
☊	• Reduction in materials used	325

Wandregal

This is an exercise in maximum functionality with minimum wastage. Rectangular cut-outs from one standard birch multiplex or white laminate-faced MDF sheet serve as the shelves while lateral sections reinforce the frame. Steel struts take the load on the shelves. Simply lean the shelving unit against a wall and load with weight to achieve full stability. Transportation is facilitated by the fact that the shelving units can be virtually flat-packed.

✏	Atelier Alinea, Switzerland	311
⚙	Atelier Alinea, Switzerland	311
📲	Birch multiplex or laminate-faced MDF, steel	
♻	• Reduction in materials used	325

Celia

All too often the glitzy world of contemporary design feeds off a limited palette of seductive technological materials. It takes the vision of the Campana Brothers to realize exciting tactile and aesthetic pleasures from humbler man-made materials. The Celia range uses the ubiquitous strandboard found on building sites the world over for temporary, disposable screens or shuttering. In the strandboard, nature is roughly thrown together and bonded with glue, heat and pressure. The Celia sideboard reminds us of the precious and unique nature of these resources and lends a quiet dignity to the material. Modest and non-elitist, this furniture hints at a quality that all members of society could aspire to and afford.

✏	Fernando and Humberto Campana,	305
⚙	Campana Studio, Brazil Habitart, Brazil	305
📲	Strandboard	
♻	• Single material	325
	• Renewable materials	325
	• Recyclable materials	324

Pocket

Most paperback books' dimensions follow international standards. This observation served as the stimulus to design minimalist shelving scaled to the depth set by these standards. Avid collectors of specific publishing brands, such as Penguin, can delight in arranging broad swathes of orange and turquoise blue spines in abstract patterns. In fact, Pocket allows users to create their very own wall art feature using lovingly thumbed copies of their favourite paperbacks. This is a minimalist altar to the joy of reading.

✏	Helena Allard and Cecilia Falk, Sweden	304
⚙	Iform, Sweden	316
🎞	Wood and laminates	
♫	• Reduction in materials used	325

Es

Nine beech wood rods are inserted into four plywood panels and locked into place using plastic rings. Grcic tests the boundaries of stability with a design that wobbles yet doesn't fall over. His design appears to fly in the face of man's desire to remove nature from the process of manufacturing, being deliberately made to look naïve and in a DIY style. The rods permit the shelving to double as a coat rack and clothes stand.

✏	Konstantin Grcic, Germany	306
⚙	Nils Holger Moormann GmbH, Germany	318
🎞	Beech wood, plywood, plastic	
♫	• Renewable materials, economically applied	325
	• Design for assembly/ disassembly	325

Compartment Man

This classic design has resided in the Röthlisberger collection for over 25 years, testimony to its timeless appeal. Borrowing from the inspiration of Naum Gabo and the Russian Constructivists, Compartment Man dares to challenge the boundary between art and design, while poking fun at both. Like any good employee today Compartment Man can multi-task – bookshelf, museum of curios, tidy-all...and more.

✎	Susi & Ueli Berger, Switzerland	304
⚙	Röthlisberger, Switzerland	320
🗑	Laminated wood, natural or black	
⏏	• Renewable materials	325
	• Multifunctional	327

Shell

Moulded aircraft-grade plywood, just 3 mm (about ⅛ in) thick, is fixed with 3D metal corner fixings to create a basic shell that can be fitted out internally as required with shelving and a clothes rail. A longitudinal hinge permits much better access than conventional wardrobes.

✎	Ubald Klug, France	307
⚙	Röthlisberger, Switzerland	320
🗑	Plywood, metal	
⏏	• Reduction in materials used	325
	• Low-energy manufacturing	325

Aman furniture range

Aman means 'believe', revealing the underlying ethos and design philosophy that clearly empathizes with Shaker, Arts & Crafts and the Japanese concept of *wabi sabi*. A unique feature of this range is a special joint called shub, which removes the need for adhesives or fastenings. The solid geometry of the oak finished with Danish oil reflects a need for 'truth to materials', which finds new expression in the shelving unit dynamically balanced against the wall.

✏	Avad, UK	311
⚙	Avad, UK	311
▤	Oak	
🎧	• Renewable materials	325
	• Durable	327
	• Reusable materials	328

Hoover

This lightweight wardrobe features stretchable side panels, which provide extra capacity for those awkwardly shaped objects. N2's minimalist philosophy excludes extraneous detail and incorporates a reductionist approach to the use of materials.

✏	Jörg Boner, N2, Switzerland	304
⚙	sdb industries, the Netherlands	320
▤	Various	
🎧	• Reduction in materials used	325
	• Improved functionality (greater capacity)	327

shelving (Startup Design)

Every designer–maker's workshop is littered with 'waste' offcuts from virgin materials. These shelves make intelligent use of short timber lengths for the uprights to improve the overall eco-efficiency of material flows through the workshop. Each shelf has its own personality, as pieces are assembled like a jigsaw in regular or irregular patterns.

✏	Startup Design, UK	309
⚙	One-off and small batch manufacturing	
🗂	Waste wood offcuts	
🎧	• Recyclable materials	324
	• Durable	327

Italic

There are just two components in this beautifully simple shelving system: bent steel rods whose ends are dipped in natural liquid gum to provide a good grip, and wooden shelves. Users can add extra components to expand their existing shelving.

✏	Lorenz Wiegand, Germany	310
⚙	Prototype	
🗂	Steel, liquid gum, wood	
🎧	• Reduction in materials used	325
	• Low-energy production	325

alata shelf system

A range of different folded steel shelf units can be combined to form bespoke shelving configurations. The manufacturing processes are kept to a minimum: cutting, bending and powder-coating the steel surfaces. This minimalist approach produces very durable, easy-to-clean and multifunctional shelving.

✏️	Lorenz*Kaz	307
⚙️	Lorenz*Kaz	307
📄	Steel, powder coatings	
♻️	• Single material	325
	• Reduced-process manufacturing	
	• Durable	327
	• Multifunctional	327

Console and shelving system

Bär + Knell demonstrate the versatility of recycled HDPE plastic board with this eclectic range of furniture. The message is unequivocal: waste is valuable and recycled waste extends the range of materials for the designer.

✏️	Bär + Knell, Germany	304
⚙️	One-offs, Bär + Knell, Germany	304
📄	Recycled plastic	
♻️	• Recycled content	324

Nomadic lifestyle

As geographic boundaries fade in a greater European Union, and citizens gain rights to working and living in EU states, there is an expectation that people movement will increase in the next decade. Recognizing that there is a new generation of cultural and economic migrants, the Nomadic lifestyle project suggests an economical solution to moving and storage. Boxes used for removal purposes can easily be converted into shelving units by the addition of internal shelves and strengthened components. Regular movers will know that half-a-dozen moves is all you can expect from conventional boxes. A little attention to detail can extend the life of this specialist corrugated cardboard box.

✎	Willem Jansen, graduate student 2003, Design Academy Eindhoven, the Netherlands	305
⚙	Prototype	
📜	Corrugated cardboard	
♻	• Lightweight strong materials	324
	• Recyclable materials	324
	• Multifunctional	327

China cabinet

Turning the concept of a cabinet to hold china inside-out, China cabinet is made from porcelain. Each piece is constructed in the same fashion but deforms during the firing to produce a series of one-off modular shelf units. The unexpected forms and unusual application of the material combine to subvert expectations of precision manufacturing. Applying traditional materials to new forms is an exciting arena for experimentation in sustainable design and is ideal for exploring the creation of local sustainable enterprises.

✎	Frederik Roije, graduate student 2002, Design Academy Eindhoven, the Netherlands	305
⚙	Small batch production	
📜	Porcelain	
♻	• Abundant geosphere resources	324
	• Modular design	327

Continua

Continua is an appropriate name for this evolving modular shelving and storage system; the design originated in 1994 but has undergone continuous evolution. The latest addition is the 'Continua glass' and 'Continua glass console' 12 mm (½ in) hardened glass shelving with underside screen prints. Core to the system concept is a series of powder-coated aluminium vertical wall rails and horizontal shelf supports. These can accommodate a diverse range of shelving, clothes hangers and storage units. Shelves are available in steel, wood and glass, according to personal taste and needs. A variety of steel drawers on wheels interlock with the fixed wall system. Continua is suitable for domestic and contract markets. Eco-efficiency is built into the concept as the original fixings can accommodate upgrades, newly designed accessories or extensions. Underpinning Continua is the pure minimalist approach of a manufacturer that understands its customers well, knowing that they want a delicate balance between delivering predictable quality and offering new aesthetic and practical choices.

✏	Pallucco Italia, Italy	319
⚙	Pallucco Italia, Italy	319
📜	Aluminium, steel, wood, glass	
🎧	• Modular	327
	• Upgradable	327
	• Multifunctional	327

eo

Colour is integral to our emotional well-being, yet most furniture designs are available only in hues that designers and manufacturers feel are in tune with overall fashion trends. This modular storage system incorporates low voltage LEDs and some clever electronics to allow users to mix blue, red and green light using a remote control. Light is projected onto the back panel of each module and reflected light is 'captured' by the matt glass panels. Individuals can create the exact hue and tone to match their emotional needs – set the modules to bright orange to pump up the (visual) volume or switch to meditation mode with calm greens if you want to chill out. Aside from eo's function as an emotional barometer, its basic modules can be fitted with internal glass or steel shelving to suit a variety of artefacts from CD players to tableware or a mini wine cellar. With such intelligent application of electronics and lighting to benefit our sense of well-being, it is only fitting that the system won the prestigious Design Award of the Federal Republic of Germany in 2002.

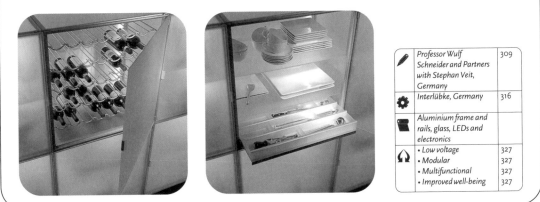

🖊	Professor Wulf Schneider and Partners with Stephan Veit, Germany	309
⚙	Interlübke, Germany	316
🗔	Aluminium frame and rails, glass, LEDs and electronics	
♻	• Low voltage	327
	• Modular	327
	• Multifunctional	327
	• Improved well-being	327

Dia

Adaptability and durability
are the two primary
prerequisites for furniture
that is intended to survive
the elements and robust
use in the garden. This
range offers a high degree
of flexibility – the chair
has an upright and a low
position, the table height
is adjustable and the
sunbed has eight possible
permutations. Polished
stainless steel and strong
fabric, impregnated with
waterproofing and UV-
stabilized, ensure a long
life. Thanks to these high-
quality materials, this
range of furniture is also
suitable for indoor use
and thus offers flexibility
and dual-functionality.

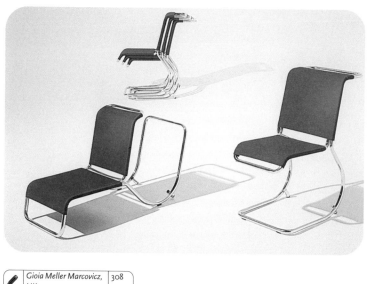

✐	Gioia Meller Marcovicz, UK	308
✿	ClassiCon, Germany	313
▤	Stainless steel, waterproofed/UV-stabilized fabric	
♫	• Durable	327
	• Multifunctional	327

Flexipal

These identical interlocking plastic modules can be articulated and held in fixed positions by tightening the adjusting screws to configure a range of furniture from tables to chairs, beds and platforms as desired. This encourages the owner to experiment with his or her own concepts and offers flexible functionality.

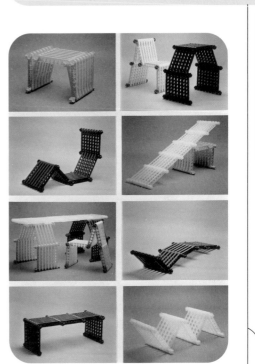

✏	J. R. Miles, UK	308
⚙	Flexipal Furniture, UK	315
🎞	Plastic	
🎧	• Multifunctional (ease of upgrading)	327
	• Single material	325

Connected chairs

This collection of furniture queries and probes at subtle connections of human interaction. The loose connectivity between the chairs confronts expectations and invites experimentation from the users – is this a sofa for three, chairs for two or a bench for one? These designs allow people to explore their own sense of design, space and dynamics.

✏	Jessica van den Heuvel, graduate 2002, Design Academy Eindhoven, the Netherlands	305
⚙	One-offs	
🎞	Wood, wool	
🎧	• Renewable materials	325
	• Multifunctional	327

Kokon

Old wooden furniture is revived by covering it with a PVC-based coating. The opportunities to create quirky new custom furniture are legion but the technique needs further refinement to find a substitute for PVC, whose environmental track record is poor. How to isolate the timber of the reclaimed furniture from intimate contact with the PVC and what to do with the items at the end of their lives are unanswered questions.

✏	Jurgen Bey, Droog Design, the Netherlands	304
⚙	Limited batch production, Droog Design, the Netherlands	313
▧	Reclaimed furniture, PVC coating	
🎧	• Use of ready-made components	326

Téo from 2 to 3. Snoozing Stool

Téo challenges the hyper-productivity of the modern era. Originating in 1998 but still produced today, Téo suggests that time out for doing nothing in particular – reading a book, taking a nap – might make us more productive when we become active again. 'Do not disturb' signals the unequivocal intention of the user and causes the viewer to think and react.

Crasset suggests that taking a nap on the Téo satisfies our need to step aside from the non-stop world. While setting a challenge for our daily work practices, Téo is also a useful multifunctional object in the home – it can also serve as an overnight guest bed or side table.

✏	Matali Crasset, France	305
⚙	Domeau & Perés, France	313
▧	Wood, double density and high resilience foam, coated cloth	
🎧	• Multifunctional • Encourages slowing down	327

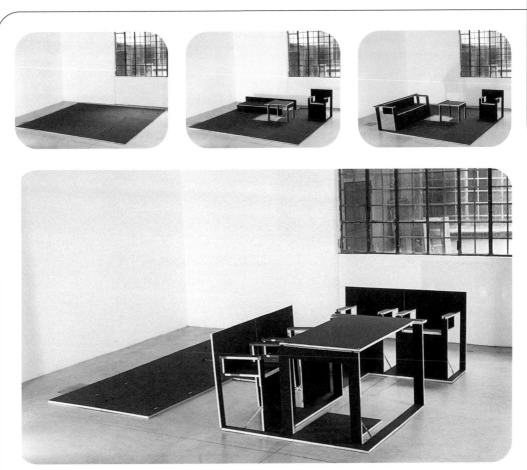

bidimensional furniture (mobilibidimensinali)

It is instinctive that we re-design the rooms we live in by re-positioning the furniture, changing the lighting and so on. Yet the dominant convention is that our furniture is a permanent, tangible object that occupies space and creates spatial dynamics by its presence. Cantono frees us from these conditions by providing a multifunctional, bidimensional floor that can be converted into tridimensional furniture as required. This design is a paradigm shift, offering the user a chance to interact and design his or her space, and offers an economical, rational and emotionally satisfying way of living in small spaces. It permits the space to fulfil many functions and creates fresh possibilities for dignified urban living without consuming vast resources per capita.

✏	Chiara Cantono, WELL-TECH, Italy	305
⚙	Chiara Cantono, WELL-TECH, Italy	305
🗍	Plywood, steel (hinges)	
♻	• Low-embodied-energy materials	324
	• Low-energy manufacturing	325
	• Multifunctional	327
	• User involvement	327

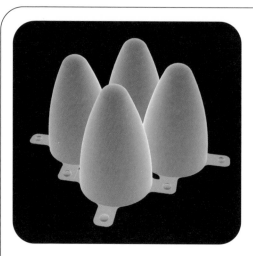

'Cones' Furniture System

Abstraction of form and function is a recurring theme of twentieth-century design. It has produced some stunning sculptural furniture – consider Oliver Mourgue's Djinn chairs, Verner Panton's modular Pantower and Salvador Dalí's Mae West sofa. The 'Cones' Furniture System is an evolution of the abstraction theme that hints at a bigger 'furniture landscape'. It consists of a series of modular geometric forms that make various organic arrangements of cones possible. The system creates a multifunctional environment for fun and relaxation, offering innumerable opportunities for enhancing our well-being.

✏	Patricia Gomes, João Cunna, Luis Temudo and Palmira Leinia, Portugal	306
⚙	Prototype	
📜	Various materials	
↻	• Multifunctional	327
	• Modular design	327
◕	IDRA award, 2000–2001	330

Hide-away

This intriguing design invites the user to take a book from the shelf, crank the handle to raise the day-bed and settle down for a good read.

✏	N2, Switzerland	308
⚙	Prototype, sdb industries, the Netherlands	320
📜	Aluminium, birch plywood, polypropylene	
↻	• Multifunctional	327

Ram

This is an experiment in perceptions, colour, spatial dynamics and form, creating a work of art within which you can sit. Its minimalist construction makes the most of the physical properties of the selected materials. Although it could benefit from an eco-redesign to reduce the environmental impacts of the chosen materials, the way it challenges socio-cultural language offers promise. It is a blank canvas to create food for the soul, a small opportunity to lift us beyond consumption of physical things to enjoy contemplative time. In the context of architecture Christopher Day called this 'spiritual functionalism', a description that could equally be applied to Ram.

✎	Felicerossi, Italy	314
⚙	Felicerossi, Italy	314
▣	Enamelled steel, Leather Lineltex fabric	
⌂	• Minimalism • Improved social well-being	327

Little Drummer

There are four sides to this intriguing percussion stool – slit drum, xylo-drum, conga and bongos – to be played by the sitter by hand or with drumsticks. This symbiotic arrangement of stool and percussion instrument is achieved by precision cutting or milling of the birch plywood percussive surfaces to 3.3 mm (about ⅛ in). The acoustic and musical elements of the stool provide a strong aesthetic personality to this wonderfully thought-out object.

✎	Sabine Mrasek and Clemens Stübner, Germany	308
⚙	Nils Holger Moormann GmbH, Germany	318
▣	Birch plywood	
⌂	• Renewable materials • Multifunctional	325 327
✦	iF Design shortlist 2003	330

Spiga

Mimicking an ear of corn, this lightweight coat rack, made of seven thin, wave-shaped, plywood cutouts attached to a metal-rod frame, is an ideal resting place for coats, hats, umbrellas, bags, newspapers and more throughout its entire length.

Hülsta furniture

Germany is undoubtedly one of the 'greenest' consumer markets in the European Union and Hülsta is a significant manufacturer of domestic and office furniture with a proven commitment to environmental performance. It was one of the first companies to reach the quality assurance standard, ISO 9001, and its entire production is certified under the Blue Angel eco-label scheme. In collaboration with Danzer, a leading veneer company, Hülsta initiated the 'veneer passport' guaranteeing that it does not originate from a tropical rainforest. Only four of their current ranges of furniture use solid wood,

again not sourced from rainforests. Particleboard or MDF is the primary material. In-house designers apply lifecycle analysis to extend the projected lifespan of products, of which most are already expected to last between thirty and forty years.

✏	Hülsta, Germany	316
⚙	Hülsta, Germany	316
📇	Veneers, solid wood, particleboard	
⏎	• Renewable materials	325
	• Blue Angel eco-label	328
	• Corporate environmental vision and policy	328

✏	Ubald Klug, France	307
⚙	Röthlisberger, Switzerland	320
📇	Plywood, metal	
⏎	• Reduction in materials used	325
	• Multifunctional	327

Kauna-phok mats & cushions

In the swamps of Manipur, India, there is a valuable local resource, the tough hard-wearing kauna-phok rush, which is harvested to make this range of cushions, pads and mats. Once worked, the material retains its inherent sponginess, making it ideal to provide a soft but firm surface. Each design is hand-made and is traded according to Fair Trade principles.

🖊	Indigenous designers, India	
⚙	Natural Collection, UK	318
📃	Kauna-phok plant materials	
↻	• Renewable and sustainably harvested materials • Fair Trade products	325

Geometric Structure Cube, Play Cube

Defying categorization these cubes are a 3D incarnation of digital craftsmanship where 'truth to materials' is realized first on a PC, then in a computerized laser-cutting machine and finally as a tangible artefact. These beautiful structures are squashy cubes that bounce back to their original shape. Decorative sculpture, seat, cushion or plaything, the cubes invite exploration. Geometric Structure Cube is 45 cm (17¾ in) on each dimension and made of EVA rubber foam. Play

Cube is larger at 60 cm (23²⁄₃ in) and is made from Plastazote, a polymer that is 100% recyclable, contains no CFCs or HCFCs and is manufactured by a safe, low-environmental-impact, nitrogen expansion technique.

🖊	Lauren Moriarty, UK	308
⚙	Batch production	
📃	EVA rubber or Plastazote foam	
↻	• Reduction in materials used	325
	• Multifunctional	327

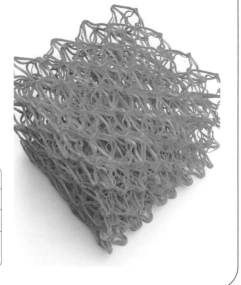

Diana

Like an origami forest of Russian Constructivist symbols, Grcic's inventive sheet steel side-tables encourage users to explore and re-interpret their functionality; a bookshelf transforms into a coffee table and then into a desk for laptop users. All steel is finished in a choice of solid powder-coated colours.

🖊	Konstantin Grcic, Germany	306
⚙	ClassiCon, Germany	313
📃	Steel, powder-coated finishes	
↻	• Single material	325
	• Multifunctional	327

do Create

Droog Design are renowned for their wry humour ('droog' translates as 'dry'), which is apparent in the project 'do Create' in 2000. Kesselskramer, a Dutch publicity company, set up a fictional experimental brand called 'do'. Ten designers were invited to participate in the Droog Design project, the outcomes of which were presented at the Milan Salone. The 'do' brand is incomplete without a ceiling light rose. This new lighting appliance parades in surreal splendour. The designer is facilitator to the consumer's imagination, resulting in whimsical designs that affirm the personality and imagination of the consumer and are a world away from the predictability of mass-produced design. This concept of designer and consumer as co-designers provides fertile ground for cementing a long-lasting relationship with the brand, an emotional mortar to create a treasured object.

✏	Marijn van der Poll and Marti Guixé, Droog Design, the Netherlands	313
⚙	One-offs	
📜	Various materials	
🎧	• User-centred design • Emotionally durable design	327

continuation of the design process by the consumer, involving a personalization or customization of the original artefact. Marijn van der Poll's 'do hit' armchair is a cube of sheet steel before being transformed into its final form. More modest physical activity is required for Marti Guixé's 'do scratch', a light box that comes to life as the owner scratches a message or drawing into the black paint. Guixé's 'do reincarnate' is a loose recipe for reviving lifeless, familiar objects by magically suspending them by an 'invisible' thread and by an extension of the wiring from

Carta

Shigeru Ban creates structures with cardboard and succeeds in elevating this humble material to a new aesthetic level. His use of cardboard tubes in projects as diverse as furniture, temporary housing for refugees and buildings for communities reveals superb understanding of the capabilities of the material. Carta is a range of furniture that makes minimal use of low-embodied-energy materials.

✏	Shigeru Ban, Japan	304
✿	Cappellini SpA, Italy	312
📜	Cardboard	
🎧	• Renewable materials	325
	• Reduction in materials used	325
	• Recyclable	324

Herz

Steel reinforcing rods, similar to those used in construction with concrete, are welded and bolted into a simple frame to which a moulded leather breastplate is attached, providing a functional, minimalist coat stand. The materials used are easily recycled and the design is both aesthetically pleasing and durable.

✏	Anthologie Quartett, Germany	304
✿	Robert A. Wettstein, Germany	310
📜	Leather, steel	
🎧	• Recyclable and compostable materials	324
	• Reduction in materials used	325
	• Low-energy manufacturing	325

Screen

A slender metal frame supports a web of interwoven plastic string. This lightweight screen can be fabricated from virgin or recycled materials, doesn't require specialist machinery to construct and is easily dismantled at the end of its life, when the materials can be salvaged. Fabrication can easily be adapted to suit locally available materials.

✏	Fernando and Humberto Campana, Brazil	305
⚙	One-off limited batch production	
🎞	Metal, plastic string	
⏎	• Reduction in materials used	325

Greensleep

Many manufacturers of mattresses claim to offer the most comfortable night's sleep but few can say that they have examined every aspect of the environmental impact of their products. The company uses only non-toxic, natural and organic materials. For the core of a Greensleep mattress, Furnature™ use a Hevea natural rubber mattress from natural rubber extracted from Malaysian plantations. This core is naturally resilient and possesses anti-bacterial and hypoallergenic properties. Outer coverings provide excellent warmth and breathability by including an organic cotton cover and quilted layers of pure wool. An untreated seasoned Canadian spruce frame from sustainably harvested forests supports rubber wood dowels on recycled plastic inserts.

✏	Furnature™, USA	315
⚙	Furnature™, USA	315
🎞	Spruce, rubber wood, recycled plastics, organic and natural textiles	
⏎	• Recycled and renewable materials	324/ 325
	• Safe, non-toxic manufacturing	327
	• Improved health	327

black 90, black 99

Black bamboo is suspended in an American black walnut frame to neatly counterbalance the delicacy of the bamboo with the solidity of the frame. In another screen, interlocking bamboo forms foldable concertina-like sections. The modernist mantra of 'truth to materials' is eloquently demonstrated here. These natural materials have an extensive history and unique colour signatures.

✏	Paola Navone, Italy	308
⚙	Gervasoni SpA, Italy	315
🗋	Bamboo, black walnut wood	
♻	• Renewable materials	325

Hut Ab

Aluminium fixings allow simple machined pieces of ash wood to articulate around a pivot to provide a multifunctional clothes and hat stand, drying rack or structure for suspending house plants. Low-energy requirements during production make this an efficient, low cost, design.

✏	Konstantin Grcic, Germany	306
⚙	Nils Holger Moormann GmbH, Germany	318
🗋	Ash wood, aluminium	
♻	• Recyclable and renewable materials	324/325
	• Low-energy manufacturing	325
	• Multifunctional	327

Lattenbet

Anyone who has moved house knows that the most cumbersome item is the double bed. Not so for this superb example, an entire double bed that can be neatly carried in its own suitcase. Bucking the trend for self-assembly furniture to be flat, stylistically drab and infuriatingly difficult to assemble, Steinmann and Schmid have devised a construction system that is not only rapid to assemble but also visually appealing.

✏	Peter Steinmann and Herbert Schmid, Switzerland	309
⚙	Atelier Alinea, Switzerland	311
🗋	Beech wood, plywood, steel, rubber	
♻	• Portable, self-assembly furniture	326
	• Reduction in materials used	325

Grito

This spun aluminium shade cleverly takes the hassle out of fitting one to a pendant light by simply slotting it over the lamp fitment. It is easy to attach, remove and clean.

✏️	El Ultimo Grito, UK	305
⚙️	Mathmos, UK	318
📩	Aluminium	
♻️	• Recyclable material	324
	• Durable	327

PO/0128

Moerel demonstrates the flexibility and beauty of her raw material in this series of designs, which are inspired by the same basic module, a ceramic sphere. There are several variants for unusual pendant lights, as well as individual candleholders. Economy of scale is possible if large-scale production can utilize different modular spheres to make a range of products.

✏️	Marre Moerel, the Netherlands/USA	308
⚙️	Cappellini SpA, Italy	312
📩	Clay	
♻️	• Abundant geosphere materials	324
	• Durable	327

Horeta

Polycarbonate is a durable, versatile polymer that can be easily recycled and is available in translucent or opaque forms. It is ideal for lightweight flat-pack, self-assembly light shades, as demonstrated here by Setsu Ito. This multilayered pendant shade lets through varying amounts of light depending on the angle of view.

✏️	Setsu Ito, Italy	307
⚙️	One-off	
📩	Polycarbonate	
♻️	• Single material	325
	• Recyclable	324
	• Self-assembly	326
	• Reduction in energy of transport	326

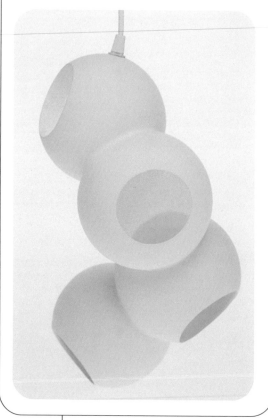

Quentin

The designers were inspired by the complex forms of folded cardboard packaging and utilitarian products such as egg boxes. Working in collaboration with a local Glaswegian manufacturer, they created a product utilizing pulp from recycled newsprint and paper-mill waste. An individual shade comprises two identical but mirror-image halves, which are formed in a mould where the pulp is vacuum-drawn. These innovative lampshades are semi-opaque, giving a unique light output.

✐	Ian Cardnuff and Hamid van Koten, VK & C Partnership, UK	310
✿	Universal Pulp Packaging, UK	322
▤	Newsprint, paper waste	
⏻	• Recycled and recyclable materials	324
	• Low-energy manufacturing	325

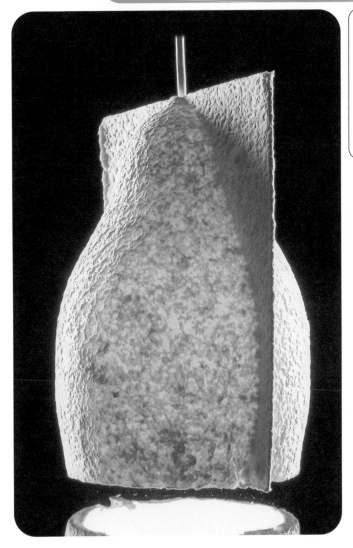

The shades can be deployed individually or in groups to form customized arrangements. At the end of its useful life the lampshade can be repulped ready for its next reincarnation. Potentially this product can sit within a closed recycling loop, ensuring maximum reuse of paper fibre and minimizing the energy required to remanufacture the product.

Agave

Initial public reaction to compact fluorescent light bulbs was as muted as the luminosity of the bulbs themselves. Happily compact fluorescent technology has substantially improved, leading to their adoption by most major light manufacturing companies to enhance their product range. Luceplan has experimented with transparency, reflection, refraction and diffusion in its latest range of light shades suitable for pendant, wall and floor lighting. Transparent methacrylate is injection-moulded into a radial structure comprising a series of arc-shaped ribs that simultaneously diffuse and 'conduct' the light. This produces a complex interplay of light quality and colour that can be further enhanced and personalized by fitting graded coloured filters (yellow, red and blue) to the shade. Spherical, elliptical and double-bodied shades add further variations to the range. Methacrylate is renowned for its distinctive optical clarity and impact resistance, which is why it is used for illuminated signs and motor vehicle lights. Unfortunately it requires acetone and cyanohydrin in its manufacture, so the best way to eco-redesign this light would be to examine how to reduce environmental impacts while retaining the brilliance of the luminosity from the compact fluorescent bulbs. Reduction in weight and alternative materials are feasible.

	Diego Rossi and Raffaele Tedesco, Italy	309
⚙	Luceplan, Italy	317
🗎	Methacrylate, compact fluorescent bulbs	
🎧	• Reduced energy consumption	327
	• Customization	327

Dawn

Reusing redundant materials or objects is a clear statement about closing the loop of resources, but it is a challenge to create new meaning and character using such materials. Dawn is one of those successful transformations where the donor material, in this case old aluminium and plastic venetian blinds, creates a new vision yet cleverly borrows from the light-filtering properties of the discarded product. The light shade is available as a flat pack, and in assembling it the consumer is, perhaps, sensitized to the design process and manufacturing.

✏	*Tiffany Threadgold*	310
⚙	*Tiffany Tomato Designs, USA, one-off and small batch production*	310
📑	*Reused mini-venetian blinds, industrial scrap*	
♻	• *Reused materials*	326

Lamp shade

This reversible shade provides a choice of two strong lighting directions depending upon whether the reflector is uppermost (for down-lighting) or on the underside (for up-lighting). This eloquent design embodies principles of minimalism and dual-functionality, both of which are very relevant to designs with reduced environmental loads.

✏	*Sebastian Bergne, UK*	304
⚙	*Radius GmbH, Germany*	320
📑	*Steel*	
♻	• *Dual-function design* • *Reduction in materials used*	327 325

Miss Ceiling light

Efficient use of a single
natural material creates a
lampshade with sculptural
characteristics, permitting
shafts of light and a warm
glow to penetrate the semi-
opaque natural plywood
and creating a dramatic
light source.

✏	Jasper Startup, Startup Design, UK	309
⚙	Startup Design, UK	309
🗋	Plywood	
♺	• Renewable single material	325
	• Reduction in materials used	325
	• Low-energy manufacturing	325

Milk-bottle light

Since the early 1990s
designers have responded
to the challenge of
considering their ethical
responsibilities to the
environment. In the
Netherlands Tejo Remy
explored the issue using
discarded plastic milk
bottles and in the UK Jane
Atfield did the same with
her RCP2 chairs using
recycled plastic sheeting.
As a consequence the
message – that modern
design must use recycled
materials – is eloquently
delivered.

✏	Tejo Remy, Droog Design, the Netherlands	313
⚙	Droog Design/DMD, the Netherlands	313
🗋	Discarded bottles	
♺	• Reuse of waste ready-mades	326

Northern Fleet chandelier

Shards of broken glass are painstakingly assembled into a cascade of light in this unique chandelier. The jagged edges of the glass make an exciting contrast with the sheer beauty of the final form and the design gently mocks at the cut-glass chandeliers of grand houses and public buildings. Quality, one-off designs may enjoy long lives, since they may attract greater custodial care than run-of-the-mill, mass-produced objects.

✎	Deborah Thomas, UK	310
⚙	One-off	
▤	Glass, wire	
↻	• Recycled materials	324

FluidSphere

This pendant light shade made from criss-crossed maple veneer strips explores an organic geometry. Beams of light escape from a cell-like structure that floats in space. Wood, that most malleable of materials, is perfectly adapted to creating new aesthetic pleasures with light.

✎	Leo Scarff, Ireland	309
⚙	One-off and small batch production	
▤	Maple veneer, steel suspension fittings	
↻	• Reduction in materials used	325
	• Renewable materials	325

Happy Blackout

Minimally packaged in a slim envelope, the wall plate and tray for the Happy Blackout is made of laser-cut brushed stainless steel attached to a greeting card. Attach the steel to the wall and insert a burnt-out light bulb in front of a standard metal-cased candle. Stand back to watch the light bulb float in mid air. A lyrical and effective solution to getting more light out of your dead light bulbs.

✏	Stiletto DESIGN VERTReiB, Germany	321
⚙	Stiletto DESIGN VERTReiB, Germany	321
🗒	Stainless steel, old light bulb, candle	
♺	• Low-energy light source	327
	• Lightweight, mono-material	324/ 325
	• Minimal packaging	326

Glühwürmchen® (Glow-Worm)

A prototype of the Glühwürmchen® plug-in lamp was designed in 1990 and it has been subject to several re-designs since then, the latest in 2002. The lamp is made almost exclusively from ready-made industrial parts except for a small aluminium adapter piece. This strategy enables small or large batches of the different models to be made efficiently and economically. This versatile lamp continues the German tradition of fitness for function and provides a long service life.

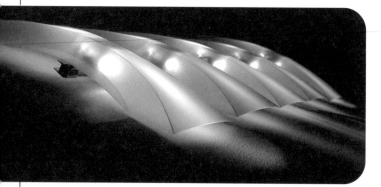

5s Light

Recycled polypropylene and aluminium are the raw materials for this sculptural, translucent and seductive wall or ceiling lighting. '5s' refers to soft, sleek, subtle, sophisticated and sustainable. Low-voltage, high-luminosity, LEDs produce negligible heat, so can be used in close proximity to the PP shades. Standard sheets of PP are cut to permit a flat pack design, which requires no adhesives and generates zero waste. A life expectancy of twenty years for the LEDs guarantees high eco-efficiency.

✏	Guy Blashki, Australia	304
⚙	Prototype	
🗒	Polypropylene (PP), LEDs, electric circuit	
♺	• Energy efficiency	327
	• Zero waste production	325
	• Lightweight, flat-pack	326
✪	IDRA award, 2000–2001	330

✏	Stiletto DESIGN VERTReiB, Germany	321
⚙	Stiletto DESIGN VERTReiB, Germany	321
🗒	Various metals, old and new Edison light bulb	
♺	• Dual-function design	327

ComeBack series

Plastic packaging waste is reincarnated as a beautiful series of shades for table, standard and pendant lamps. The diversity of colour of the original waste source is reflected in the random, mosaic-like arrangement in the manufactured sheeting that is the base material for the shades.

✏	*Bär + Knell, Germany*	304
⚙	*Bopp Leuchten GmbH, Germany*	312
📷	*HDPE waste*	
🎧	*• Recycled material*	324

GS3 Light and Noodle Block Light

Low-energy light bulbs are encased in a protective plastic casing surrounded by a beautiful geometrical network of laser-cut EVA rubber foam. Is this twenty-first-century digital craft, where nimble fingers on the computer using 3D modelling, CAD and laser-cutting software replace the manual dexterity of craftsmen? This is an exciting area waiting for more experimentation at the digital–craft interface where the virtual can inform the real and vice versa. If used wisely, laser-cutting can result in more efficient use of raw materials and embraces principles of 'lightness' as advocated by Adrian Beukers and Ed van Hinte.

✏	*Lauren Moriarty, UK*	308
⚙	*Batch production*	
📷	*EVA rubber foam, plastic, low energy light bulbs*	
🎧	*• Reduction in materials used*	325

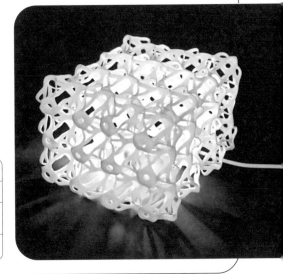

Flamp

This wooden-based table lamp is dipped in phosphorescent coating so that it absorbs the energy from sunlight and re-radiates it for up to twenty minutes. An ideal 'emergency' light after sunset.

✏	Martí Guixé, Spain	306
⚙	Small batch production	
▤	Phosphorescent paint, wood	
↻	• Solar-powered non-electric light	327

Clips

A simple stainless steel frame clips over a discarded drinks can and supports a polypropylene shade. As your favourite brand of drink changes you can dispose of the old one (at a can bank) and insert a can that held the flavour of the month.

✏	Bernard Vuarnesson, France	310
⚙	Sculptures-Jeux, France	310
▤	Polypropylene, stainless steel	
↻	• Encourages reuse of ready-mades	326
	• Reduction in materials used	325

Fake Lamp

Guaranteed to add a surprise dimension to any room, the Fake Lamp is a large 171 cm- (67½ in-) wide sheet of white MDF that challenges our perceptions of domestic lighting.

✏	Sophie Krier, the Netherland	307
⚙	Moooi, the Nethelands	318
▤	MDF	
↻	• Zero energy use	327
	• Improved well-being	327

floor lamp

An update on the spun-fibre tripod lamps of the 1950s, the low-energy fluorescent strip light is surrounded with an opaque acrylglas shade. Slender wooden supports evoke a contemporary feel to the tripod. Detailing and materials used are kept to a minimum, mirroring trends fifty years ago as Europe recovered after the War. If substantial reductions of resource flow are to become a reality then all designers and manufacturers must make efficient use of materials. Some have already started.

✏️	Robert A. Wettstein, Germany	310
⚙️	Robert A. Wettstein, Germany	310
📜	Acrylglas, wood fluorescent light fitting	
♻️	• Reduction in materials used	325
	• Low-energy light source	327

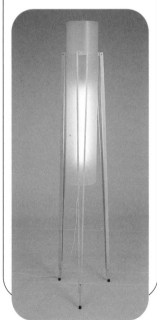

Aquarius

Matt stainless steel cylinders interlock to form this versatile table and floor vase/lamp. The tall cylinder is 122 cm (48 in) high while the smaller ones are 55 cm (21½ in). An internal tempered thick glass vessel accommodates the flowers. Fitments are available for 230V and 110V but all variations use low-wattage fluorescent bulbs (11W).

✏️	Filippo Dell'Orto, Italy	305
⚙️	Pallucco Italia, Italy	319
📜	Stainless steel, glass fluorescent bulbs	
♻️	• Multifunctional	327
	• Durable	327

Parallel Universe Series

Translucent fluted polycarbonate provides the channelling to hold stripes of recycled coloured glass in designs personalized to customers' requirements. This detailing can be applied to free-standing light screens, table lamps or pendant fixtures. Aluminium and wood-pulp corrugated board is used for the ends and closures, enabling a future reordering of the coloured stripes at a later date.

✏️	David Bergman, Fire & Water, USA	304
⚙️	Fire & Water, USA	304
📜	Polycarbonate, recycled glass, aluminium, corrugated board	
♻️	• Recycled and recyclable materials	324

bolla S, M, L, XL

Mimicking seed pods hanging from a tropical tree, these biomorphic rattan floor lamps fulfil many functions and needs. The designs embed what Christopher Day, the British architect and writer, called 'spiritual functionalism'. He was referring to the ability of buildings to lift the human spirit by their form, spatial dynamics and sensory palette but the term can also be applied to products. Sodeau's lyrical use of woven rattan makes this much more than just a light. It evokes many senses, from an appreciation of the aesthetic form to the tactility of the material and, in doing so, it generates emotional and mental reactions that aim to improve our well-being.

✏	Michael Sodeau, UK	309
⚙	Gervasoni SpA, Italy	315
🗌	Rattan, MDF	
🎧	• Renewable materials	325
	• Retention of local craft skills	324
	• Improved sense of well-being	327

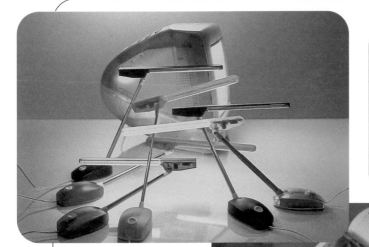

✏	*Artemide SpA, Italy*	311
⚙	*Artemide SpA, Italy*	311
🗐	*Synthetic polymers, low-voltage lamp*	
🎧	• *Reduction in consumables used*	325
	• *Energy conservation*	327
	• *Easily repaired and disassembled*	327
🔍	*Design Sense awards, Shortlist, 1999*	

e-light

The e-light integrates a number of technological improvements over conventional desk lamps. The lifetime of the lighting filament is twenty times greater than that of an incandescent bulb and two to three times that of a fluorescent bulb, and it uses one-fifteenth as much mercury as the latter. Creating a light spectrum similar to daylight, it is five times brighter than a tungsten bulb. As the e-light produces negligible thermal emissions, the need for heat-resistant materials is significantly reduced. Components can easily be separated, facilitating recycling and reuse. Reversible joints and compact design provide flexible lighting configurations and a small footprint.

Mini Desk Lamp

Good ideas are often recycled, but the Mini Desk Lamp goes one step further by reusing part of an iconic design of the late 1950s, British Leyland's Austin Mini car, designed by Alec Issigonis. Original Austin Mini parts for the distinctive sidelights are rehoused in a plastic body, fitted with an automatic on/off tilt switch and painted in the original body paintwork colours. The rebirth of a mini Mini classic?

✏	Paul Topen, Designed to a 't', UK	310
⚙	Designed to a 't', UK	313
🗋	Steel, plastic, electrical wiring	
☊	• Use of ready-made components	326

LD1

Bright white light at low voltages and without the generation of heat was the dream of lighting technicians for many years. The arrival of light emitting diodes has opened up many possibilities for experimentation and design. This one from Ligne-Roset uses 117 LEDs against a reflector mounted on a lacquered aluminium stand. LD1 brings a new geometry to light diffusion.

✏	Ligne-Roset, France	317
⚙	Ligne-Roset, France	317
🗋	LEDs, electronic circuitry, aluminium	
☊	Low-energy light	327

Tube

The familiar fluorescent light gets the minimalist treatment from Christian Deuber. A slender synthetic tube protects the light with steel- and rubber-footed closures, allowing the light to be placed wherever it is required. The use of fluorescent bulbs, in this case 58W, which are much more efficient users of energy than incandescent sources, adds to the versatility of this product.

✏	Christian Deuber, N2, Switzerland	305/308
⚙	Pallucco Italia, Italy	319
🗋	Fluorescent light, steel, rubber	
☊	• Multifunctional light • Reduction in materials used • Low energy consumption	327 325 327

Golden Delicious

Illuminance from one 60W spherical opaque bulb is diffused by twenty-one bulbs placed in the polycarbonate bowl. This whimsical idea is beautiful in its simplicity and the shade acts as a repository for spare or expired bulbs.

✏	Ralph Ball, UK; Ligne-Roset, France	317
⚙	Ligne-Roset, France	317
🗋	LEDs, electronic circuits, aluminium	
☊	• Multifunctional	327

PO98 10/10C, 11/11C, 12/12C

In a clever extrapolation of scale, the table lamp becomes a floor or standard lamp. These lightweight constructions combine visual stimulation and humour in an economical design.

✏	Marcel Wanders, the Netherlands	310
✿	Cappellini SpA, Italy	312
🗒	Wire, polymer	
🎧	• Reduction in materials used	325

Pharos floor lamp

The designer has succeeded in transforming a garden cane with a cylindrical papyrus shade into an elegant, minimalist standard lamp. Natural variability within the papyrus paper creates a range of unique textures and light patterns, mimicking the spun-fibre shades of the 1950s.

✏	Jasper Startup, Startup Design, UK	309
✿	Startup Design, UK	309
🗒	Bamboo, papyrus paper	
🎧	• Renewable materials • Low-energy manufacturing	325 325

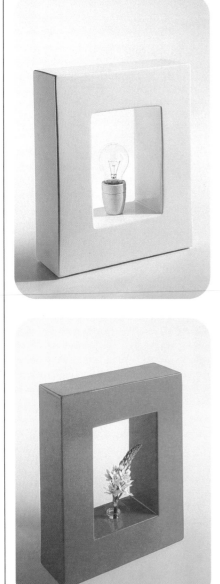

PO/9902C-D

Framing the bulb not only focuses the viewer on the light source but also provides a protective package during distribution and retailing.

✏	Jeffrey Bernett, UK	304
⚙	Cappellini SpA, Italy	312
📄	Cardboard, lampholder	
↻	• Reduction in materials used	325

PO/0001, 0006

Special clays are fashioned into durable lighting units. Ceramics are traditionally used for bases for table lamps but here the material forms the base and the shade.

✏	Marre Moerel, the Netherlands	308
⚙	Cappellini SpA, Italy	312
📄	Earthenware, porcelain	
↻	• Abundant geosphere materials	324

Post It Lamp

A strong cardboard tube with plastic end caps arrives through the post, the contents are extracted and within minutes it is assembled into a compact but functional table light. While not particularly robust, the Post It Lamp minimizes transport emissions and costs, as the packaging is used to form the lamp base. This is a neat idea that can be extended to other design applications.

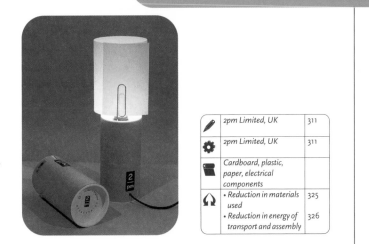

✏	2pm Limited, UK	311
⚙	2pm Limited, UK	311
📜	Cardboard, plastic, paper, electrical components	
♻	• Reduction in materials used	325
	• Reduction in energy of transport and assembly	326

Merlino

Magically Merlino transforms from a mirror into a decorative image and then a wall light. This is achieved by a series of luminous sections and screens that are either painted or pierced. The mirrored surface gives way to a printed image as the viewer pulls the first string and then new effects emerge as the back-lit light is switched on by the second pull string. Items that normally vie for attention suddenly merge into a multifunctional object and, in doing so, expand the range of expression of each individual function.

✏	Jacopo De Carlo, Andrea Gualla with Raffaella Godi	305
⚙	DeCarloGualla Studio, Italy, prototype	305
📜	Glass, metal, paint, light fittings	
♻	• Multifunctional	327

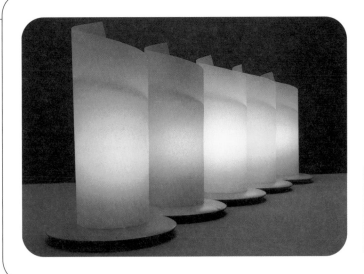

Sailbuoy Canvas

Four coloured filters can be fitted to the polypropylene-paper laminate shade to alter the lighting mood. A wooden base is fitted with a lampholder for a 10-watt compact fluorescent lamp (CFL), ensuring low energy consumption and heat output.

✏	Neil Wilson, UK	310
⚙	Lampholder 2000 plc, UK	317
📜	Polypropylene, paper, wood	
♻	• Multifunctional task or mood lighting	327

The Eye of the Peacock

Plastic bottles are shredded and reconstituted to form a fascinating melange of colour and texture, the original bottle tops and sealing rings further enhancing the texture and variety of this wall panel. Illuminated from behind with fluorescent lighting, this illustrates the capacity of new materials to create a visual stimulus.

✏	Bär + Knell, Germany	304
⚙	One-off	
📃	HDPE and LDPE bottles	
♺	• Recycled materials	324

Light-pot

These polypropylene table or floor lamps, 18 cm or 35 cm (7 or 13⅘ in) in diameter, make the most of this versatile and recyclable material as the illuminance grows out of its containment. Terracotta pots, with their ready-made drainage holes, are an obvious source of light shades for the DIY enthusiast if Light-pot fails to enter mass production.

✏	Juan Benavente Juanico, Spain	307
⚙	Prototype	
📃	Polypropylene	
♺	• Single material	325
	• Recyclable	324

wall light

Redundant stainless steel washing-machine drums are readily transformed into lighting shades. While large-scale production is beset with difficulties, not least obtaining a steady stream of used drums, this design approach can be extended to a wide variety of well-made mass-manufactured objects. Individual consumers should enjoy experimenting with their own consumer junk.

✏	Aki Kotkas, Finland	307
⚙	Small batch production	
🗎	Reused washing-machine drums, metal, bulb and wire flex	
↻	• Local reused objects	326

Light columns

Plastic packaging waste offers a wonderful palette of colours and graphical shapes when recycled and reconstituted into thin, semi-opaque sheeting. Suitable for one-off, small-batch and high-volume production, these long cylinders of plastic recyclate illuminated with fluorescent lamps create an eclectic range of decorative lights.

✏	Bär + Knell, Germany	304
⚙	One-off	
🗎	Discarded plastic packaging	
↻	• Recycled materials	324

REVOLVER

A galvanized wire socket connects virgin or burnt-out light bulbs in the circular holder but only one bulb is alive. This ingenious arrangement creates a dual-function design: it is a repository for light bulbs, and these bulbs create a unique shade (for the pendant version) or light diffractor for ceiling or wall-mounted units. Your spare Edison bulbs need no longer languish in a cupboard but can contribute to mood lighting and decoration.

✏	Stiletto DESIGN VERTReiB, Germany	321
⚙	Stiletto DESIGN VERTReiB, Germany	321
🗄	Various metals, old and new Edison light bulb	
🎧	• Dual-function design	327

power glass®

A sandwich of conductive material, which is completely transparent, is embedded between layers of ordinary glass. Single or multilaminate conductive glass affords different power-carrying capacities, so this patented technology can be used in a range of applications for lighting, switches, electronic displays and so on, especially for low-voltage applications.

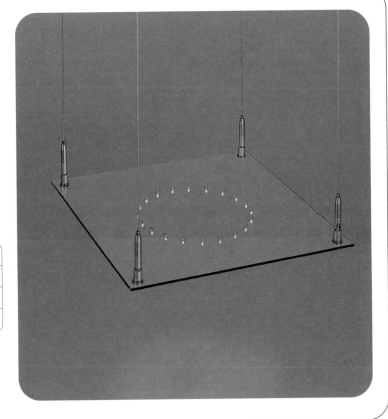

✏	Glas Platz, Germany	315
⚙	Glas Platz, Germany	315
▤	Transparent conductive material, glass	
↻	• Multifunctional	327

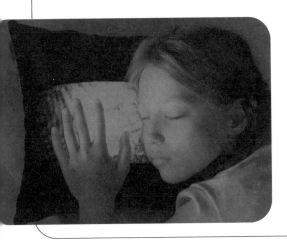

Pillow Light

Switch off your room lights! There are times when a soft glow in the dark is a comfort. This black hemp fabric cushion contains internal illumination creating an interference-light effect by using red LEDs operated by a 6V battery on a twenty-minute timer switch. The low-voltage LEDs ensure long battery life but maybe a future model could include a recharge or wind-up battery facility.

✏	Stiletto DESIGN VERTReiB, Germany	321
⚙	Stiletto DESIGN VERTReiB, Germany	321
▤	Hemp fabric, LEDs, battery and switch	
↻	• Low-energy light source	327
	• Renewable fabric	325

SugaCube

Individual LEDs are encased in click-together two-tone acrylic blocks, enabling consumers to create complex patterns of blocks for bespoke lighting. This playful concept combines the technological flexibility of LEDs with the fun of children's Lego™ building bricks. Users continue the design process by customizing the final lighting product.

✏	Studio Jacob de Baan and Frank de Ruwe, the Netherlands	309
⚙	Conceptual prototype	
📷	Polymer, single LED	
🎧	• Low-energy lighting	327
	• Modular	327
	• Multifunctional	327

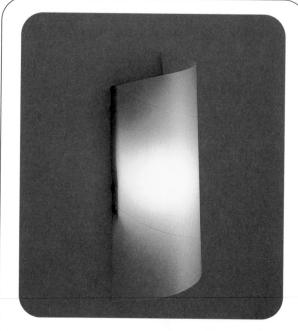

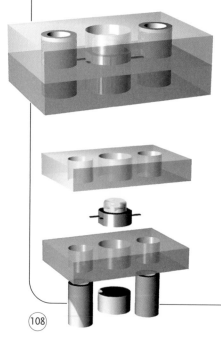

Wall bracket

Stripped down to its bare essentials, this wall bracket is an economical design. The electronic ballast and compact fluorescent lamp (CFL) holder sit on a simple pressed-metal bracket to which a curved sheet of polypropylene is attached.

✏	Lampholder 2000 plc, UK	317
⚙	Lampholder 2000 plc, UK	317
📷	Metal, polypropylene	
🎧	• Reduction in materials used	325
	• Low-energy lighting	327

Pod Lens

Most lighting is static, irredeemably rooted to the a building's electric cabling. Pod Lens is a modular system of a polycarbonate pod unit with bulb and flex and a series of bases for standard or floor lighting. For indoor or outdoor use, the pods provide flexible and decorative lighting at the whim of the user.

✏	Ross Lovegrove, UK	308
⚙	Luceplan, Italy	317
📜	Polycarbonates, electrical components	
⟲	• Multifunctional lighting system	327
	• Upgradable and repairable	327

Moonlight MFL

A robust, weatherproof, semi-translucent, polyethylene material is moulded in four sizes and fitted with different sockets to enable the low-wattage lamps (5-23 watts) to be fixed into the earth or used on hard surfaces. Feel a mood swing coming on? Simply change the coloured bulb filter, choosing from up to 250 colours. Moonlight MFL is a versatile, low-energy, 'mood and colour', indoor/outdoor lighting system.

✏	Moonlight Aussenleuchten, Germany	318
⚙	Moonlight Aussenleuchten, Germany	318
📜	Polyethylene	
⟲	• Multifunctional	327
	• Low energy consumption	327

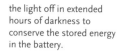

Tsola

Most outdoor, solar-powered lights are above-ground installations, which makes them vulnerable to the elements, accidental damage and vandalism. Tsola is designed to be installed flush with the ground and can be walked or driven upon without damage. This low-maintenance light is equipped with a timer that automatically switches the light off in extended hours of darkness to conserve the stored energy in the battery.

✏	Sutton Vane Associates, UK	309
⚙	Light Projects Ltd, UK	317
📕	Photovoltaics, heavy-duty glass, stainless steel, battery	
♫	• Solar power	327

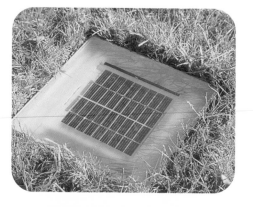

Solar Bud

A photovoltaic panel generates energy from sunlight, stores it in a battery and releases it to three low-voltage, red LEDs, all in a self-contained unit that is placed in the desired position by pushing it into the soil or other suitable medium. Ideal for garden decorative or safety lighting, the Solar Bud would also be at home in the window box of an urban bedsitter.

✏	Ross Lovegrove, UK	308
⚙	Luceplan, Italy	317
📕	Metals, photovoltaics and light emitting diodes (LEDs)	
♫	• Solar-powered lighting	327
	• Very low-energy LED bulbs	327

Sherpa/Sentinel

Just thirty seconds' winding on the handle linked to the AC alternator gives thirty minutes of light from the 3.3V 1565mA high-efficiency Xenon dual-filament bulb on normal beam setting. An LED charge level indicator tells you the optimal winding speed. Alternatively plug it into the mains supply using the AC/DC adaptor. A fully charged NiMH 1000mAh battery gives five hours of light on full beam. Ideal for everyday or emergency use, the Sherpa is available in the USA under the Coleman Sentinel brand.

✎	Freeplay Energy Ltd, South Africa & Europe	315
✿	Freeplay Energy Ltd, South Africa & Europe	315
▧	Various materials	
♻	• Renewable energy options	327

StarLed® Light

Candles remain a potent sign of human faith. Here, perhaps, if we aspire to a more sustainable future, is the twenty-first-century technological equivalent. StarLed® is a portable lamp using a single bright LED given jewel-like prominence by its transparent or metalized methacrylate body and prismatic head. A dedicated electronic circuit and three rechargeable AA nickel metal hydride (NiMH) batteries sit in the base plate of the candle. A single charge generates four hours of light and it is recharged by placing the candle on a mains recharger that fits to the base. This design does much to focus on issues of lighting energy, and creates a special mood lighting, but only a full lifecycle analysis would reveal how many candles you could burn to equate to the embodied energy of manufacturing, distributing and retailing each StarLed®. However, over time, as the embodied energy is negated by recharging the product, it will become a 'zero energy' device, providing that the mains electricity emanates from renewable sources. Future evolution of this product may consider an alternative to the methacrylate. This is a high-embodied-energy polymer with excellent optical qualities but the manufacturing process is chemically complex and involves hydrocyanins. Glass, glass composites and other translucent or transparent materials may provide acceptable substitutes.

✎	Alberto Meda and Paolo Rizzatto, Italy	308/309
✿	Luceplan, Italy	317
▧	Various technosphere materials	
♻	• Reduced energy consumption	327
	• Improved well-being	327

Bubble Light

Crossing over between portable light and anti-stress object, the Bubble Light is a squeezable silicone globe containing a switch, LEDs and an integral rechargeable battery. Squeeze for 'on' and the globe emits a blue, green or orange glow. Is this the candle to light you to bed, a personal mood light or a 'must have' gizmo? Potentially the technological development embedded in this low-voltage lamp is ideal for off-grid domestic renewable energy systems, although at present demand is more likely to be from urban loft dwellers.

✏	Aaron Rincover, Mathmos Design Team, Mathmos, UK	318
⚙	Mathmos, UK	318
▤	Silicone, LEDs, rechargeable battery	
⤵	• Rechargeable light source	327

Aladdin Power

The hand-wound electricity generator is not a new idea. The Russian army has supplied its conscripts with a robust, hand-powered torch since the 1940s, and plastic-bodied, hand-cranked torches have been available since the 1970s. But Nisso Engineering's design uses lighter, modern materials to improve the efficiency of the design and possibly make this an attractive option for powering other hand-held electronics such as mobile phones. However, these hand-cranked torches cannot store energy in a battery or in a wind-up mechanism as featured in Freeplay Energy's products (p. 111).

✏	Nisso Engineering, Japan	318
⚙	Nisso Engineering, Japan	318
▤	Polymers, electronic components, dynamo	
⤵	• Renewable energy source	327
	• Reduces consumables (battery)	327

JAR, JAM, JEL

Various ready-made polyethylene containers are transformed into a portable light (JAR), a light-cum-pouffe (JAM) and an illuminated table (JEL). Existing technology and production capacity are diverted to provide a range of new objects offering flexibility to their new owners.

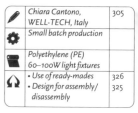

✏	Chiara Cantono, WELL-TECH, Italy	305
⚙	Small batch production	
▤	Polyethylene (PE) 60–100W light fixtures	
⤵	• Use of ready-mades	326
	• Design for assembly/ disassembly	325

Jigsaw

Prior to the arrival of electricity and the tungsten bulb, oil-lamps were a primary source of light. Today they are rarely used in Western Europe despite widespread availability of suitable fuel. The Jigsaw oil-burner is a beautiful glass receptacle with cotton wick. It is functional, aesthetically pleasing and economical to operate. Grouped together or as solitary lamps they cast a warm comforting light and are suitable for mood lighting in bars, restaurants and the dining table at home.

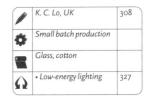

✏	K. C. Lo, UK	308
✿	Small batch production	
▥	Glass, cotton	
↻	• Low-energy lighting	327

SL-Torch

An 80% reduction in materials used is achieved by making the battery into the handle in this neat torch design. Insert the battery into a housing, which holds the bulb, and twist to turn on the torch.

✏	Antoine Cahen, Les Ateliers du Nord, Switzerland	305
✿	Leclanché, Switzerland	317
▥	Battery, bulb, plastic	
↻	• Reduction in materials used	325

Solaris™ lantern

Two hours of sun provide one hour of light for this lantern, which is capable of functioning at -30°C (-20°F) and altitudes in excess of 7,000 m (23,000 ft). Fully charged, the NiMH battery, which is free of mercury, cadmium and lead, will provide light for six hours, but if the battery discharges 90% of its capacity a low-voltage disconnect is automatically triggered. This saves battery life and ensures it will last for up to a thousand recharges.

✏	Light Corporation, USA	317
✿	Light Corporation, USA	317
▥	Photovoltaics, plastics, NiMH battery	
↻	• Renewable power source	327
	• Avoidance of hazardous substances in the battery	325

Ecotone Ambiance Slimline

As the process of shrinking the compact fluorescent lamp (CFL) continues, Philips have developed a matt-glass, candle-shaped bulb with an E-14 thread. This encourages use of CFLs in a wider range of light fittings and so can help save energy. Available in 6W, 9W and 11W, the lamps are manufactured with a minimal amount of mercury and lead and have a projected lifespan of six years or so.

✎	Philips Lighting BV, the Netherlands	319
⚙	Philips Lighting BV, the Netherlands	319
🎞	Glass, metals including mercury and lead	
🎧	• Encouraging energy conservation	327
◉	iF Ecology Design Award, 2000	330

Mini-Lynx Ambience

Although this CFL looks like a conventional incandescent bulb and is of the same size, it uses 80% less energy. Making CFLs this size has been a considerable challenge, since the electronic ballast and convoluted fluorescent tubes produce bulbs that often protrude beyond conventional light fittings. Sylvania have extended the market reach by shrinking a standard CFL into the recognizable shape of an incandescent bulb and offering the bulb in white and, unusually, apricot and rose.

✎	Sylvania Design Team, Switzerland	309
⚙	SLI Lighting, UK	321
🎞	Glass, electronic ballast, plastic	
🎧	• Energy conservation	327

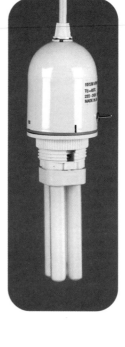

Lampholder 2000

Compact fluorescent lamps (CFLs) with standard bayonet fixings include their own electronic ballast to 'kick-start' the light. Lampholder 2000 is a light fitting (in pendant, batten-holder, flush-mounted or down-light formats) with an integral electronic ballast suitable for direct usage of four-pin-connector CFLs. After three changes of a four-pin CFL it is more cost-effective than the standard bayonet CFL and reduces materials used.

✎	Lampholder 2000 plc, UK	317
⚙	Lampholder 2000 plc, UK	317
🎞	Various	
🎧	• Energy conservation • Reduction in materials used	327 325

LED® DecorLED

Light emitting diodes (LEDs) operate on low voltages and are very efficient. LEDtronics offers a range of standard Edison base-fitting AC light bulbs in pear, globular and spot shapes. A bulb fitted with 17 LEDs provides full-spectrum white light equivalent to the illumination provided by a 25-watt conventional tungsten bulb but consumes only 1.7 watts. Typically these LED bulbs generate very little heat, 3.4BTUs/hr, compared with 85BTUs/hr for an equivalent tungsten bulb. Aside from offering huge energy savings, LEDs last up to ten times longer than CFLs and 133 times longer than tungsten bulbs.

✏	LEDtronics, USA	317
⚙	LEDtronics, USA	317
▤	LEDs, metal, glass	
↻	• Energy conservation • Extended lifespan	327

Philips Ecotone Ambiance

Although compact fluorescent lamps consume up to 80% less energy than incandescent tungsten bulbs, the early CFL designs included a U-shaped tube, which often protruded when used in conventional fittings. Philips have created a satisfying compromise by reshaping the CFL to conventional lines and size, while still accommodating the electronic ballast in the base. Bulbs to 5W, 9W or 11W output are to the same basic design. The only downside of this energy-efficient story is that small amounts of mercury are required to manufacture each lamp.

✏	Fons Baohm and Patrick van de Voorde, Philips Lighting BV, the Netherlands	319
⚙	Philips Lighting BV, the Netherlands	319
▤	Glass, plastics, electronics, mercury	
↻	• Low energy consumption	327
	• Universal design for standard fittings	324
	• Long life	

FRUIT and ICE

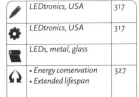

Low-energy light bulbs suffered an image problem in the late 1990s with their utilitarian configurations of thin fluorescent tubes. The Lamps of Desire project by Studio Jacob de Baan in cooperation with major lighting companies, including Sylvania, NOVEM, Osram and Philips, examined a more emotive response to encourage consumers to be green. FRUIT encases the fluorescent elements in an evocative, yet reductionist, fruit-like form, reinterpreting a 'green' product as a sophisticated modern design. Further emotional ambitions are realized in the quirky forms of ICE. There's no doubt that the Lamps of Desire project galvanized lighting manufacturers to be more adventurous in promoting low-energy lighting.

✏	Studio Jacob de Baan with Martijn Wegman and Marije Franssen, the Netherlands	309
⚙	Sylvania, NOVEM, Osram and Philips	309/319
▤	Glass, fluorescent low-energy lights, electronics	
↻	• Low-energy lighting	327

iChef

Induction technology uses magnetic fields to transfer heat from a wire coil, generated by electrical energy, into the contents of a cooking vessel. Almost 95% of the electrical energy is converted into heat, which is claimed to be five times more efficient than a gas ring and significantly faster than a conventional electric hotplate. Glass is used to separate the wire coil from the cooking pan or kettle but, as all the heat is transferred to the vessel, the glass top remains cool, providing a safety advantage over conventional cookers. Precise temperatures can be selected by using a digital display.

🖊	Induced Energy, UK	316
⚙	Induced Energy, UK	316
🗐	Metal, glass	
↻	• Energy efficient • Improved safety	327 327

Aga

Aga-Rayburn cookers are a symbol of durable, classic design, with over half a million units supplied to the UK and exported worldwide. 'Agas', as they are fondly known, have been made near the iron-making town of Telford, Shropshire, since the beginning of the Industrial Revolution. Indeed, the Coalbrookdale foundry, which supplies castings for the cooker, made sections of the world's first iron bridge in 1779 at Ironbridge. Scrap and pig iron are the raw ingredients to create the hand-made cast iron sections that form the basic components of the cooker. Poor-quality castings are simply recycled. Originally a solid-fuel cooker, the Aga-Rayburn has been improved over the years to accept oil, natural or propane gas or off-peak electricity, and since the 1940s can also provide domestic hot water and central heating needs. Agas are a coveted status symbol in any domestic kitchen, representing 'good middle-class taste'. There may be more energy-efficient designs on the market but few can match the lifespan of an Aga, which, of course, can be refurbished or recast in the future.

🖊	Dr Gustaf Dalen, Swedish physicist, 1920s	
⚙	Aga-Rayburn, UK	311
🗐	Cast iron	
↻	• Recycled and recyclable materials • Durable	324 327

Café Duo HD 1740/42, Cucina Duo

This slimline one- or two-cup coffee maker combines aesthetic flair with Philips's high eco-standards. In-line drip filters are easily removed from the top, the product does not consume any electricity in stand-by mode and all parts are marked for recycling.

🖊	Philips Design, the Netherlands	319
⚙	Philips Electronics, the Netherlands	319
🗐	Metal, thermoplastic, electrical components	
↻	• Energy efficient • Design for disassembly and recycling	327 325
✪	iF Design Award, 2000	330

Metal Cube HD 4603

Philips have borrowed from the design of 1930s and 1950s teapots to create a twenty-first-century form for the kettle. As its name suggests this cordless kettle has a cubic form comprising a polished stainless steel body with insulated ceramic handle, lid and base. The spout is wide to facilitate filling directly from the tap. It is equipped with standard contemporary features, such as a water-level indicator, but includes extras such as a three-level automatic safety system, a quick re-boil function and a 360-degree rotational base. It has a powerful 3100W stainless steel concealed element that can boil 200 ml water in under thirty seconds and rates in the middle to top of the range in terms of energy efficiency. All parts are easily disassembled for recycling and plastic parts are clearly identified for the correct waste stream.

✏	Philips Design, the Netherlands	319
⚙	Philips Electronics, the Netherlands	319
📜	Stainless steel, ceramic, electrical circuitry	
🎧	• Improved ergonomics	327
	• Durable	327

Solar cooker

Cardboard coated with a special reflective surface focuses the sun's energy on to a dark-coloured cooking pot. In subtropical and tropical regions it is possible to save the equivalent of 30% of the annual firewood consumption of a typical household using this cooker. This device provides people in the developing world who face fuel wood shortages with the means to sterilize water and have hot food.

✏	Bernard Kerr and Pejack Campbell, USA	307
⚙	Solar Cookers International, USA	321
📜	Cardboard, reflective foil	
🎧	• Solar power (passive)	327

POLTI Vaporetto 2400

Super-heated steam is an
effective agent in cleaning
and sterilizing carpets,
mattresses and upholstery,
which obviates the need to
use strong and toxic
chemical cleaners. Steam
is also a safer method than
insecticides of killing dust
mites and other insects.
Tap water is heated in a
pressurized steel container
and dispersed via the outlet
hose, which can be fitted
with a range of brushes
and hose-ends.

POLTI Ecologico AS810

For those who suffer
from asthma this vacuum
cleaner is a boon. Its water-
based filtration system
removes 99.99% of dust
up to 0.3 microns including
pollen, dust-mite faeces
and cigarette smoke. Paper
bags are replaced with
a removable water filter,
which is emptied after use
and has an lifetime of
approximately six months.

✎	Polti, Italy	319
⚙	Polti, Italy	319
▤	Various plastics, stainless steel	
♺	• Reduction in use of consumables	327
	• Improved health environment	327

✎	Polti, Italy	319
⚙	Polti, Italy	319
▤	Various plastics, metals	
♺	• Reduction in use of consumables	327
	• Improved health environment	327

Dyson Dual Cyclone range

Hoover and other major manufacturers of vacuum cleaners watched their global market share decrease as James Dyson's bagless version, using two ultra-fast-spinning centrifugal chambers, grabbed a significant market share in the 1990s; the DC02 became the best-selling upright cleaner in 1993. Unlike conventional cleaners, in which suction decreases as filters and bags become full, the cyclone system maintains 100% suction and the troublesome task of removing a full bag is replaced with emptying the main chamber. Dyson Appliances now produce a range of upright and horizontal vacuum cleaners and the cyclone system has been established worldwide as other manufacturers introduced similar machines to their existing ranges. Today the principle of centrifugal vacuum design is well established, eliminating the wastage of vast quantities of paper bags. Current Dual Cyclone models include the original DC03, DC04 and DC05 but the range has been expanded with Root Cyclone models DC07, DC08 and DC11 with more power.

✎	James Dyson and Dyson Appliances, UK	314
⚙	Dyson Appliances, UK	314
📋	Plastics, motor, electrics	
🎧	• Avoids use of consumables (paper bags)	327
	• Improved user funtionality with greater cleanliness	327
	• Design for easy maintenance	327

Lavamat Lavalogic 1600

Building on the success of its Öko-Lavamat range, AEG now produce a 6 kg wash-capacity machine using just 39 litres of water and 1.02kWh of electricity per wash to meet the EU Energy Label 'A' Class for washing machines. It has LCD displays and twenty-four wash programmes including an'Energy saving programme' and 'Hand wash programme' for wood, silk an delicate fibres. Spin speeds range from 400rpm to 1600rpm. Loading has been facilitated by widening the porthole to 30 cm (12 in) and at full operation the machine remains quiet at only 46dB (A).

✏	AEG, Germany	311
⚙	AEG, Germany	311
🗌	Steel, metals, rubber, electronics, motor and pump	
♻	• Energy and water conservation	327/ 328

Pedal-Powered Washing Machine

Appropriate levels of technology are important when local communities are faced with limited resources, manufacturing equipment and cash. The Industrial Design and Sustainable Development Group, known as GDDS, at the Federal University of Campina Grande, has designed this human-powered washing machine for manufacturing in local 'hole in the corner' workshops. The same recumbent bicycle-power unit can be fitted to other activities such as driving a water pump. This technology considers the needs, skills and cultural make-up of the community and delivers a reliable product.

✏	Dr Luiz Guimarães, GDDS, Federal University of Campina Grande, Brazil	306
⚙	Conceptual prototype for local workshops	
🗌	Various locally available materials	
♻	• Appropriate technology	324
	• Self-help design	326

TIME 1200

Reviving memories of an earlier era in domestic laundry appliances, the TIME 1200 Top Loading washer is the ideal machine for singles, couples or small families. Rated 'A' on Energy and Wash EU labels even at 40° Celsius, the 4.5 kg wash-capacity machine has an auto half load, 'Hand Wash' and 'Sportswear' programmes. It features a variable spin dial with speeds up to 1200rpm. Weighing just 63 kg, this represents a weight saving of between 10 kg to 15 kg over most horizontal drum machines. Hoover led the market in the early 1990s with its New Wave range of machines, the first to obtain the EU 'A' Energy Rating. Now most manufacturers offer at least one 'A' Energy Label machine, so perhaps Hoover have targeted a niche market with its new top loader. Lifecycle studies suggest that single-occupancy home owners or tenants might still be better off taking their laundry to the local laundrette where there's usually a free conversation on offer too.

✏	Hoover, UK	316
⚙	Hoover, UK	316
📄	Steel, metals, rubber, electronics, motor and pump	
♻	• Energy and water conservation • Reduction in materials used	327/ 328 325

Staber Washer

Unlike horizontal-axis-driven front-loading washing machines, the Staber Washer offers a top-loading machine into which the stainless steel basket of laundry is loaded. Energy-saving features include the use of a variable-speed motor. Easy access to the internal components can be gained by lifting the front panel and fitting a self-cleaning filter, thus facilitating maintenance.

The manufacturers claim reduced energy, water and detergent consumption.

✏	Staber Industries, USA	321
⚙	Staber Industries, USA	321
📄	Stainless steel, steel, resin and various other materials	
♻	• Energy and water conservation • Energy efficient to Energy Star guidelines	327/ 328 328

Supercool™

Traditional refrigerant manufacturing involves the use of chlorofluorocarbons (CFCs, HCFCs) as coolants but Supercool AB have exploited the Peltier effect of a doped bismuth telluride thermocouple, which avoids using any of the ozone-depleting gases. A thermoelectric panel operated on a low-voltage system (12V, 24V) consumes a modest 10W to provide sufficient cooling for a small hotel minibar. A further advantage is that the mechanism operates silently, unlike the familiar hum of conventional coolant systems.

✏	Supercool AB, Sweden	321
⚙	Supercool AB, Sweden	321
📃	Thermoelectric module, bismuth telluride, plastic	
🎧	• Reduction in energy consumption	327
	• Non-toxic refrigeration system free of CFCs or HCFCs	326

Vestfrost BK350

Vestfrost is one of the world's largest manufacturers of refrigerators and freezers and took an early lead in showing environmental responsibility by removing all CFCs and HFCs from its model range in 1993.

Using the alternative 'Greenfreeze' refrigerants, Vestfrost remains the only manufacturer in Europe holding the EU Eco-label for this category of appliances.

✏	David Lewis, UK	307
⚙	Vestfrost, Denmark	322
📃	Metal, plastics, rubber, electric motor and compressor	
🎧	• Low energy consumption	327
	• Clean production	325

Supercool™ box

A panel of thermocouples of doped bismuth telluride is capable of pumping heat and provides the cooling for this transportable refrigeration box suitable for commercial or domestic use at 12V or 24V.

✐	*Supercool AB, Sweden*	321
✿	*Supercool AB, Sweden*	321
📃	*Thermoelectric module, bismuth telluride, plastic*	
🎧	• *Reduction in energy consumption*	327
	• *Non-toxic refrigeration system free of CFCs and HCFCs*	326

Planet DC

This DC refrigerator/ freezer of 0.33 cu. m (11³⁄₅ cu. ft) capacity, operates from 12V or 24V and can be sustained with any small, domestic, renewable-energy system using deep-cycle batteries, for example a photovoltaic module capable of generating 150W.

✐	*Planet, USA*	319
✿	*Planet, USA*	319
📃	*Various*	
🎧	• *Low-voltage device for domestic renewable-energy systems*	327

Wind

In the Industrial Revolution iron and steel usurped natural materials, so it is refreshing to see the process cleverly reversed in the housing of this electric fan, in which woven rattan replaces the conventional pressed sheet steel or plastic. At the same time the fan is transformed from an object of cold functionalism to one of playful character. Most of the materials can be recycled or composted.

✐	*Jasper Startup, Startup Design, UK*	309
✿	*Gervasoni SpA, Italy*	315
📃	*Rattan, steel, electrics*	
🎧	• *Recyclable and compostable materials*	324

Felt 12 x 12

Be your own fashion designer using Fortunecookies's felt squares backed with Velcro: assemble a jacket, trousers, wedding dress or any other garment in your own personalized style. Bored with the look? Deconstruct your design and start again. Fashion is placed back in the hands of the consumer.

✏	Fortunecookies, Denmark	306
⚙	One-off, Fortunecookies, Denmark	306
📜	Felt, Velcro	301
↻	• Modular system for reuse of components	327
	• Renewable material (felt)	325

✏	Hussain Chalayan, Amaya Arzuaga, Gaspard Yurkievich, Pret à Porter	
⚙	Acordis Fibres Ltd, UK	311
📇	TENCEL®	
♺	• Renewable, compostable materials • Cleaner production	324/325 325

TENCEL® fashion garments

Three internationally renowned designers reveal the versatility of a man-made fibre using natural cellulose derived from managed forests. TENCEL® is from renewable resources and is manufactured in a closed-loop clean production process. With good drapability and a wide choice of surface finishes and weaves, TENCEL® fabrics offer the convenience and feel of modern synthetics and have a reduced impact on the environment. All this is proof that today's levels of comfort and style can be maintained without sacrificing the environment. TENCEL® is one of the modern success stories of the global textile industry.

AW03 and AW04

Ciel is a new collection of luxury sport and street outerwear for women, reflecting the solid eco-design principles and supply chain management of Sarah Ratty's pioneering company Conscious Earthwear. All fabrics are certified to Oeko Tex or other recognized low environmental impact standards. Organic linen and Azo dye-free fabrics are used for the new A/W 2004 collection and new items use organic alpaca knitwear in natural origin colours. Recycled or recyclable sources are sought for any man-made fibres used. All Ciel garments can be washed on a low heat cycle of 30° Celsius to reduce energy consumption by 75%. As part of the working group for Pesticide Action Network (PAN) and the UK Soil Association Textile

Certification working group, Ratty knows how rigorous you have to be to design and manufacture fashion that takes its environmental responsibilities seriously. Her eco-intelligent designs attract many well-known fashionistas, including Zoë Ball, Macy Gray, Denise van Outen and Asian Dub Foundation.

✎	Sarah Ratty, Ciel, UK	309
⚙	Ciel, UK	313
📜	Organic textiles	
↻	• Renewable materials • Avoidance of toxic substances	325 325

Levi's Engineered Jeans®

Jeans are firmly rooted in popular culture and have been the workwear garment of choice for millions over the last century. Cotton fibre has been the favoured raw material for all jeans manufacturers but cotton fabrics carry a significant raft of environmental burdens from pesticide applications and other toxic agents used in the textile production process. Today Levi Strauss, one of the world's largest clothing manufacturers, uses a special natural cellulose-based, low-environmental-impact fibre called TENCEL® for its new line of Levi's Engineered Jeans.

✎	Levi Strauss, Inc., USA	317
⚙	Levi Strauss, Inc., USA	317
📜	TENCEL®	
↻	• Renewable, compostable materials • Clean production	324/ 325 325

Sensor sportswear

Today's fascination with fashion and electronic technology find a meeting point in these prototype garments by Philips Design/Philips Research. Street-cred denim is embedded with an audio system in Audio Streetwear, and in-flight communications are integrated into the stretch wool Imaginair-Airline Workwear outfits for cabin crew. Sensor Sportswear includes electronics to monitor bodily functions, while snowboarders intent on going off-piste can do so in the knowledge that the in-built global positioning satellite will stop them from getting lost. Water and electronics don't generally mix, so presumably the electronic elements can be detached before the clothing is consigned to the laundry bin.

✏	Philips Design, the Netherlands	319
⚙	Prototype	
📦	Textiles, electronics	
🎧	• Improved functionality	327

Mit Scarf

Fusion, as the name suggests, is a design agency that creates mutant designs by combining two functions into one. The Mit Scarf does away with all those sad single gloves often seen absentmindedly abandoned by their owners. Keeping neck and hands warm comes together in the Mit Scarf as a glove is thoughtfully stitched in each end of the scarf. Simple cutting patterns make for minimal wastage.

✏	Fusion, UK	315
⚙	Fusion, UK	315
📦	100% polyester fleece	
🎧	• Dual-function design	327
	• Reduction in materials used	325

Emilianas

This prototype is the result of an investigation to find maximum synergy between form and function by using one material, one piece and one seam. Made from 90% felt, these slippers embrace the feet like a little carpet. They invite you to put them on and comfort yourself with their warmth and tactile surface.

✏	Ana Mir and Emili Padrós, Emiliana Design Studio, Spain	308
⚙	Prototype	
📄	Felt	
🎧	• Reduction in materials used	325
	• Renewable materials	325

Typically Dutch

Today, symbols of national identity are often subsumed into a global cultural milieu as marketing and advertising agencies corporatize design and its semantics. Here is a refreshing reinterpretation of identity that builds positive associations of Dutchness and 'otherness' by incorporating influences from other cultures. Felt liners and Islamic carving and relief texture on the traditional Dutch clogs question dogmatic interpretations of what it is to be Dutch, proving good design can happily leap across political and cultural boundaries.

✏	Nine Geertman, graduate student 2003, Design Academy Eindhoven, the Netherlands	305
⚙	Prototype	
📄	Wood, felt	
🎧	• Renewable materials	325
	• Innovation of traditional skills	325

Soul Mates

Disposable footwear might seem an odd design strategy but these slippers are made from pulped cellulose fibre that is easily returned to the recycling loop. This footwear is vacuum-formed in aluminium moulds then dried in the sun to create a surprisingly rigid corrugated sole. Intended for use in domestic, hotel or light manufacturing environments, these slippers are expected to last for up to one week before they need recycling or composting.

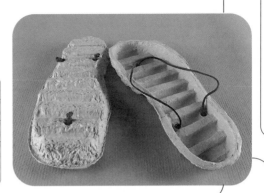

✏	Vikram Mitra, National Institute of Design, India	308
⚙	Prototype	
🎒	Pulped cellulose fibre, cotton thong	
♻	• Recyclable, compostable	324
	• Low-energy manufacturing	325

OURO

These shoes combine efficiently reused, recycled and offcut materials to produce a fashionable yet fashionless result. Old car tyres, industrial production scrap and post-consumer shoe soles are reincarnated as new soles by reconstituting and moulding the recycled chips. Chrome-free vegetable tanned leather is used for the upper, and offcuts from the moulded leather are used as well. This is a tough, no-nonsense shoe for the green consumer.

✏	Lea Bogdan, USA	304
⚙	Small batch production	
🎒	Re-ground used car tyres, virgin, scrap and post-consumer leather, recycled shoe soles	
♻	• Reused, recycled and biodegradable materials	324
🏆	IDRA award, 2002–2003	330

Modular Hinge Sandal

Most manufactured shoes are not easily disassembled. This sandal comprises a frame, an upper cover and a lower cover, each layer being attached to the others by snap bolts. Should any layer wear, or the user want to give the sandals a fresh look, it is easy to undo the snap connectors and replace a layer. This feature improves its lifespan by allowing people to customize and repair their sandals, and facilitates recycling of mono-material components. Disassembly also offers manufacturers opportunities for developing new models and making repeat sales of 'spare parts', facilitating closing the loop – you can't have your new part unless you bring back the old.

✏	Arvind Gupta, USA	306
⚙	Prototype	
🎒	Thermoplastic resin Nylon 66	
♻	• Design for disassembly	325
	• Upgradable, repairable	327
🏆	IDRA award 2002–2003	330

A-button

Designers often neglect important aspects of everyday living. Not so Antonia Roth, whose elegant button design takes a fresh look at something that people with limited dexterity struggle with daily. The elongated shape makes it easier to hold and to pull through the button-hole, so it offers improved independence for children, the elderly or anyone with arthritis or limited hand mobility.

✏	Antonia Roth, Fachhochschule Hannover, Germany	309
⚙	Prototype	
▤	Polymer	
↻	• Universal design	327

The Body Shop range

Since its formation in the 1970s the ethos of The Body Shop has been to provide a holistic, natural approach to body care and hygiene, with due consideration to the environmental, ethical and social responsibilities of the business. That approach still drives what has become the role model for an ethical international business. Encouraging recycling of packaging materials, such as the HDPE bottles used for many formulations, is an integral part of the day-to-day business. Having used up the product, the user is encouraged to take the bottle back either for a refill or for recycling. Product information is generally printed directly on to the bottles to eliminate the need for stick-on labels and to facilitate recycling.

✏	The Body Shop, UK	312
⚙	The Body Shop, UK	312
▤	Natural oils, conditioners, soaps and recyclable HDPE	
↻	• Reusable and recyclable containers • High natural-content ingredients	327

Simply

This fashion bag takes the iconic form of the ubiquitous folding paper bag but uses vegetable tanned cowhide instead and comes in a range of bright colours or natural hide. This lightweight but durable bag is ideal for making a trip to the local store or corner shop without making an overt statement by using a branded supermarket 'bag for life' made of less durable low-density polyethylene (LDPE).

	BREE Collection, Germany	312
⚙	BREE Collection, Germany	312
🗋	Vegetable tannins, cowhide	
🎧	• Renewable materials	325
	• Avoidance of toxic substances	325

purses, bags

End-of-line and faulty carpet stock is transformed into an innovative range of neck and hand purses, clutch bags and idiosyncratic mini-bag designs. Applying heat and pressure to special moulds transforms the existing carpets into waterproof, smooth, hardened surfaces. Exterior textiles, decorations, embellishments, fastenings and fittings use textured and plain rubber, net and felt.

	Kelly Atkins, Carpet-Burns, UK	304
⚙	Kelly Atkins, Carpet-Burns, UK	304
🗋	Reused carpet, rubber, net, felt	
🎧	• Reused and reycled materials	324/ 326

Citymobil C1

In May 1998, at the annual BMW Day of Technology, the Citymobil C1 was announced as a new-concept vehicle specially designed to offer individual mobility in cities and metropolitan areas. Single-occupancy cars account for almost 80% of urban journeys. The C1, launched in April 2000, provides an alternative for urban car drivers who are frustrated with congested roads. Conscious of some of the drawbacks of two-wheeled transport, BMW have striven to improve safety standards and comfort. Crash bars and fairings and a stiff frame, resistant to twisting, provide front-end collision protection similar to that of a small car. An ABS option ensures skid-free braking, and a unique curved windshield arches over the driver, providing better weather protection than conventional motor scooters. Fuel consumption is a respectable 2.9 litres per 100 km (about 97 mpg) and emissions are reduced by a three-way catalytic converter. Can the C1 capture a new audience with its lower road tax and operating costs, reduced fuel consumption and lower emissions? The big question is whether the city suits and trendsetters will be lured out of the comfort of their existing vehicles.

✏	BMW, Germany	312
⚙	BMW, Germany	312
📃	Alloy frame, rubber wheels, various plastics – some recyclable and recycled	
↻	• Fuel economy • Alternative mode of transport to single-occupancy cars	327 326
💬	Design Sense awards, 2000; iF Design Award, 2001	330

Ecobasic

Fiat understands the requirements of the low-cost, small-car market. From the Topolino and the Fiat 500 series in the late 1940s and early 1950s to the Panda and Cinquecento in the 1980s and 1990s, Fiat has always kept true to its vision of economical products to ensure that the 'freedom of the road' reaches a mass market.

Ecobasic is the prototype for a production model capable of yielding about 35 km per litre (100 mpg) at a price of £3,300. This minimalist design incorporates a steel frame to which coloured plastic panels and a one-piece polycarbonate tailgate are fixed. A 1.2-litre, sixteen-valve, four-cylinder Fiat JTD turbodiesel engine is coupled with a five-speed manual gearbox. Selection of the 'econ' transmission mode automatically cuts the engine if the car is stationary for more than four seconds but the engine restarts as soon as the accelerator is pressed. This 2000 conceptual prototype has contributed to the evolution of Fiat's standard models.

✏	Fiat Auto, Italy	314
⚙	Prototype, Fiat Auto, Italy	314
🗜	Steel, thermoplastics, various other materials	
♫	• Fuel economy • Lightweight, modular construction	327 325

Peugeot H2O

Peugeot could be accused of mixing their metaphors with this hydrogen-powered fire-fighting concept vehicle named 'H2O'. Of course it alludes to the fact that the combustion of hydrogen from the on-board fuel cells produces emissions of harmless water vapour, although in this concept the fuel-cell is only an auxiliary power source to the main battery-powered electric engine. This electric vehicle was first trialed in a Taxi PAC (pile à combustible; fuel cell) in June 2001. PSA Peugeot Citroën developed the concept further in the Peugeot H2O by adding an auxiliary fuel-cell power unit providing electrical energy to run pumps, smoke extractors, electrical sockets and communications systems.

Any idea that the fire engine can generate its own water for putting out fires should be quickly dispelled. This 1.7 tonne vehicle's primary purpose is for reconnaissance in difficult-to-reach areas of cities and towns. It carries only sufficient equipment to deal with minor emergencies, but is manoeuvrable in confined spaces. It is fitted with proximity sensors in the bumpers and a radar system in the front panel to enable it to work in dense smoke. Should the smoke inhibit sufficient oxygen up-take for combustion of the hydrogen, an on-board oxygen tank takes over. Hydrogen is actually generated on-board from an aqueous solution of sodium borohydride and a catalyzer.

✏	PSA Peugeot Citroën, France	320
⚙	PSA Peugeot Citroën, France	320
📜	Various man-made materials and systems, electric motor, auxiliary hydrogen fuel-cell	
🎧	• Reduced emissions • Improved emergency operational capability	326

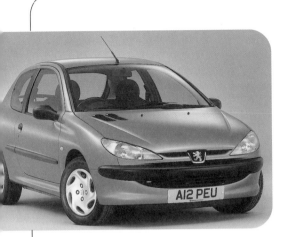

Peugeot 206 1.4 HDi

Ask most people if they would be happy travelling on a motorway at an average speed of 97 km/h (60 mph) on just 7.44 litres (1.64 gallons) of fuel, and they'd probably all say, yes! This was the result of a test by a London *Evening Standard* journalist as he drove the 1.4 litre HDi diesel-engined Peugeot 206 around the entire 193-km (120-mile) length of the M25 motorway around London. Fuel consumption was an efficient 26 km per litre (73.3 mpg), even better than the combined fuel economy figure of 23.4 km per litre (65.7 mpg) quoted by Peugeot. This equates to a CO$_2$ emissions of 113 g/km. Over the last thirty years the design of diesel engines has significantly improved fuel economy, reducing average fuel consumption from 10 litres to less than 6.5 litres per 100 km, an overall efficiency gain of 35%. The test above delivered 3.85 litres per 100 km, showing how efficient the HDi engine can be on long runs. Improvements in fuel economy have flattened off in recent years. Designers will have to examine ways of making cars lighter, without sacrificing strength and safety, reducing drag and transmission or frictional losses, in order to register improved fuel economies. Of course, even greater gains could be had if drivers were happy to travel at slower speeds. The Peugeot 206 1.4 HDi generates 68bhp at 4,000rpm, accelerates from 0 to 100 km/h (62 mph) in 15 seconds and has a top speed of 166 km/h (104 mph), signifying a lot of excess power. Until the consumer clamours for slower, more economical cars or a fossil-fuel crisis forces governments to act, manufacturers are unlikely to lead the process of change. However, whoever comes up with the concept of the 'slow car' could create a new market sector overnight.

✏	PSA Peugeot Citroën, France	320
⚙	PSA Peugeot Citroën, France	320
▤	Various man-made materials	
☊	• Fuel economy	327

G90 concept car

Innovative weight-saving and excellent aerodynamics give fuel economy, which is further improved by the three-cylinder ECOTEC engine developed for the Corsa by Vauxhall Motors Ltd in the UK.

✏	General Motors, USA, with Vauxhall Motors Ltd, UK and Opel, Germany	315 322
⚙	General Motors, USA, with Vauxhall Motors Ltd, UK and Opel, Germany	315 322
▤	Various	
☊	• Fuel economy	327

EV1

Launched in December 1996, the EV1 was General Motors's first production vehicle for non-commercial use to be entirely powered by an electric engine. It was an attempt to deal with the effects of state legislation to reduce vehicle emissions, such as the 1985 Californian law that set a 2% sales volume of zero-emission cars by 1998. Capable of achieving 100 km/h (62.5 mph) in 9 seconds from a standing start and with a top speed of 129 km/h (82 mph), the EV1 was no slouch, but its performance was constrained by the 532 kg (1,172 lb) of batteries needed to move the vehicle's mass of 817 kg

(1,797 lb). Fortunately developments in battery technology have ensured that such power-to-weight ratios are now a thing of the past.

✏	General Motors, USA	315
⚙	General Motors, USA	315
🗎	Various	
♻	• Zero emissions (if recharged from a renewable energy supply)	326

Insight

Claimed to be the world's most efficient petrol-powered car, the Insight is actually a hybrid petrol/electric car featuring Honda's integrated motor assist (IMA™). The IMA combines a high-efficiency 995cc petrol engine with a 10kW ultra-thin, in-line, brushless DC electric motor to achieve about 30 km per litre (83 mpg) and 80 g/km (4.5 oz/mile) CO_2 emissions (less than half the EU 2000 limit). Connecting directly to the crankshaft, the electric motor draws power from a 20 kg, 144V nickel metal-hydride battery via an electronic power control

unit (PCU) when the car is accelerating, to provide 'Motor Assist'. This improves power output and low-speed torque. During deceleration the batteries are recharged, so the car is independent of external electricity sources. Reduction in kerb weight, which is just 835 kg (1,837 lb), is achieved by a lightweight aluminium body (borrowing design principles from the Honda NSX sports car).

✏	Honda, Japan	316
⚙	Honda, Japan	316
🗎	Various	
♻	• Fuel economy • Lightweight construction	327 325

NECAR 5

Four years ago the Mercedes-Benz A-Class compact car fitted with a 75hp (55KW) fuel cell, called the NECAR 4 (New Electric Car), proved that a high-performance, long-range, hydrogen-powered car was a practical reality. NECAR 5 continues the company's evolution of prototype hydrogen cars, and in May 2002 one drove 5,000 km (3,000 miles) across the USA to enter the record books. Refuelling every 500 km (300 miles) at pre-arranged depots for Methanol, the primary fuel for the fuel cell, the journey took sixteen days. As their only emissions are water vapour, hydrogen-fuelled cars can offer many benefits, especially in reducing aerial pollutants. The environmental ramifications of converting petroleum and diesel cars to hydrogen are not fully computed yet. Methanol can be synthesized from plants, such as sugar cane, and hydrogen can be produced by utilizing energy from renewable power plants (solar, wind, water). The real complexity lies in constructing a reliable, safe and comprehensive distribution system for hydrogen-generating fuels. Building such a system will require lots of energy and such energy is likely to come from fossil fuels. The hydrogen economy story is just beginning.

✐	Mercedes-Benz/Daimler Chrysler, Germany	313
⚙	Mercedes-Benz/Daimler Chrysler, Germany	313
🎞	Various materials	
🎧	• Reduction in emissions • Potential renewable energy car	326 327

VW Lupo 3L TDI

In Berlin on 16 May 2000 a Volkswagen Lupo 3L TDI commenced a journey to circumnavigate the world. It was fitted with a three-cylinder, 1.2-litre (¼-gal) turbo direct injection (TDI) engine and the objective was to break existing records by making this the most fuel-efficient car to undertake this challenge. A total of 53,333 km (33,333 miles) was travelled in eighty days across five continents at an average speed of 85.6 km/h (53.5 mph) at a remarkably economical fuel consumption of 2.38 litres per 100 km (0.84 gallons per 100 miles). Production models of the Lupo achieve 2.99 litres per 100 km (1.05 gallons per 100 miles) under MVEG cycle tests. The designation '3L', for 'three-litre', refers to its fuel economy rather than to the more traditional indication of engine size. This economy translates into a 1000-km (625-mile) journey on one full tank of fuel and carbon-dioxide emission levels below 90 g per km (5.1 oz per mile). An efficient engine is complemented with lightweight construction materials, especially aluminium, in body, chassis and running gear components. An automated five-speed, direct-shift gearbox and automatic stop-start system help maximize fuel economy. Why can't all manufacturers produce such fuel-efficient compact cars?

✐	Volkswagen AG, Germany	322
⚙	Volkswagen AG, Germany	322
🎞	Various	
🎧	• Fuel economy	327

ICVS
(Intelligent Community Vehicle System)

The realization that transport systems need an urgent rethink has prompted industry, government and academia to examine so-called intelligent transport systems (ITS). One of the more advanced concepts is Honda's ICVS (Intelligent Community Vehicle System), whose working components are three electric vehicles and an electric/manual bicycle, which operate within a defined geographical area and under shared community usage. The City Pal four/five-seater compact electric car is a multi-user option and the other versions are for a single user. Two personal electric vehicles, the 'Stepdeck' and the 'Monopal', offer short-distance mobility suitable for inner-city business users, commuters and shoppers. Finally, the electric-assisted bicycle, 'Racoon', is an all-purpose, utilitarian mode of transport. In the ICVS, cars and bicycles function as complementary tools in an urban ecosystem.

✐	Honda, Japan	316
⚙	Prototypes	
▧	Various	
☊	• Community ownership	326
	• Reduction in air emissions	326
	• Possibility of powering from renewable energy sources	327

reduction of 25% in carbon dioxide in the exhaust emissions. Further reductions are achieved in the Multipla Hybrid by combining a petrol engine with electric motor system. In hybrid operating mode 6.8 litres (1.5 gals) of fuel plus 3kWh are consumed per 100 km and in electric mode power consumption is about 30kWh per 100 km (62 miles). The car produces about half the emissions of a conventional petrol-engined version.

Multipla and Multipla Hybrid Power

The Multipla is capable of carrying six passengers in an upright steel frame. Power unit options include a dual-fuel, 1.6-litre, sixteen-valve, four-cylinder engine capable of using petrol or methane. Methane is a clean fuel, free from benzene and particles, and gives a

✎	*Fiat Auto, Italy*	314
⚙	*Fiat Auto, Italy*	314
▭	*Steel, composite panels*	
☊	• *Reduction in emissions (air pollution)*	326
	• *Hybrid power*	327

P2000 HFC Prodigy

Ford unveiled the fuel-cell technology for the hybrid hydrogen-fuelled electric Prodigy family sedan at the Geneva Motor Show in 1999. In a joint programme with Ballard Power Systems and Daimler Chrysler, Ford has developed an advanced power unit that is capable of delivering 75kW (100PS) from four hundred hydrogen cells in a three-stack proton exchange membrane (PEM) weighing 172 kg (378 lb). The electric induction motor, which delivers 120PSA, weighs 91 kg (200 lb). An on-board capacity of 1.4 kg (3 lb) of hydrogen gives a range of around 160 km (99.3 miles) with a combined EPA-cycle fuel economy equivalent to 28.5 km per litre (80 mpg) of petrol.

With a total kerb-weight of 1,514 kg (3,330 lb), acceleration is about 14 seconds from 0 to 100 km/h (62 mph) with a top speed of 145 km/h (91 mph). These HFC cars are now being tested in a three-year programme in California. Other variants of the P2000 range include a lightweight version being developed at the Ford Forschungszentrum Aachen (FFA) in Germany. Weighing just 900 kg (1,980lb) and powered by an experimental 1.2-litre DiATA compression-ignition, direct-injection engine, fuel consumption is 26.2 km/litre (74 mpg).

✎	*Ford Motor Company, USA*	315
⚙	*Ford Motor Company, USA*	315
▭	*Various*	
☊	• *Zero emissions*	326
	• *Fuel economy*	327

New Prius

Since 1997 the Prius has sold over 130,000 units, making it the world's most successful hybrid car. Building on this success, the New Prius incorporates the Toyota Hybrid Synergy Drive®, a sophisticated combined system of energy management that finds the best fit between the drive power of the petrol engine, the power of the electric motor and balancing power delivery to wheels with generation of energy by regenerative brake control. The key to this system is the 'power split device' that delivers a continuously variable ratio of petrol engine and electric motor power to the wheels. At start-up the electric motor works alone but as the accelerator pedal is pressed the engine and motor provide power. Hard acceleration is achieved by drawing additional power from the batteries. Deceleration and braking causes the high-output

regenerative breaking system is then stored in the high-performance battery. What does all this mean in terms of the car's performance? Combined fuel consumption is 23.4 km per litre (65.7 mpg), resulting in a respectable 104 g/km CO_2 emissions. Particulate matter emissions are almost non-

respectively than required by EURO IV regulations for petrol engines. Thanks to this low pollution footprint the Prius is subject to lower rates of Vehicle Excise Duty in the UK and is exempt from the Congestion Charge in London. Toyota continues to commit substantial resources to

overall fuel efficiency – 29% to the Prius's Hybrid Synergy Drive's 28%. However, Toyota has set targets of overall efficiency of 42% for evolving generations of FCHV vehicles.

motor to act as a high-output generator driven by the car's wheels. Kinetic energy recovered as electrical energy by this

existent and CO_2 and NOx levels are very low. In fact, the hydrocarbon and nitrogen oxide emissions are 80% and 87.5% lower

develop improvements in other types of engines and hybrid technologies. In December 2002 a limited number of the Toyota FCHV (Fuel Cell Hybrid Vehicle), running on high-pressure hydrogen, were released. Early indications are that the FCHV route offers similar

✏️	Toyota, Japan	322
⚙️	Toyota, Japan	322
📷	Various materials	
♻️	• Reduced emissions	326
	• Hybrid power	327

Sparrow

Those who recall the infamous C5 electric car designed by Sir Clive Sinclair in the early 1980s in the UK, regarded by many as a liability in fast-moving traffic, might just

consider this new one-seater electric car. With body styling reminiscent of the 'bubble' cars of the 1960s, the Sparrow can achieve speeds of up to 112 km/h (70 mph) from the thirteen batteries that

store the electrical energy. A full charge gives a range of 96 km (60 miles), making it extremely economical. With side-door access and full weather protection, the Sparrow might tempt people who are unwilling to venture out on a motor scooter or motorcycle. When it is being charged up, however, it has to be remembered that the vehicle is only as green as the mains electricity supply to which it is connected.

✏️	Corbin Motors, USA	313
⚙️	Corbin Motors, USA	313
🗞️	Various	
♻️	• Zero emissions (when using renewable energy supplies)	326
	• Alternative mode of one-person commuter transport	326

Smart Car

In October 1998 the Smart Car, jointly developed by the German manufacturer Daimler Chrysler/Mercedes Benz and Swatch, the renowned manufacturer of colourful modern watches, passed prolonged safety tests and was launched in the European market. This super-compact car looks as though it has been driven straight out of the pages of a comic strip. Measuring a mere 2.5 m (8 ft 2 in), it is half the length of a standard car (reducing materials consumption) and uses one-third less

petrol (on average 21 km per litre, or 60 mpg) than its market rivals, the Ford Ka and Volkswagen Polo. Although the car is designed as an urban runabout, the turbo-charged petrol engine provides rapid acceleration (60 km/h, or 37.5 mph, in

six to seven seconds) and a top speed of 130 km/h (82 mph) to meet the high expectations of today's motorists. The occupants are protected by a strong monocoque steel frame and there is a range of modular panels and interior elements allowing

personal customization, catering for changing fashions and facilitating repair. This could herald an important development in the car industry in which interchangeable components contribute to extending the lifespan of a vehicle.

✏️	Daimler Chrysler, Germany	313
⚙️	Daimler Chrysler, Germany	313
🗞️	Various	
♻️	• Fuel economy	327
	• Improvements in upgradeability and repairability	327

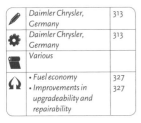

Segway Human Transporter

The Segway Human Transporter (HT) was launched in the USA in 2001 and received massive media attention. It's easy to see why; this self-balancing, personal transportation device combines practicality, versatility, mobility and carrying capacity while

being a visual curiosity too. Solid-state gyroscopics, tilt sensors, high-speed microprocessors and powerful electric motors link up to create a machine that senses minute shifts in the operator's balance or in the terrain, and immediately adjusts to restore balance. These adjustments happen one hundred times a second when travelling in a straight line or manoeuvring in confined spaces. This intuitive balance is complemented by an equally intuitive stop/start/move system. There is no conventional accelerator, brake or steering: lean forward and the machine

moves forward, straighten up and it stops, lean backwards and it begins to reverse. The steering grip on the fixed handlebar is rotated using your wrist, using the right hand to go to the right and vice versa. It is possible to turn in situ within the maximum dimension of its footprint, 64 cm (25 in). This sophisticated machine combines innovations from a number of companies: Delphi Electronics provide the controller boards on the main platform with a Texas Instruments digital signal processor; the balance sensor assembly with five gyroscopes is supplied by Silicon Sensing Systems; the 1.5 kilowatt (2hp) wheel motor, which operates at up to 8,000rpm, is produced by Pacific Scientific (Danaher); and the two-stage 24:1 reduction gearbox is a joint venture between Segway and Axicon Technologies. There are currently two models, the iSeries, with larger tyres and wider profile for variable terrain, or the smaller and more portable pSeries, ideal for commuting that involves multiple transport modes. Each model is equipped with a series of 64-bit security keys set to pre-

coded governors: the black key limits maximum speed to 9.6 km/h (6 mph) for beginners; yellow to 12.9 km/h (8 mph); and, for the 'open road', 20 km/h (12.5 mph). A single charge of the battery gives a range of 13–19 km (8–12 miles). Segway see this mobility product operating in a variety of environments and meeting the needs of everyone from commuters and students on campus to warehouse personnel. While the remarkable levels of innovation are plain to

see, it is too early to say whether this intriguing vehicle will really meet society's needs or remain a vehicle for niche activities. For local travel it certainly offers considerable emissions savings over motor scooters or motorbikes.

✎	Segway, USA	321
⚙	Segway, USA	321
🔋	Various materials, electronics, motors, battery	
♎	• Energy reduction • Improved choice of mobility mode	327 326

Voloci

Weighing just 36 kg (79 lb), including a 9 kg (19 lb 11 oz) 36-volt metal nickel hydride (NiMH) battery, this is claimed as the lightest electric motorbike on the market. An aluminium frame accommodates the battery in the diagonal. A high-efficiency microprocessor-controlled gearless drive system operates off a quiet brushless motor. Just twist the throttle and go. With a fully charged battery the Voloci has a range of up to 32 km (20 miles), a top speed of 48 km/h (30 mph) and acceleration of 0–32 km/h (20 mph) in five seconds. A second battery can double the range and a sealed lead acid battery option is also available, although it ups its weight to 47 kg (103 lb 6 oz).

✐	NYCEwheel, USA	319
⚙	NYCEwheel, USA	319
🛢	Aluminium, NiMH or lead acid battery, motor, rubber, various metals	
♻	• Reduction in materials used	325
	• Energy-efficient mobilty product	327

Hunter

This lightweight, 54 kg (119 lb), moped is powered by a hydrogen fuel cell (the H2 Stack) mounted under the rotationally moulded polyethylene frame. With an output of 3.0KW the cell generates sufficient power to operate two 48-volt, 1.5KW motors mounted in each wheel hub, and achieve speeds of up to 45 km/h (28 mph). Easy manoeuvrability and nippy acceleration of 0–45 km/h in four seconds makes it an ideal commuter or fun two-wheeler. A five litre, 5 kg (11 lb), hydrogen cylinder fits comfortably under the seat and ensures a range of up to 100 km. This H_2 bike is aimed at a mass market with an affordable price tag. It is a well-researched prototype that challenges the motorbike industry to find low-cost options and help car commuters convert to low-impact water-emission hydrogen-powered bikes.

✐	Peter Jaensch, Behind-the-wheel Product Design, Germany	307
⚙	Prototype	
🛢	Various materials	
♻	• Renewable energy source (if hydrogen is generated using wind, sun or water power)	327
	• Zero emissions	326

✏	Students, professors and alumni, Massachusetts Institute of Technology, USA	308
⚙	Students, professors and alumni, Massachusetts Institute of Technology, USA	308
📃	Photovoltaics, lightweight metals and composites	
↻	• Renewable energy • Zero emissions	327 326

Daedalus 88

Weighing just 31.4 kg (69 lb), the Daedalus 88 aircraft set a new endurance distance record for a human-powered aircraft of 199 km (129 miles) over 3 hours 54 minutes, from Crete to the island of Santorini, Greece. The feasibility of man-powered flight is beyond doubt but translating the technological advances into everyday transport provides a significantly greater challenge.

Gossamer Albatross

Powered flight was made a reality by the Wright Brothers but it was Paul McCrady of AeroVironment who pioneered human-powered flight with the Gossamer Albatross aircraft that made the crossing of the English Channel in 1979. Made of lightweight synthetic materials, this strange craft collates expertise in materials technology with advanced aerodynamics. This was a welcome invention but it has proved difficult to design larger human- or solar-powered passenger-carrying aircraft.

✏	Paul McCrady, AeroVironment, USA	308 311
⚙	Prototype, AeroVironment, USA	311
📃	Carbon-fibre and graphite resins, Kevlar	
↻	• Zero emissions	326

Independence® 3000 iBOT™ mobility system

Conventional wheelchairs, whose design has largely remained static for several centuries, offer only limited mobility. iBOT enables disabled people to navigate rough, uneven surfaces, to 'stand' up and to climb and descend stairs. This is achieved by gyroscopic articulations of the frame and three sets of wheels.

✎	DEKA Research and Development and Independence Technology, USA	305 316
⚙	Independence Technology, USA	316
▤	Various	
⮎	• Improved functionality and mobility for disabled people	327

Biomega bicycles

Freedom and vitality are synonymous with that downhill rush on a bicycle. Most people don't need the eighteen to twenty-one gears of a mountain bike as most usage is urban or suburban with the occasional countryside foray. What they need is a well-made, reliable, appropriately geared and beautifully designed machine that will provide excellent service and long-life. Enter Biomega, a Danish company that, harnessing the skills of some top designers, has transformed expectations for the cyclist. Marc Newson's MN01 Extravaganza and MN03 Relampago are based on a strong, rigid super-formed aluminium monocoque chassis, which is simpler to produce than the normal method of welding metal tubes. Ross Lovegrove has created the aesthetically more conventional RL01, which has an unusual frame of bamboo and aluminium. Beatrice Santiccioli's Amsterdam is a shaft-drive bicycle for women.

🖋	Various designers including Marc Newson, Ross Lovegrove, Beatrice Santiccioli	
⚙	Biomega, Denmark	312
📑	Super-formed, aluminium, bamboo, metal alloys, rubber	
🎧	• Human-powered • Improves users' health and well-being	327 327

Hybrid bike

Prestigious car manufacturers are entering the fray in the electric-assist bicycle market. This offer from Mercedes-Benz provides about 30 km (18.75 miles) of assisted travel on a full battery charge, which takes five-and-a-half hours. Build quality matches the brand name, so we can expect this bicycle to be durable.

🖋	Daimler Chrysler Japan Holding, Japan	313
⚙	Daimler Chrysler Japan Holding, Japan	313
📑	Various	
🎧	• Electric and human power • Zero emissions (if recharged from renewable sources)	327 326

Delite

A standard frame fitted with a range of modular components can suit either touring or racing bicycles or a hybrid. Suspension is provided by front fork dampers and a damped rear sub-frame assembly.

✏️	riese und müller GmbH, Germany	320
⚙️	riese und müller GmbH, Germany	320
📃	Various	
🎧	• Modular design	327

Chameleon

Despite the well-proven rigidity and ride quality of small-wheeled folding bikes, there are many who prefer the more conventional ride of 24-inch wheels. Airnimal's Chameleon is a re-examination of folding options to accommodate the larger wheels to deliver higher levels of ride stability and comfort without sacrificing portability. This is a high-spec machine comprising a 7005-T6 alloy tubing frame, bladed carbon fibre front forks and 27-speed triple chainring Shimano 105 groupset gears with STI shifts. The ovalized monotube doesn't have any folding components and provides a rigid backbone to the rear triangulated suspension sub-frame and seat post. Weighing about 10 kg (22 lb) without pedals this is a 'packable' rather than a quick-folding bike. Three levels of packing are possible. Level one ensures you can put two bikes in a conventional car boot and requires removal of the seat post and front wheel using quick releases. Level two requires removal of both wheels and the handlebar in order to pack it into a 610 x 610 x 280 mm (24 x 24 x 11 in) suitcase. In level three, suitable for air travellers, you can fold the Chameleon frame and accessories neatly into a carry-on, hand-luggage bag 560 x 360 x 200 mm (22 x 14 x 19 in), although wheels and any saddle bags will have to join the rest of your luggage in the cargo hold. Probably more suited to the long-distance or weekend recreational cyclist than the daily train commuter, the Chameleon is a versatile bicycle. A modified racer version is available.

✏️	Airnimal Designs Ltd, UK	311
⚙️	Airnimal Designs Ltd, UK	311
📃	Various alloys, carbon fibre, rubber, steel	
🎧	• Lightweight design • Portable	326

Strida 2

Unfolding the Strida 2 takes ten seconds and immediately reveals its radical triangular frame, a departure from the typical arrangements in other folding bicycles. It weighs in at just 10 kg (22 lb), the tubes being of aluminium and the wheels and other components made from glass-reinforced polyamide, a strong, durable, lightweight polymer. A conventional chain is replaced by a belt drive over low-friction polymer cogs, making for an oil-free and low-maintenance bicycle. Tyre and belt repairs are facilitated by the offset frame-wheel arrangement. Apparently it takes time to master the ride as the frame is not as torsionally stiff as other folding bikes, but thereafter the rider is guaranteed an intrigued audience as he or she sails by car-bound commuters.

✏	Mark Sanders, Roland Plastics, UK	309
⚙	Roland Plastics, UK	320
▤	Glass-reinforced polyamide, aluminium, rubber, stainless steel	
♫	• Reduction in materials used	325
	• Ease of maintenance	327

Windcheetah

A cruciform frame enables the rider to adopt a low centre of gravity, which, when coupled with carbon-fibre fairing, provides very efficient aerodynamics. Pinpoint accuracy of steering is achieved by means of a unique joystick system that gives good stability in cornering. Lightweight materials and precision engineering make this the Rolls Royce of recumbents. The efficiency of the design has attracted interest from courier and local delivery companies who wish to develop zero-emissions transport policies for urban areas.

✏	Advanced Vehicle Design, UK	311
⚙	Advanced Vehicle Design, UK	311
▤	Metal alloys, rubber, carbon fibre, Kevlar	
♫	• Human-powered	327

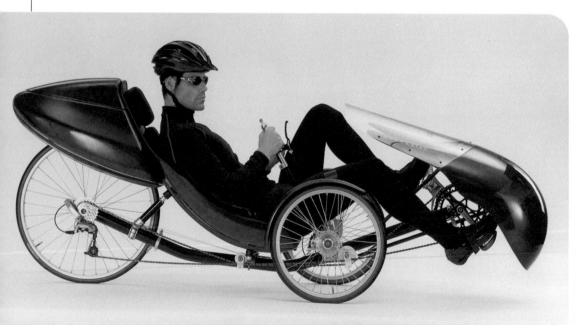

ZEM cycles

Let's face it, the tandem bicycle is not conducive to nurturing long-term relationships, so it is opportune that ZEM have a more modern take on the socio-cultural aspects of cycling. There's a two-cycle ZEM where the occupants sit side by side or the four-cycle ZEM where everyone is a proactive contributor but in a car-like configuration. Suitable for recreational cycling or serious work-outs, both models are robustly constructed, the two-cycle weighing in at 42 kg (92 lb) while the four-

cycle is a sturdy 96 kg (211 lb), and they can take an allowable gross weight of 220 kg and 480 kg (484 and 1,056 lb) respectively. Key design features of the two-cycle include 46 cm (18 in) wheels, two independent seven-speed hub gears, SRAM S7 with reverse gear for all speeds, an ARMOR disc brake system, parking brake, four-point Elastomer shocks and Hella/Basta lighting. Each bucket seat can be individually adjusted for maximum comfort. Fitments are similar for the four-cycle but there

are four sets of gears and GRIMECA brakes in two independent systems and 36 cm front and 40 cm rear wheels (14 and 16 in). Interestingly the two-cycle is classified as a bicycle worldwide but the four-cycle version requires a special licence in Japan

and Switzerland. Both ZEMs offer an alternative choice for those contemplating taking to a bicycle and encourage participation by family and mixed-age groups.

✏️	ZEM, Switzerland	323
⚙️	ZEM, Switzerland	323
🗃️	Various materials, 6060 aluminium frame	
🎧	• Human-powered vehicle	327

Brompton

Folding bicycles are not a new invention but Brompton has manufactured durable products over the last three decades and is probably one of the most popular brands in the UK. The robust 'full-size' steel frame can be

folded within twenty seconds to make a compact package that weighs less than 12 kg (27 lb) and measures only slightly bigger than the 40-cm (16-in) wheels. Currently the range includes two three-

speed (L3, T3) models, one four-speed (L5) and one five-speed (T5), all with rear carrier and dynamo. The T types are supplied with lights too. Optional extras allow customization but the design remains fundamentally little changed since its inception, making it less prone to the whims of fashion. Owning a folding bicycle allows you to cut overall journey times by combining cycling with public or other private transport. Folding bicycles make a viable contribution towards a more integrated and sustainable transport system.

✏️	Brompton Bicycle Ltd, UK	312
⚙️	Brompton Bicycle Ltd, UK	312
🗃️	Rubber, steel, plastic	
🎧	• Multifunctional • Durable	327 327

Gemini

Ironically 'Mountain' or 'All-terrain' bikes are the dominant type of bicycle in many cities and urban areas. Yet these bicycles are rarely suited to the demands of transporting young children or heavy loads. Gemini (denoting the 'Twins' in astrology) includes riese und müller's innovative Transport System that completely re-designs the handling and balance points for a loaded bicycle. It includes an integrated rear frame carrier and special mounting points for child seats and luggage baskets. The front child or load carrier is positioned between the handgrips and above the centre of the turning axis for the steering. The frame re-design and 61-cm (24-in) wheels give much better stability and safer handling, to reduce the risk of accidents. With a carrying capacity of 40 kg (88 lb) excluding the rider, the Gemini can operate with two children or other heavy loads. This is an intelligent re-appraisal of our cycling mobility needs.

✎	riese und müller, GmbH, Germany	320
⚙	riese and müller, GmbH, Germany	320
📃	Various metals, rubbers, plastics	
🎧	• Multifunctional load carrier	327

Xootr Cruz

Skateboard culture meets the bicycle in this resurrection of the old push scooter. Lightweight aluminium frame, cast wheels and a low-slung laminated birch wood deck ensure manoeuvrability and stability. This vehicle is very portable, weighing just 4.5 kg (10 lb) and folding to a package less than 800 mm (31 in) long.

✎	Nova Cruz Products, USA	319
⚙	Nova Cruz Products, USA	319
📃	Birchwood, aluminium, polyurethane	
🎧	• Human-powered	327

Electric Shopper, Eurobike, Commuter Folding, Powatryke

Powabyke is a range of electric-assist bicycles to suit all ages, commuters and recreational users. The range without pedal assist varies from 21 to 48 km (13–30 miles) according to the exact model. Batteries are 14-amp, 36V, sealed lead acid, which reach full charge over eight hours and drive a 150W or 200W front- or rear-hub-drive DC motor. A folding version offers commuters an easy, less energetic start to the day.

✎	Powabyke Ltd, UK	319
⚙	Powabyke Ltd, UK	319
📖	Various	
🎧	• Hybrid human-/electric-powered transport	327

Shimano wheel system

In the West huge new markets for mountain, all-terrain and road bicycles have been created by a resurgent interest in cycling as a recreational pursuit. Sadly, many of these bicycles are unlikely to last more than five to ten years as manufacturing quality is poor. Like the automobile industry, bicycle manufacturing is extremely competitive and is dominated by a fashion-led agenda. Fuel consumption and carbon dioxide emissions data are part of the sales package for cars. An equivalent measure for bicycles would be a measure of the human energy needed to cover specific distances and gradients, measured in milliJoules per kilometre (MJ/km) or something similar. With such a measure the consumer would really be able to assess the efficiency of each bicycle. That efficiency would be dependent on the weight and aerodynamic characteristics of the bicycle, and it is here that the Shimano wheel system would score highly. This is an innovative wheel with the spokes anchored in the side wall of a lighter-than-average carbon-fibre rim. Strong rim-to-hub triangulation, with an unusual lateral crossover pattern, creates great lateral rigidity. Combining all these innovations gives a very strong lightweight wheel that improves acceleration. Carbon fibre ensures toughness, durability and aerodynamism to deliver a smooth ride. This is a bicycle wheel designed to last a lifetime, at least for the average recreational user, and so offers improved eco-efficiency.

✏	Shimano, Inc., Japan	321
⚙	Shimano, Inc., Japan	321
📜	Carbon fibre, metal	
🎧	• Improved eco-efficiency • Durable	327
🔍	iF Design Award, 2002	330

SRAM 9.0 sl

SRAM manufactures brakes, gears and gear shifts to high standards of aesthetics and functionality, using between 30 and 50%-recycled content for many of the sub-components, which can be disassembled for pure-grade recycling in the future.

✏	SRAM Corporation, USA	321
⚙	SRAM Corporation, USA	321
📜	Part recycled content - rubber, metal composites	
🎧	• Recycled content • Design for disassembly	324 325
🔍	iF Ecology Design Award, 2000	330

backpack

One of the problems of being a cyclist commuter is juggling assorted briefcases and bags while maintaining composure, style and functionality. This backpack provides voluminous storage while converting into a saddlebag and daypack. There are numerous design challenges to overcome before cycling becomes a favoured mode of transport for the daily commute. The entire journey from home to office is ripe for design innovation, from clothing to safety devices, bicycles, cycleways and workplace facilities.

Leggero Cuatro 1 / Cuatro 2

Safety features of this bicycle trailer for children include a low centre of gravity, seat belts, a protective plastic shell and a warning flag. All-weather protection allows flexibility of use and ensures that the children have a good view and can feel the breeze.

🖊	Christophe Apotheloz, Switzerland	304
⚙	Brüggli Produktion & Dienstleistung, Switzerland	312
📇	Various	
🎧	• Human-powered transport for the family	327
🏆	if Design Award 2003	330

🖊	Aran Hartgring, graduate student 2003, Design Academy Eindhoven, the Netherlands	305
⚙	Prototype	
📇	Various	
🎧	• Multifunctional	327

Curraghs

Imagine making a lightweight, versatile rowing or sailing boat using traditional methods for less than £150. Look no further than the Irish curragh 'skin-boats', whose pedigree stretches back to the early Neolithic period. All curraghs are built upside-down, with the willow withies or hazel rods stuck into holes in a wooden gunwale. The sides are shaped by weaving in a pattern called French randing then the withies are bent over and tied with tarred hemp twine to form the keel-less base of the boat. Today, the frame structures are covered with heavy calico (canvas) tacked to the gunwale and covered with tar for waterproofing. Until the nineteenth century the traditional covering was cowhide. At least twelve different types of curragh are found on the Atlantic seaboard of Ireland, varying from Bunbeg and Dunfanaghy curraghs of 3–6 m (10–20 ft) length, to the great 8 m- (26 ft-) long, four-man, Kerry naomhóg capable of taking loads of nearly two tonnes. Usually the curraghs are rowed: each pair of oars are thin-bladed with a wooden counterweight inserted over a wooden pin. Occasionally a simple lugsail is hoisted on a detachable mast. Curraghs are reliable sea-going boats but are also at home in estuarine waters or inland lakes. Designs evolve as makers innovate with local materials and respond to local conditions. Such boats are a good example of economic, environmental and social sustainability.

✏	Traditional and hobby designer–makers, Ireland and UK	
⚙	Serial one-offs	
📋	Willow, hazel, pine, deal, canvas, tar, hemp twine	
🎧	• Renewable materials	325
	• Biodegradable	324
	• Local sustainable design	325
	• Retention of traditional craft skills	324

Soundgarten

Playing with sound is a learning experience with the Soundgarten. Individual cones, each with a different icon and unique sound, can be arranged within the central resonating body. Tactile skills to arrange and 'play' the cones help create a unique sound environment whose volume and array can be modified and recorded. Aimed at children three to eight years of age, it is intended to support musical education and offers an audio experience for everyone irrespective of musical ability.

✏	Michael Wolf, Germany	310
⚙	Prototype	
📋	Various materials	
🎧	• Interactive design	327
	• Tool for education	326
🔍	iF Design Award, 2003	330

Synchilla® Snap T®

In 1991 the outdoor clothing manufacturer Patagonia declared in their catalogue that 'Everything we make pollutes'. This was the beginning of the company's process of reducing its environmental impacts by switching to organically grown cotton and by manufacturing fleeces derived from recycled plastic bottles. Post-consumer recycled (PCR) Synchilla® fleece was developed with Wellman, Inc., in 1993. Each garment saves twenty-five two-litre PET bottles from landfills.

✎	Patagonia, USA	319
⚙	Patagonia, USA	319
🗎	Synchilla® fleece, Supplex® nylon	
♻	• Recycled materials • Reduced emissions (compared with virgin PET fibre)	324 326

Veloland

Veloland is an information provider and service company responsible for a network of over 6,300 km (4,000 miles) of national and regional cycling trails in Switzerland. Maps, guides and a website provide wide access to information nationally and beyond. Trails have been linked with public-transport networks and bicycle rental at SBB railway stations, and the whole system is covered with consistent, standardized signage.

✎	Veloland Schweiz, Switzerland	310
⚙	Various	
🗎	Various	
♻	• Encourages alternative modes of transport (integrating cycling with public transport)	326

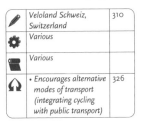

NIGHTEYE®

This lightweight headlamp is fitted with the patented Ultralight using a low-energy Xenon bulb with a special reflector to provide a quality light source. A rear red LED personal safety light fits to the back of the headband. All components, including the polycarbonate casings, clip together and so can be separated for recycling.

✎	PROFORM Design, Germany	309
⚙	Nighteye GmbH, Austria	318
🗎	Polycarbonate, elastic, LEDs, Xenon bulb	
♻	• Low energy consumption	327
🏅	iF Design Award, 2000	330

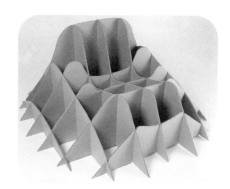

Terra Grass Armchair

A subtle merging of man and nature is embodied in this witty outdoor seat reminiscent of some mini Bronze Age burial mound. The structural framework is provided by corrugated cardboard to which locally sourced soil is added and grass seed applied. Just a few weeks later succulent grass covers your very own green throne for the garden.

✏	N Fornitore, Italy	318
⚙	Purves & Purves, UK	318
🗦	Corrugated cardboard, grass seed, soil	
♻	• Renewable, compostable and locally sourced materials	324/325

du feu

In temperate Western Europe most people's idea of outdoor cooking is burnt barbecue food in the summer months. While the taste of the food can be extremely variable, the bonhomie of cooking al fresco usually more than compensates. Liefting has built on this communality with an elongated table that combines gas rings with preparation and eating surfaces. Solid wood surfaces and steel fixtures bring an industrial confidence and durability to this design.

✏	Mander Liefting, graduate student 2002, Design Academy Eindhoven, the Netherlands	305
⚙	Prototype	
🗦	Wood, steel	
♻	• Durable materials • Communal facility	324

Terracotta incense burner

Technology from the Middle Ages is not often revered these days, but many design solutions from then have yet to be bettered. This terracotta incense burner originating from France traditionally burns cade wood incense from the small juniper-like shrub found in southern France. Cade is a natural air purifier and insect repellent that also actively absorbs bad aromas. Pure cade wood is easily mixed with other fragrant powders or flower parts such as lavender, rosemary or orange. These mixtures are completely hazard-free, unlike deadly aerosol insecticides. Two versions of the burner are produced, an indoor one just 110 mm (4½ in) diameter, or a larger outdoor burner 255 mm (10 in) diameter. Either version is an ideal way of 'transporting' oneself to the verdant aromatic slopes of Les Alpes Maritimes or Provence.

	Indigenous makers, France	
⚙	Natural Collection, UK	318
📖	Terracotta, herbs	
♻	• Abundant geosphere and renewable materials	324
	• Safe, non-toxic	327

E-tech

The E-tech two-stroke motor for chainsaws and strimmers combines efficient power production with a new catalytic converter, ensuring that these motors meet the world's strictest standard for emissions for motorized hand-held garden and forestry equipment, the 1995 California Air Resources Board (CARB) standard. Electrolux, the world's largest manufacturer of chainsaws, has also forged partnerships with petroleum companies to develop fuels that reduce emissions. For example, the Finnish company Raision offers a vegetable chain oil for chainsaws through Husqvarna.

	Husqvarna/The Electrolux Group, Sweden	316
⚙	Husqvarna/The Electrolux Group, Sweden	316
📖	Metals, plastics	
♻	• Reduction in emissions	326
	• Reduction in consumables	327

FSC Fencing

A decade ago most consumers had not heard of the Forest Stewardship Council, the organization dedicated to encouraging sustainable management practices in forests, timber producers and manufacturers. Today the FSC endorse a wide range of products. These individually designed fencing panels are manufactured from planed, pressure-treated European timber from FSC certified wood. Pressure-treating wood by injecting preservatives is the industry's standard way of extending durability, since these products are made from fast-growing softwoods, mainly conifers. In the longterm, the pressure-treatment cycle could be omitted if larger quantities of hardwood, such as oak (*Quercus spp.*) and sweet chestnut (*Castanea sativa*), are grown in small roundwood plantations that are regularly harvested or coppiced. Sweet chestnut, often used as fencing posts, is renowned for its durability.

	Natural Collection, UK	318
⚙	Natural Collection, UK	318
📖	Timber	
♻	• Certified timber from sustainably managed forests	324

Can-O-Worms

A compact self-assembly series of nested circular trays, made from 100% post-consumer recycled plastic, is supplied with a coir fibre block. This block is moistened and broken up, then placed in the bottom tray to provide 'bedding' for a colony of native composting worms. As a tray is filled with household or garden waste, another is added to build up the stack. The worms migrate up and down the stack through the mesh in the bottom of each tray, digesting waste and turning it into compost.

✏	Reln, Australia	309
⚙	Reln, Australia, with Wiggly Wigglers, UK	322
▦	Plastics, live worms	
⏏	• Encourages local biodegradation of waste	328
	• Recycled and renewable materials	324/ 325

Compost converter

Manufactured from 95%-recycled plastic, the 220-litre (48-gal) capacity Compost converter is a strong, rigid bin with a wide top aperture, which facilitates disposal of biodegradable domestic and garden waste and its conversion into compost, and a wide hatch at the bottom for removing mature compost. Blackwall make a range of compost bins from 200 litres to 708 litres (44–156 gals) capacity for domestic use, from injection-moulded, flat-pack, wood-grain-effect, recycled thermoplastic bins to blow-moulded, cylindrical bins mounted on a tubular galvanized-steel frame, permitting aeration of the compost by regular inversion or tumbling. Bins are guaranteed for at least ten years.

✏	Blackwall Ltd, UK	312
⚙	Blackwall Ltd, UK	312
▦	Plastics, tubular galvanized steel	
⏏	• Encourages local composting	328
	• Recycled materials	324

adjustable spades and fork

An adjustable telescopic shaft provides a variable length of 105 cm to 125 cm (41–49 in) to accommodate people of different heights. Anyone who has done any serious gardening will know that inappropriate tools can cause back and posture problems. These well-made garden tools ensure efficient and healthy work practices.

✏	Fiskars Consumer Oy, Finland	314
⚙	Fiskars Consumer Oy, Finland	314
▦	Metal, plastics	
⏏	• Universal design	327
	• Improves health	327
🔍	iF Design Award 2003	330

MicroBore

Porous piping placed in or on the soil surface provides a means for the precise delivery of water direct to the plant root zone. The flexible, rubber-based hose made from shredded, recycled tyres is perforated with holes, allowing water to trickle into the soil. Pipe diameters vary from 4 mm or 7 mm ($\frac{1}{6}$ or $\frac{1}{4}$ in) for the MicroBore, which is ideal for watering window boxes and office plant displays, to 13 mm, 16 mm or 22 mm ($\frac{1}{2}$, $\frac{1}{3}$ or $\frac{9}{10}$ in) for the HortiBore and

ProBore, which are for commercial horticulture and landscaping. System accessories allow you to customize the irrigation system and permit the use of stored rainwater with in-line filters, taps, tee-junctions and end-stops.

✏	Porous Pipe, UK	319
⚙	Porous Pipe, UK	319
📄	Recycled tyres	
♻	• Water conservation	328
	• Recycled materials	324

Solar mower

In 1997 Husqvarna combined robotic technology with solar power to create a unique lawnmower capable of autonomously maintaining an area of up to 1,200 sq. m (1,440 sq. yds). By 1998 the Auto Mower offered another robotics-driven option able to recharge itself from a mains electricity supply.

✏	Husqvarna/The Electrolux Group, Sweden	316
⚙	Husqvarna/The Electrolux Group, Sweden	316
📄	Photovoltaics, plastics, metals, motor, battery	
♻	• Solar-powered	327
	• Zero emissions	326

Milan

Composting efficiency is often improved if the raw garden waste is shredded first. Meeting all European safety standards and qualifying for the German Blue Angel ecolabel, the Milan sets the standard for garden shredders. Gloria developed a new cutting system, Vario-Cutbox, which is easier to use and reduces noise.

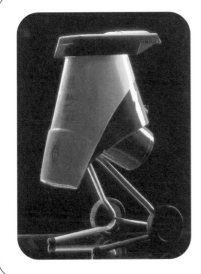

✏	Gloria-Werke, H. Schulte-Frankenfeld, Germany	315
⚙	Gloria-Werke, H. Schulte-Frankenfeld, Germany	315
📄	Plastics, steel, rubber motor	
♻	• Encourages local composting	328
	• Safe operation	327
⚘	Blue Angel ecolabel	

Glass Sound

Chunky speaker cabinets are redundant in this ultimate minimalist sound system. Suspended from stainless steel wires, which carry the signal, a thin glass diaphragm emits sound waves. This system uses NXT technology to deliver the sound.

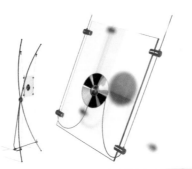

✏	Christopher Höser, Designteam, Glas Platz, Germany	315
⚙	Glas Platz, Germany	315
📜	Glass, stainless steel	
♻	• Reduction in materials used	325
🏆	iF Ecology Design Award, 2000	330

Xenaro GDP 6150/1

Multimedia standards are becoming the battleground for market domination, so it is refreshing to find an electronics company that recognizes the need of consumers to use the format that most suits them. The Xenaro DVD multimedia player is a small-footprint device in a tough metal case, capable of playing DVD, VCD, S-VCD and audio CDs. It can also record CD-R/RW in MP3, MPEG1 and MPEg2 formats. Such cross-format capability is to be welcomed as it ensures one device can do the job of many. While this can result in loss of sound quality, this is not the case with the Xenaro as it features Dolby Virtual Surround.

✏	Grundig Produkt Design, Germany	316
⚙	Grundig AG, Germany	316
📜	Various metals, polymers, electronics	
♻	• Multifunctional (cross-media compatibility)	327
🏆	iF Design Award, shortlist 2003	330

MUJI CD player

Consistent with MUJI's philosophy of reductionism and minimalism, this CD player has a pared down form while simultaneously creating an almost surreal piece of wall sculpture. This intriguing object does some semiotic gymnastics. Pull the power cord and watch the CD spin like a fan of an air-conditioning unit. But it is more than a whimsical design. Speakers are integral to the lightweight, 550 g (19 oz), player so every effort has been made to minimize the consumption of materials.

✏	IDEO Japan, Naoto Fukasawa and MUJI, Masaaki Kanai, Japan	307
⚙	MUJI, (Ryohin Keikatu Co. Ltd), Japan	318
🗎	Various polymers, metals and electronics	
⌾	• Lightweight design • Reduction in materials used	325 325
⬤	iF Design Award, 2002	330

iPod

With the dust of the storm surrounding the copyright issues over the downloading of MP3 music files still swirling around, Apple Computer took the industry by surprise and launched the iPod MP3 player. Available for both Mac and PC users, the iPod is actually the front end of a complete system for downloading, cataloguing and playing music from licensed copyright sources. The iPoD is dematerialization in action. Weighing just 182 g (6.5 oz) the device has a storage capacity of either of 15, 20 or 40 Gigabytes (with the new, even smaller, iPod Mini having a storage capacity of 4 GB), representing up to 10,000 music tracks, Apple claim. It takes only minutes to download tracks from the Internet, and the ten-hour Lithium Polymer battery recharges via the computer's power source. Imagine the equivalent music archive on CD; it would be a veritable stack of hundreds of CDs with their plastic cases and all that embodied and transportation energy to get the stock to the retail point of sale. The lightweight polycarbonate/ABS top clicks into a polished stainless steel case, providing ergonomic perfection and an intuitive communication interface. Most importantly, it is a highly portable accessory and, to all intents and purposes, supplants the need for other kinds of portable personal stereos. An object lesson in dematerialization, the iPod suggests possibilities for future product-service-systems (PSS). However, it is worth remembering that even today less than 10% of the world is actually connected to the Internet, so the benefits of dematerialization are not universal.

✏	Apple Industrial Design Team, USA	311
⚙	Apple Computer, USA	311
🗎	Polycarbonate, ABS, stainless steel, electronics	
⌾	• Dematerialized product-service-system • Reduction in materials and energy consumption	324 325
⬤	iF Design Award, 2003	330

Freeplay Ranger/Coleman Outrider

This dual-band, AM/FM Freeplay Ranger radio is compact but rugged. The wind-up handle on the back of the radio provides about thirty-five minutes' playing time at normal volume for each thirty-second wind and a small LED indicator lets you know the best wind speed. In direct sunlight a small photovoltaic panel on the top allows continuous playing without recharging but if the NiMH battery ever runs down you can wind up or recharge it from the mains using the AC/DC adapter. A fully charged battery can give up to twenty-five hours playing time. The radio is available in the USA as the Coleman Outrider.

✏	Freeplay Energy Ltd, South Africa & Europe	315
⚙	Freeplay Energy Ltd, South Africa & Europe	315
▤	Various materials	
🎧	• Renewable energy options	327

Summit

Evolving from the S360 model is the new Freeplay Summit, a top-end radio with all the self-sufficiency energy features a user could want – an AC alternator, a 4.2V 34mA rated solar panel and a 100mA AC/DC mains travel adaptor, all linked to a rechargeable NiMH battery pack. LCD Digital tuning provides thirty pre-selected stations but the four wave bands, FM/MW, SW and LW, mean plenty of choice. For eco-travellers anywhere, the Summit treads lightly wherever it goes.

✏	Freeplay Energy Ltd, South Africa & Europe	315
⚙	Freeplay Energy Ltd, South Africa & Europe	315
▤	Various materials	
🎧	• Renewable energy options	327

SmartWood guitars

Gibson Guitars are renowned for the quality and sound of their acoustic and electric guitars. This model uses hard maple, Honduras mahogany and Chechen woods certified under the SmartWood and FSC schemes and supplied by EcoTimber, Inc. There's good evidence that Gibson guitars are cherished by their owners and accordingly have an in-built longevity. Use of certified woods for high-value products doubly reinforces the message about designing for longevity using materials from sustainable sources.

✏	Gibson Guitars, USA	315
⚙	Gibson Guitars, USA	315
▤	Various woods, metal, electronics	
⌂	• Renewable materials • Certified SmartWood, FSC timber	325 324

Tykho

Tough thermoplastics predominate in casings for electronic goods but Marc Berthier demonstrates that bucking convention produces a new sexy look for his VHF radio. Rubber also confers benefits over plastics by offering some shock resistance and weatherproofing.

✏	Marc Berthier, France	304
⚙	Lexon Design Concepts, France	317
▤	Rubber, electric components	
⌂	• Renewable and synthetic material	325

Terracell

Miniaturization is a given for personal electronic products: small is sexy and saleable. Terracell is a prototype mobile phone for a more holistic view of miniaturization, minimizing the embodied energy of the entire lifecycle – minimal material, manufacturing processes, transport, power supply and disposal energy. Cradle to cradle thinking is facilitated by design for disassembly at end-of-life for materials recycling. A methanol-base battery provides a long-term rechargeable power source.

✏	Michael Lemmon, USA	307
⚙	Prototype	
▤	Recycled aluminium and plastics, lead-free system board, LCD screen	
⌂	• Design for disassembly • Lyfecycle analysis	325 324
✪	IDRA award, 2002–2003	330

Savvy

This feature-packed mobile phone provides a 'joystick' central control to access menu options, which include a calculator, clock with stopwatch and games. Customers can choose from a range of coloured plastic components to

customize the look of their phone. This same feature ensures parts are easily disassembled to update as fashion dictates or repair or recycle. With low power consumption and a 30% reduction in the number of components, Philips are striving to give today's modern icon of communication a green conscience.

✏	Philips Design, the Netherlands	319
⚙	Philips Electronics, the Netherlands	319
📑	Recyclable plastics	
♻	• Design for disassembly and recycling	325
	• Low energy consumption	327

Xenium™

Philips Electronics are continuously developing their policy of designing products with reduced environmental impact, as is evident in this range of mobile phones. With a casing of punched metal, a 35% reduction in components, reduction in energy consumption and use of smaller recharge batteries, this phone has a smaller ecological footprint than previous models. At the end of its life it is easily disassembled into pure-grade materials or material groups. And it blends functionality with simple good looks.

✏	Philips Design, the Netherlands	319
⚙	Philips Electronics, the Netherlands	319
📑	Metal, electronics	
♻	• Energy reduction	327
	• Reduction in consumables	327
🔍	iF Design Award, 2000	330

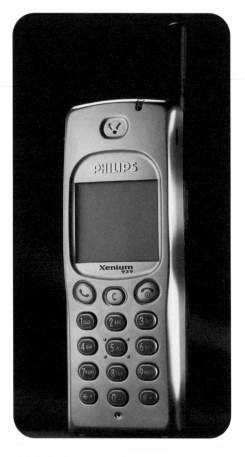

P800 Multimedia mobile phone

The rapid penetration of mobile phones into contemporary culture mirrors the phenomenon of the Sony Walkman in early 1980s, so it is not surprising to see Sony challenging the technological and cultural boundaries of the mobile phone market. The key advantage of the P800 is its ability to multi-task thanks to a range of functions that an average hand-held computer would be proud of. A 128Mb Duo memory stick and integrated MP3 player mean that the P800 is not only a mobile phone but also provides an essential role as a PDA (Personal Digital Assistant) to link to a PC, a digital camera and Internet access. A colour display and stylus ensure a tactile and usable interface. Multifunctional electronic products ensure users get the most from the rest of their hardware and software. The big question is how long do such products maintain their currency before obsolescence rears its ugly head? As a member of the Dow Jones Sustainability Indexes, Sony have embedded lifecycle thinking and sound environmental management into their day-to-day business. It is reasonable to expect that the P800 will be superseded by newer models but when such equipment meets the end of its useful life Sony Ericsson should take responsibility for recycling materials and safe disposal of any waste. Electronic equipment must meet stringent recycling targets in the next few years under the European WEEE Directive as it is enshrined in member states' legislation.

✏️	Nya Vattentornet, Sony Ericsson, Sweden	321
⚙️	Sony Ericsson, Sweden	321
📖	Polymers, metals, electronics	
🎧	• Multifunctional	327

Canon IXUS range

A new film format, Advanced Photo System (APS), was introduced worldwide by the photographic industry in the mid-1990s. APS encouraged new camera designs since the film area is 24% smaller than standard 35 mm film, resulting in much smaller cameras than conventional SLRs. Other advantages include multiformat-frame option (normal, intermediate or panoramic), automatic mid-film rewind/reuse and a strip on the edge of the film that records exposure details to improve printing results. In short, APS offers significant reductions in the consumption of raw materials, film and prints in the consumer market. Launched in 1996, the IXUS camera quickly became popular and was probably responsible for converting many newcomers to APS, with its James Bond looks and sharp prints. The range includes the IXUS III, Z65, 250 , M–1 and X–1.

✏️	Yasushi Shiotani, Canon, Inc., Japan	312
⚙️	Canon, Inc., Japan	312
📖	Stainless steel, polycarbonate, ABS	
🎧	• Reduction in materials and consumables used	325/ 327

B. M. vase, fruitbowl and spoon

B. M. refers to the Bell Metal Project, 1998–2000, for which Pakhalé applied and refined the *cire perdue* or lost wax casting technique popular in central India for hundreds of years. Wax harvested from jungle trees is purified and extruded to produce round profile strings, which are wound around a sand/clay/dung core. Once the desired shape is attained the whole wax structure is encased in more sand/clay/dung mixture, to make a mould, which is then fired. Molten scrap brass, bronze and copper are poured into the mould and allowed to cool. Once extracted from the mould the cast piece is hand-cleaned and brushed. Pakhalé breathes new life

into these time-honoured techniques by sympathetically and skillfully creating contemporary forms. The B. M. objects elicit a renewed joy in design, reclaiming its purpose to satisfy physical, mental, emotional and spiritual needs rather than just deliver profits and cultural status.

✏	Satyendra Pakhalé, the Netherlands and India	308
⚙	Atelier Satyendra Pakhalé, the Netherlands	308
📓	Brass, bronze, copper scrap	
🎧	• Recycled metals	324
	• Low-energy manufacturing	325
	• Innovation of traditional technology	325

Pinch

In homage to the beauty and functionality of the clothes peg, the artist Jos van der Meulen brings our thoughts full circle – from the twigs in the forest to an artefact in the service of man. Like little statues these unique branches stand silently waiting for something to happen, for something to appear in their little sprung jaws – a keepsake, photo or other object to become momentarily a point of personal reference, a focus of conversation, or more.

✏	Jos van der Meulen, the Netherlands	310
⚙	Goods, the Netherlands	315
📓	Wood, steel	
🎧	• Renewable and recylable materials	324/ 325

Bob

This cold-construction concrete vase challenges the dominance of ceramics and searches for new expressions of that most modern of materials – concrete. Only a detailed lifecycle analysis will reveal whether ceramics, with their high-energy requirement to fire the clay, are more or less benign than concrete, which also requires energy and results in emissions to the air during the manufacture of the cement.

✏	Goods, the Netherlands	315
✿	Goods, the Netherlands	315
📱	Concrete	
🎧	• Abundant geosphere materials	324

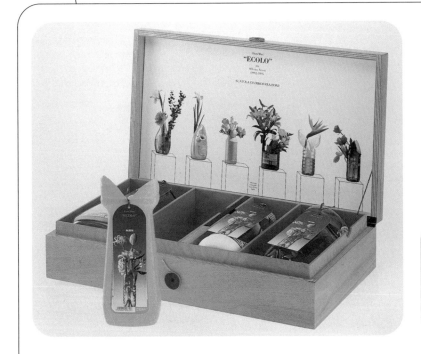

Ecolo

A booklet conceived and written by Enzo Mari and published by the renowned Italian manufacturer Alessi inspires the reader to transform the effluent of consumer culture into delicate, beautiful objects. A shampoo bottle is transformed, like a butterfly emerging from its pupa, from useless spent object to graceful flower vase. Design is taken out of the hands of the specialists and returned to the masses.

✏	Enzo Mari, Italy	308
✿	Alessi, Italy	311
📱	Post-consumer containers	
🎧	• Reuse of waste objects	326
	• Interactive design	327

TR112

Rosewood, wenge, zebrano, walnut or maple veneers form the woven strips of these hand-made Californian curved square trays, perfect for holding fruit or decorative objects. The veneer needs only an occasional wipe with oil or wax to bring out the beautiful tones and hues of the wood.

✏	Lian Ng, Publique Living, the Netherlands	320
⚙	Publique Living, the Netherlands	320
▤	Wood veneers	
♺	• Durable	327
	• Renewable materials	325

How Fusion

This series of desk-top or table-top knick-knacks are made from polished cast aluminium. The range includes pen holders, a joss-stick holder and a variety of small vases, bowls and containers. Each object has its own personality and tactility, and will endure the ravages of time much better than mass-produced plastic objects.

✏	Setsu and Shinobu Ito, Italy	307
⚙	Nava, Italy	318
▤	Aluminium	
♺	• Single material	325
	• Recyclable	324
	• Durable	327

Wagga-Wagga

This piece balances harmony and tension, using readily available materials. The principle can be extended to a wide range of materials using cold construction and/or heat deformation.

✏	Camille Jacobs, Singapore	307
⚙	Limited batch production	
▤	Float glass, bamboo	
♺	• Reduction in materials used	325
	• Recyclable materials	324

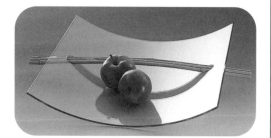

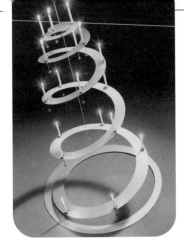

Spiralbaum

This flat-pack, laser-cut, plywood square unfolds as a helix when suspended. It is the ultimate in minimalist Christmas trees, is easily stored away for the next festive season and saves another Sitka spruce from being consigned to the landfill site each New Year.

✏	Feldmann & Schultchen, Germany	306
⚙	Prototype for Werth Forsttechnik, Germany	322
▤	Plywood, steel	
↻	• Reusable product	324

Sponge vase

Marcel Wanders commandeers nature's own manufacturing, adds his own porcelain tube and sets vase design on a new course. Designers should actively seek opportunities for 'harvesting' nature's products, which, with minimal energy input or modification, can be re-manufactured into new objects. Can we look forward to specialist 'product farms' where sponges are 'bio-manufactured' in neat rows, bamboos grow to EU regulation size in specially built moulds and bio-plastics spontaneously grow to predetermined forms?

✏	Marcel Wanders, Droog Design, the Netherlands	310
⚙	Droog Design / Moooi, the Netherlands	313/318
▤	Natural sponge, porcelain	
↻	• Renewable material	325

Timber

Weather-beaten wooden boats hold a certain fascination and the marks of Father Time remind us of pleasure boating, salt air and the inevitable conquest of the sea. Davids scours ship- and boatyards for salvaged planking and scrap wood and lovingly transforms these resources into individual objects.

Damaged paint, nail holes and wear and tear create a canvas of expression, histories of a previous life.

✏	Rik Davids, the Netherlands	305
⚙	Goods, the Netherlands	315
▤	Salvaged boat timber	
↻	• Reused materials	326

transform from everyday to special, from industrial production to unique one-offs, from the ordinary to the beautiful. Moooi challenge the concept of mass production.

Salvation Ceramics

Celebrating everyday mass-produced things, the sum of the parts creates a greater whole as each 'new' Salvation Ceramic comes to life. The Boyms design a range of basic shapes that are assembled by stylist Rebecca Wijsbeek. Inviting us to examine the original components, these objects

✏	Constantin and Laurene Boym, USA with Recbecca Wijsbeek, the Netherlands	305
⚙	Moooi, the Netherlands	318
📜	Secondhand ceramics, adhesive	
⟁	• Reused objects	326
	• Low-energy manufacturing	325

The Soft Vase

Challenging our perceptions about polymers and the way we use them, Hella Jongerius experimented with flexible rubberized polyurethane to create a traditionally styled vase. It provokes us to question how we value plastics. Objects made from plastics can be highly valued in the case of 'designer' objects or regarded as a throwaway item in the case of the ubiquitous plastic bag. Instead of using hard durable ceramics or tough shiny ABS, the traditional materials for a vase, Jongerius has chosen soft flexible polyurethane, thus encouraging the user to experiment with altering the shape of the vase.

Fragility

Porcelain has a tactile quality that sets it apart from stoneware, terracotta and other ceramics. It is therefore well suited to delicate applications, something that van den Heuvel has exploited in her decorative designs. Lace and other fine woven fabrics are dipped in porcelain slip to emerge as fragile, impractical

✏	Jessica van den Heuvel, graduate 2002, Design Academy Eindhoven, the Netherlands	305
⚙	One-offs	
📜	Textile, porcelain	
⟁	• Abundant geosphere materials	324

objects. Their function is aesthetic – to give pleasure and engage the curious. Decoration for its own sake.

✏	Hella Jongerius, JongeriusLab, the Netherlands	307
⚙	Droog Design, the Netherlands	313
📜	Polyurethane, elastomers	
⟁	• Improved user-friendliness	327
	• Durable	327
	• Recyclable	324

Attila

Consumers' voracious appetite for convenience drinks will ensure that the humble steel or aluminium drinks can will be a feature of the twenty-first-century landscape. While recycling of these cans improved significantly during the 1980s, any device that actively encourages people to recycle more is a good thing. Attila is a durable crusher that is a pleasure to use: simply place your can in the bottom of the translucent column and enjoy that satisfying crumpling noise as the 'anvil' crushes the can with the downward push of the arms.

✏	Julian Brown, Studio Brown, UK	305
⚙	Rexite SpA, Italy	320
🗎	Injection-moulded ABS, polycarbonate, Santoprene	
↻	• Encourages recycling	326

Cricket

Consumption of bottled water and soft drinks contained in PET plastic bottles has risen dramatically in the last decade, so any device that facilitates recycling is to be welcomed. This witty bottle crusher makes recycling fun and improves storage capacity of containers for collecting waste bottles.

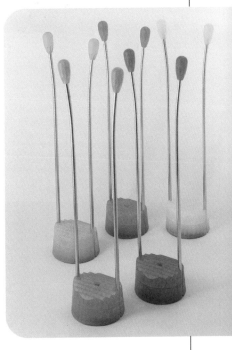

✏	Julian Brown, Studio Brown, UK	305
⚙	Rexite SpA, Italy	320
🗎	Steel, plastic	
↻	• Encourages recycling	326

LINPAC environmental kerbside collection box

Since the introduction of the LINPAC Environmental kerbside collection box in 1996 to the city of Sheffield, UK, over twenty million plastic bottles have been diverted from landfill sites to recycling plants where the plastic is reused to create yet more boxes. This robust box, with high-impact and -deformation characteristics, encourages greater recycling by local authorities and private contractors.

✐	LINPAC Environmental, UK	317
⚙	LINPAC Environmental, UK	317
▤	Used HDPE bottles	
♻	• Recycled materials	324
	• Encourages recycling	326

Zago™

Recycling domestic waste has an image problem, so anything that can elevate this activity into fun is welcome. Three Zago™ triangular rubbish bins made from flat-pack, recycled cardboard neatly sit together to form a functional separator for different waste streams. The photographic exteriors clearly indicate each particular waste stream and reinforce the message that waste is a valuable resource.

✐	Benza, Inc., USA	312
⚙	Benza, Inc., USA	312
▤	Recycled cardboard	
♻	• Recycled materials	324
	• Encourages recycling	326

Bottle stopper and opener

Oxo have a reputation for excellent attention to detail and ergonomics for their hand tools. This easy-to-use device combines two functions and thus improves on conventional products.

✏	Human Factors in co-development with Oxo International, USA	319
⚙	Oxo International, USA	319
📑	Hardened rubber	
🎧	• Dual-function	327

GreenWare

This curvaceous, ergonomically formed cutlery is made of soy bean biodegradable plastic. Production waste has been minimized by creating a universal, standard handle to which the individual functional elements (knife, fork, spoon) are attached. These utensils are not single-use items but can be washed and reused just like normal cutlery until they reach the end of their useful lives, at which point they can be composted.

✏	Julia Carlson, USA	305
⚙	Prototype	
📑	Soy Bean™ biodegradable plastic	
🎧	• Renewable and compostable material	324/325
◉	IDRA award, 2002–2003	330

JOY-STICK

Much of the culture of Asian countries has been absorbed into everyday European life, especially in the proliferation of Asian restaurants and take-aways, yet chopsticks, or eating with the fingers, remain an elusive skill to some Europeans. The designers refer to this hybrid multifunctional object as being an 'Object Genetically Modified'. It is, indeed, a combination of design types that offers fun and functionality but its future use remains untested.

✏	Samuel Accoceberry and Antonio Cos, Italy	304
⚙	Prototype	
📑	Beech wood, steel	
🎧	• Low-energy materials and manufacturing	324/325
	• Multifunctional	327

Cutlery tool

This could be the prototype for a 'universal' cutlery design as it cleverly combines the functions of knife, fork, spoon and teaspoon all in one piece.

✏	Nina Tolstrup, UK	310
⚙	Prototype	
📑	Plastic	
🎧	• Multifunctional	327
	• Reduction in materials used	325

kitchenware

Carefully selecting the most unusual, old, stainless steel washing-machine drums, Kotkas gives them new forms as dish racks and fruit bowls. Recycled or virgin timber from Finland is shaped to take the cut perforated stainless steel drums, then treated with flax oil. It is intended that these designs will be produced in UUSIX workshops in Helsinki where long-term unemployed people are given an opportunity to learn new skills.

🖊	Aki Kotkas, Finland	307
⚙	Small batch production	
📜	Reused washing machine drums, timber	
♻	• Renewable resources • Reused objects • Local employment	325 326 324

Drinking glass

Clever cutting of two PET bottles enables two sections to be rejoined; an original screw top is used as the clamp to form a new glass. It remains unclear whether the offcuts are recycled or can be used to generate other products such as napkin rings.

🖊	Aki Kotkas, Finland	307
⚙	Limited batch production	
📜	PET bottles	
♻	• Reused objects	326

Lavabowl

Buying tableware and kitchen ceramics is always fraught with a catalogue of dilemmas. Do you choose this season's fashionable colour, go for impractical but funky styles or settle for a 'classic' design? Other dilemmas arise around issues of price and quality. Lavabowl avoids all these difficult decisions by being solid and dependable, modern yet traditional, neutral but not boring, and it also manages to produce a porcelain-like 'sound'. De Leede developed a special clay with Royal Tichelaar Makkum by mixing crushed gravel into the clay. This prevents the thick walls of the bowls from cracking but also imparts its own texture and colour flecks to the high-temperature fired bowls.

🖊	Annelies de Leede, the Netherlands	305
⚙	Goods, the Netherlands	315
📜	Clay, gravel	
♻	• Abundant geosphere materials	324

Porcelain tableware

Brenda Fee aims to stimulate much more than our taste buds with her tactile, sculptural and functional porcelain tableware. Culinary skills aside, there's no doubt that food tastes better when served on the finest china platters. Small and Large Weave Plates with Bowl hint at tempting morsels of Eurasian food eaten with chopsticks and fingers. Each design is textured by hand and so each plate is subtly different. Limited editions of sets of three Satin platters extend the sculptural boundaries of tableware. Concealed holes in the salt-and-pepper shakers reveal the objects

functionality at the last minute, sitting happily on the dining table as an aesthetically pleasing function abstract artefact.

	Brenda Fee, UK	306
⚙	Small batch production	
▤	Porcelain	
🎧	• Abundant geosphere material	324

Bone China Line

Just a generation ago the best china tableware would sit in the sideboard waiting for special visitors and guests before it made its appearance. Today the idea of 'best china', and the sense of occasion and pleasure that this activity brought, have faded. Everyday ceramic tableware often doesn't do justice to the food or drink being served but Bodum have sought to change that in their Bone China Line by combining the concepts of 'everyday' and 'best' with modern eating habits. This seven-part dinner service comprises dinner and dessert plates, soup/pasta/rice bowls, a mug, coffee cup and saucer, and espresso cup and saucer. Subtle design detail emerges on inspection, the surfaces of each piece forming an inner and an outer parabola. Tough and durable, bone china is also fired at relatively low temperatures compared to other ceramics, such as porcelain.

	Carsten Jørgensen	307
⚙	Bodum, Switzerland	312
▤	Bone china	
🎧	• Durable	327
	• Abundant geosphere materials	324
🏆	iF Design Award, 2003	330

Basic cookbook

It's difficult to imagine how to cook without using the sense of sight, yet this imaginative cookbook is full of useful tips for those whose tactile, audio, olfactory and taste senses are honed to perfection yet who can't see. The basic cookbook is full of recipes in Braille and the written word. It contains useful tips for visually impaired or blind people, such as how to know when butter fat in the frying pan is turning brown – use margarine with a high water content, it stops sizzling just when it begins to turn colour.

Living water project

There's a generation of young people who believe that potable water arrives in blue-tinted plastic bottles and is carried around as an emblem of urban chic. This perception hides some uncomfortable truths. Many European water-bottling companies are pumping from ancient subterranean aquifers. These aquifers are replenished only slowly and some not at all, so to all intents and purposes these

water resources are finite. The energy involved in extraction, purification, bottling, distribution and disposal of the waste bottles is significant compared with a similar quantity of tap water. And, as enforcement of European legislation regarding pollution and water quality has improved in the last decade, so has the quality of water supplied to consumers. The Living water project attempts to re-engage consumers

with the pleasures of having potable water on tap. Water tumbles through these carafes and is energized and oxygenated to dispel the 'flat' nature of the product emerging from the tap. These carafes are striking pieces mixing the semantics of the wine carafe with scientific distillation equipment. Perhaps the concept can be transferred to portable water bottles in order to rid us of the mountains of PET waste and massive energy consumption associated with brand bottled waters.

	Judith Eurlings, graduate 2002, Design Academy Eindhoven, the Netherlands	305
	One-off	
	Textiles, vinyl	
	• Universal design	327
	• Tool for communication	326

	Anouk Omolo, graduate 2002, Design Academy Eindhoven, the Netherlands	305
	Prototypes	
	Glass	
	• Water conservation	328
	• Reduction of embodied and transportation energy	326

Green Map Atlas

Not only does the Green Map System encourage greener forms of consumption but it fosters a new view of the totality of the local environment – the people, the shops, the diverse businesses and culture – that nurtures the long slow journey towards more sustainable ways of living. Originated by Wendy Brewer in New York, the Green Map System has rapidly expanded around the globe. Its success is not only in the core concept and graphical icons but in its willingness to let local design and local solutions have their voice. In a natural evolution of the system, the new Green Map Atlas brings together this 'localness' so that ideas can be shared and exchanged. Consistent with

 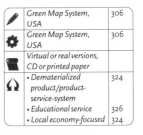

the philosophy of the organization, the atlas can be accessed by individuals according to their preferences and financial means – it is a dematerialized product (as PDFs), a digitized product (CD-ROM) or a locally printed edition.

Ananda

These totemic bookmarks represent the iconographic language of Satyendra Pakhalé , full of energy, life and vitality. While most bookmarks are reduced to rectilinear nonentities, the Ananda almost will the user to pick up a book, if only to have the tactile experience of inserting them in between the pages. Here's one bookmark that won't be consigned to the bin but becomes a personal token, a reminder of one of the true values of objects in our lives.

	Satyendra Pakhalé, the Netherlands and India	308
	De Vecchi, Italy	313
	Silver	
	• Precious material	324
	• Durable	327

	Green Map System, USA	306
	Green Map System, USA	306
	Virtual or real versions, CD or printed paper	
	• Dematerialized product/product-service-system	324
	• Educational service	326
	• Local economy-focused	324

Signlingo

From January to May 2003 at the Danish Design Center in Copenhagen the work of the textile designer Vibeke Rohland challenged public perceptions about the nature of textiles under the exhibition banner 'Signlingo'. Producing silkscreen, hand-printed one-offs on green cotton or flax fabrics, Rohland realizes subtle questions about our social condition with her vibrant graphic designs. The exhibition included *Standstills* and *Target*, modern ying-yang signs as a reaction against the 'overuse' of colour, designs and objects; *Barbie be babe* celebrating Barbie Doll's thirty-fifth birthday; and *Drugstore* and *Pill Stills* reflecting on our modern drug dependency. This graphicacy surprises, lingers and suggests an interesting line of communication for issues around sustainability. *Standstills* is going into commercial production with Hay Cph for a range of covers, bedlinen, towels and other home products.

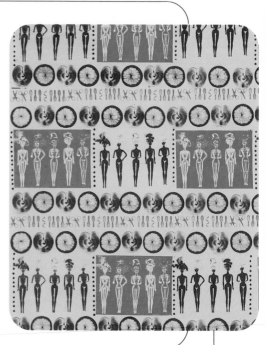

✏	Vibeke Rohland, Denmark	309
⚙	One-offs	
🧻	Cotton, flax, reactive pigments	
🎧	• Renewable materials • Communication design	325 326

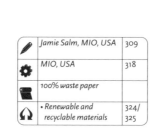

✏	Jamie Salm, MIO, USA	309
⚙	MIO, USA	318
🧻	100% waste paper	
🎧	• Renewable and recyclable materials	324/325

Tangent™ 3D wallpaper

A core philosophical premise at MIO is the idea of 'responsible desire', that is creating attractive, desirable products using readily available urban resources. Tangent™ is manufactured entirely from waste paper in a series of tiles that can be configured in different patterns and tinted with water-based paints. MIO transform a humble material into a visual, tactile wall sculpture adding a new dimension, literally, to the concept of wallpaper. Bored of last year's wallpaper? Strip off the Tangent™ and dispose it at your local recycling point.

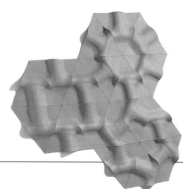

Lifeline Radio

Although not sold commercially, this radio reflects the philanthropic roots of the Freeplay Foundation, an organization dedicated to educational and humanitarian projects based on the idea of communication as a means of empowerment. The Lifeline Radio is a unique multi-band AM/FM/SW radio for families and children in remote rural locations. It is operable by wind-up generator and a small solar panel. It is available to support broadcasting projects for children, youth or other humanitarian causes operated by aid and donor agencies. This is just the arena in which the original Freeplay BayGen by inventor Trevor Bayliss was designed to operate.

✎	Freeplay Energy Ltd, South Africa and Europe	315
⚙	Freeplay Energy Ltd, South Africa and Europe	315
▤	Various materials	
⋒	•Renewable power	327
	• Durable	327
	• Tool for communication/ education	326

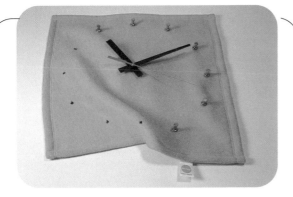

Pin Up clock

Reviewing default uses of materials for specific products encourages experimentation. Rigid materials are the norm for clock faces but Benza has been inspired to use wool fleece pinned to the wall.

✎	Giovanni Pellone and Bridget Means, Benza, Inc., USA	308
⚙	Benza, Inc., USA	312
▤	Wool fleece, pins, clock mechanism	
⋒	• Renewable material	325

doormat

Building on the success of their award-winning carpets from hemp and paper at the Swiss Design Prize awards, 1999, Teppich-art Team have incorporated plastic and rubber waste into a range of doormats. Not only do the waste materials add a visual highlight but they increase the surface texture to facilitate removal of dirt and dust. As eco-efficiency of manufacturing increases there will be an increased demand for products to be manufactured from waste streams. The trick is to add value to these waste materials as they are re-born, to present them as 'new'. Attention to detail, as exhibited here, is important if designers and manufacturers are to succeed.

✎	Teppich-art Team, Switzerland	310
⚙	Anstalten Thorberg, Switzerland	311
▤	Hemp, paper, recycled plastics and rubber	
⋒	• Renewable and recycled materials	324/ 325

Durex Avanti

Natural or synthetic latex has long been the preferred material for manufacturing condoms but the material still suffers from an image problem. Latex produces its own distinctive odour and, owing to the thickness of material required to ensure full protection during intercourse, can result in a lack of sensitivity to the wearer. It also produces an allergic reaction in some people. After considerable research a version of polyurethane proved itself in tests. It is as strong as latex but 40% thinner, is odourless and almost transparent. Add a little flavouring – do you fancy tangerine, strawberry, spearmint? – and here is a little self-help device guaranteed to assist in population control and the fight against the spread of sexually transmitted diseases including AIDS.

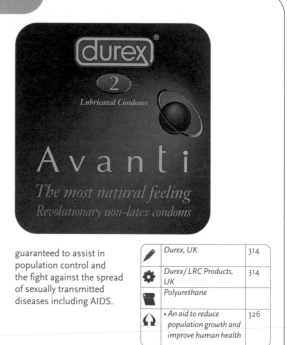

✐	Durex, UK	314
✿	Durex/ LRC Products, UK	314
▤	Polyurethane	
♺	• An aid to reduce population growth and improve human health	326

Earthsleeper™

Made entirely of Sundeala board from recycled newsprint with wood corner joints and wood nuts and bolts, these coffins are highly biodegradable and make less environmental impact than conventional wooden coffins. Coffins are available ready-assembled or as flat-pack, self-assembly units.

✐	Vaccari Ltd, UK	322
✿	Vaccari Ltd, UK	322
▤	Sundeala board	
♺	• Recycled materials	324
	• Compostable	324

Victor Mobilae

Despite years of denial by political lobbyists in many of the G8 nations, there is growing tacit recognition that climate change is a global phenomenon that is

already changing our lives. Manufacturing industries are slowly addressing their responsibilities. White-goods manufacturers in Europe are legally bound to label their washing machines and refrigerators using the EU Energy Label, and the use of CFC and HCFC refrigerants has largely been phased out. Car manufacturers have recently started listing European Emissions Standard data by giving the carbon dioxide emissions in grammes per kilometre. It is, however, unlikely most consumers will be able to put these figures into perspective – Victor Mobilae perfectly encapsulates the message of climate change. This

technically advanced mobile garden moves 167 cm (65¾ in) further north each day, representing the speed at which vegetation is being forced to adapt as our climate warms up. The sole function of this design is socially responsible communication.

✏	René van Corven, graduate student 2003, Design Academy Eindhoven, the Netherlands	305
⚙	One-off	
▤	Galvanized steel, rubber, motor, soil, plants	
↻	• Design as tool for communication • Socially responsible design	326

Mirror mirror

Utilizing existing manufacturing capacity by using ready-mades can be a well-considered eco-design strategy, but shifting perceptions can be a challenge. This design gets it right as the life-buoy gets a makeover.

✏	Startup Design, UK	309
⚙	One-off and small batch manufacturing	
▤	Polyethylene, mirrored glass	
↻	• Durable • Recyclable materials • Use of ready-mades	327 324 326

Tree-hugger

Aimed at those who live in small spaces, the tree-hugger is a device that is easily secured to existing structures using a Velcro fastening. Once in place it enables large quantities of washing to be hung and dried. Johnston's secondary aims were to encourage people to air-dry their clothes by attaching the tree-hugger to balconies, trees and fence posts, rather than resort to using an energy-demanding tumble drier. Traditional or pyramidal washing lines are partially dematerialized as the product is lightweight and flat-pack. It encourages consumers to use their own ingenuity in order to make the device fully operational.

✏	Eimeir Johnston, Ireland	307
⚙	Small batch production	
▤	Various materials	
↻	• Reduction in materials used • User involvement	324 327

Stokke Xplory™

After the success of Stokke's Tripp Trap adjustable children's chair (p. 29) it was only a matter of time before the company tackled new design challenges in their range of products. Xplory™ includes an adjustable footrest, maintains a low centre of gravity, and has four wheels to ensure stability at all times. The central spine is the key to the innovative range of positions in which the child seat can be fixed, so the angle of recline can be chosen to suit the child's requirements. Air bubbles inside the 'solid' rubber tyres and a shock absorber between the back axle and the frame provide a comfortable ride. Accessories are also attached to the spine and include a changing mat. Stokke have also thought long and hard about the ergonomics in relation to the adult carer. Handle height is adjustable from 67–116 cm (26–46 in) above ground and the U-shaped rear axle means that a relaxed walking posture is easily maintained.

✎	Bjørn Refsum, Stokke Gruppen, Norway	309
⚙	Stokke Gruppen, Norway	321
▤	Aluminium profiles, polymers, polyester fabric	
↻	• Multifunctional	327
	• Improved ergonomics	327

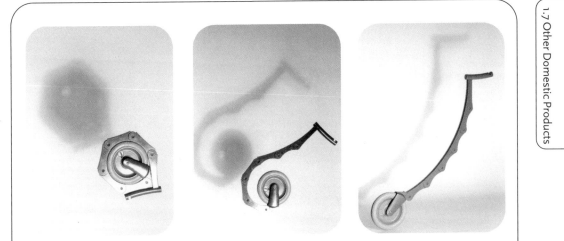

Kango

Here's a means of cutting down on the number of car journeys to local supermarkets. This mono-wheel trolley is capable of carrying a week's worth of groceries, is easily manoeuvred and, after use, is folded up into a handy package. Ideal for regular or casual use, for leisure or travel purposes, the Kango leaves all those ugly, two-wheel, tartan leatherette trollies in the shade.

Feldmann & Schultchen, Germany	306	
Patented prototype – Feldmann & Schultchen, Germany	306	
Cordura fabric, rubber, fibre-reinforced plastic		
• Encourages energy-efficient shopping		

Muscle Power toothbrush

Muscle Power products raise consumer awareness at the point of purchase by posing the question, 'Do I really need to buy a battery-powered tooth brush?' The consumer is confronted with a choice of energy sources for the simple preventative health-care task of brushing his or her teeth. The solution is provided by designing a wind-up toothbrush, which can be used in static mode, like a conventional toothbrush, or can deliver a rigorous massage. Schreuder found that thirty seconds is the average time people spend cleaning their teeth, but that up to two minutes is really required for effective action of fluorides. Fully wind up the mechanism and exactly two minutes of power are delivered. A daily drudge becomes less taxing, especially for children, in whom habit-forming hygiene needs to be induced.

✏	Hans Schreuder, the Netherlands	309
⚙	MOY Concept & Design, the Netherlands	318
📜	Plastics	
↻	• Improved health	327
	• Human-powered	327

DURAT® design collection

Those with minimalist and modernist aspirations should look no further than the DURAT® design collection of tens of different washbasins, bathtubs, shower trays and kitchen worktops with integral sinks. Individual designs are available in forty-six standard colours but bespoke patterns, edge designs and other finishes are possible. DURAT® is a smooth, hard polyester-based material that contains 50% recycled plastics and is 100% recyclable. It is warm to the touch and suitable for a wide variety of applications in any 'wet zone' in the home or work environments. The manufacturing technology permits jointless customized panels, worktops and one-off pieces.

✏	Tonester, Finland	322
⚙	Tonester, Finland	322
📜	Recycled and virgin polyester-based plastics	
↻	• Recycled content	324
	• Recyclable	324

Preserve®

Everything about the Preserve® toothbrush and its packaging suggests that Recycline is a company that takes its environmental responsibilities seriously. A clear, rectilinear box is made from a cellulose polymer derived from trees; the removable cap is HDPE and is intended as a carrying case rather than disposable packaging. A paper insert in the box carries the brand information. It is printed with soy inks and is made from 100% recycled paper fibre including 50% post-consumer waste. The toothbrush handle and head are ergonomically designed with soft bristles on the outside to protect the gums and stiffer bristles to penetrate between the teeth. Although the bristles are virgin nylon, the shaft is made from 100% recycled plastics (minimum 65% from recycled yogurt pots). When the bristles are worn out the used brush can be returned to Recycline in a postage-paid envelope available at the retailer or direct from Recycline. This is responsible closed-loop manufacturing where the manufacturer and consumer form a partnership in the lifecycle of the product.

✏	Recycline, USA	320
⚙	Recycline, USA	320
📜	Recycled plastics and paper, cellulose-based polymers, nylon, HDPE, soy bean inks	
↻	• Recycled, recyclable and renewable-source plastics	324/325
	• Product take-back	328
	• Lifecycle analysis	324

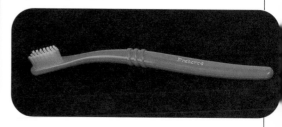

Gelapeutic Bath

Bathing is a ritualized experience, immersion in water to renew and purify. Generally washing one's body is a solitary activity but the gelapeutic bath offers opportunities for individual or communal experiences. Envisaged as a central focus of the habitat, the 'bath' comprises capsules of gelatine-like material containing substances for cleaning, toning and protecting the skin. As a living component the gel is capable of self-regeneration and self-cleaning. Warming or cooling the gel provides a range of body sensations to stimulate or relax. Immersion in the gel is unfamiliar yet embracing, a sort of womb-like humid environment in which to blend mentally, emotionally and physically. Gelapeutic Bath is part of a larger malleable concept in an 'active society', a society based upon the development of individuals with less anxieties. Malleable refers to an individual's ability to adjust and mould his or her private and pubic domains. These domains become an experiential laboratory encouraging reflection and transformation, simultaneously a mirror and a catalyst.

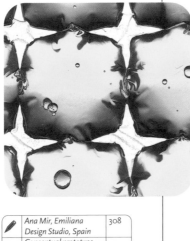

✒	Ana Mir, Emiliana Design Studio, Spain	308
⚙	Conceptual prototype	
▬	Gelatine-based substance	
♺	• Experiential design • Improved sense of well-being	327

Aquaball®

Place two pellets inside the spiky plastic ball, the Aquaball, and put it with your dirty washing into your machine. The pellet ingredients include fatty alcohol polyoxyethylene ether, higher alky sulphate, CMC, sodium carbonate, EDTA, fragrance (derived from essential oils) and FWA (fluorescent whitening agent), which 21st Century Health claim are free from harsh chemicals and therefore safer for people prone to allergies, eczema and sensitive skin. The pellets dissolve faster than detergents, increasing the level of ionized oxygen and raising the pH, so the ingredients get to work early by lifting dirt from fabrics. Rinse cycles can be shortened by up to thirty minutes for an average wash, so saving lots of water. There are fewer residues left over than with conventional washing powders and a very low phosphorus content, so there is a comparative reduction in water-borne pollutants per wash.

✒	Aquaball®	304
⚙	21st Century Health, UK	311
▬	HDPE casing, ionic crystals and salts	
♺	• Reduced consumables • Reduced water consumption and pollution	327 328

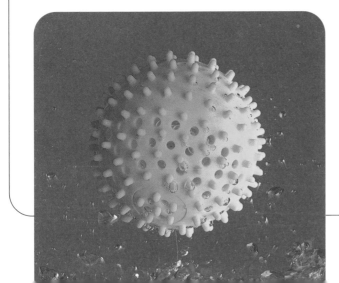

Power Grip

Unlike 'pistol grip' cordless drills and screwdrivers the Power Grip has a very compact but ergonomic handle, ensuring it is easily used with one hand and is much more versatile in confined spaces.
A reduction in weight prevents user fatigue without loss of power. This design demonstrates significant reductions in materials used compared with the established leading designs for this market.

✏	Design Tech, Germany	305
⚙	Metabo, Germany	318
🎞	Various materials	
♺	• Reduction in materials used	325
	• Improved ergonomics	327
✪	iF Design Award, 2003	330

Fingermax

These finger brushes offer creative opportunities to those who find holding a conventional paintbrush difficult. Universal fitting is achieved by moulding a thermoplastic resin polymer in a spiral shape with an elliptical cross-section.

✏	Büro für Form, Germany	305
⚙	Fingermax Gbr, Germany	314
🎞	Polymer	
♺	• Universal design	327
	• Design for need	326
✪	iF Design Award, 2000	330

Bob Pen

An average primary school has significant flows of consumables as part and parcel of everyday teaching. Careful consumption can form a core part of the curriculum and, more importantly, encourages good habits for life. This reusable, refillable art marker pen is made from a minimum number of parts and uses non-toxic inks.

✏	Linda Heammerle, Unitec, New Zealand	306
⚙	Prototype	
🎞	Various materials, non-toxic inks	
♺	• Reduction in consumables	327
	• Safe, non-toxic	327
✪	IDRA award, 2000–2001	330

TULIP

Waste scraps of acryl-resin sheets are laminated and shaped by lathing to create new tactile, decorative door knobs. Recycled aluminium is cast and machined for the base plate. This is an example of a design that the consumer perceives as new yet is really recycling by stealth. The design adds considerable value to the reused and recycled materials.

✏	Tyson Atwell, USA	304
⚙	Prototype	
🎞	Acryl-resin wastes, recycled aluminium	
♺	• Reused and recycled materials	324/ 326
✪	IDRA award, 2002–2003	330

Habitatio naturalis

Blurring the boundaries between inside and outside, Habitatio naturalis asks renewed questions about our relationship with the living world. Grass and flower seeds embedded in the woven texture of the blind germinate and grow when subject to the occasional spray of water and exposed to sunlight. This poetic emergence of nature out of context is unexpected and, if not maintained, ephemeral.

External turf roofs and walls are becoming familiar signatures of bioclimatic architecture, but living interiors is a new challenge. Health-giving properties of house plants as air purifiers are well known. Extend the range of plants from the pot to the walls and new scenarios emerge.

✎	*Krista van der Knijff, graduate 2002, Design Academy Eindhoven, the Netherlands*	305
⚙	*Prototypes*	
🗒	*Polyester, growing medium, plant seeds*	
☾	• *Renewable materials*	325
	• *Improved health*	327

charcoal hanger

Waste pieces of wood from sustainably harvested timber are bent and moulded to shape then carbonized by firing in a furnace. Charcoal dust is washed off the surface to prevent any damage to clothing. Carbonization of the wood ensures that excess humidity and odours are absorbed, giving the hanger additional clothing care qualities. Beyond the useful life of the hanger the charcoal can be returned to the soil as a nutrient or used for water filtering to complete the lifecycle of the product.

✎	*Koji Takahashi, Japan*	309
⚙	*Small batch production*	
🗒	*Charcoal*	
☾	• *Renewable and sustainably harvested material*	325
✹	*IDRA award, 2002–2003*	330

Pet Pod™

This quirky design makes a comfortable shelter and living space for a cat or small dog. The papier mâché gives insulation, so the Pet Pod is the ideal solution for pets housed in unheated buildings.

✎	*Vaccari Ltd, UK*	322
⚙	*Vaccari Ltd, UK*	322
🗒	*Papier mâché*	
☾	• *Recycled materials*	324
	• *Compostable*	324

Clean House, Clean Planet Kit

Five empty plastic bottles are supplied with 112 g (4 oz) of Dr Bronner's Peppermint Pure-Castile Soap and printed recipes for the user to make and fill the bottles. Recipes include 'It's a Lotsa Polish' (olive oil and vinegar furniture polish) and 'EarthShaker Kitchen Cleaner' (scented baking soda). This self-help system reduces reliance on harsh chemicals found in many commercial cleaners by using readily available household consumables.

✎	Dr Bronner's Magic Soaps, USA and 21st Century Health, UK	313/311
⚙	Dr Bronner's Magic Soaps, USA and 21st Century Health, UK	313/311
▤	Plastic containers, paper soap	
↻	• Dematerialized product	324
	• Reduced consumables	327
	• Reduced water consumption and pollution	328

U-Box

This multipurpose polypropylene box offers several compartments and permits boxes to be stacked. It provides a versatile storage unit for home, workshop or office.

✎	Hansjerg Maier-Aichen, Authentics artipresent GmbH, Germany	308/311
⚙	Authentics artipresent GmbH, Germany	311
▤	Recyclable polypropylene	
↻	• Recyclable materials	324
	• Multifunctional	327

Paperbag bin

The printing industry generates vast quantities of material that is never actually used for the purpose intended. Excess print runs, abandoned promotional literature, pulped magazines and books – the volume of waste generated is high. Forty-eight-sheet advertising posters are extremely difficult to recycle because of high concentrations of coloured inks and special waterproofed papers. These latter characteristics are ideal for certain new products, as Goods demonstrate with their eye-catching waste paper bins made out of these advertising posters. Creative ideas from a brainstorming session often end up in the bin; now even the best ideas, with a wonderful ironic twist, become the bin!

✎	Goods, Netherlands	315
⚙	Goods, Netherlands	315
▤	Unused billboard advertisements	
↻	• Reuse of redundant componants	326
	• Low-energy manufacturing	325

eco-ball™

Those sensitive to today's chemically based washing powders have an alternative method available in the form of the eco-ball™. This is a plastic ball that contains ionic powder that releases ionized oxygen into the water and so facilitates penetration of water molecules into fabrics to release the dirt. A little washing soda helps deal with very dirty washing but it is claimed that a set of three balls will help clean the equivalent of 750 washes before losing their activity.

✎	eco-ball, UK	314
⚙	eco-ball, UK	314
▤	Plastic, ionic powder	
♻	• Reduction of consumables	327

Ecover®

The name Ecover, like The Body Shop, needs little introduction to those who became green consumers in the 1980s. Established in 1979, Ecover has always espoused a business policy that recognizes that economics must be in harmony with ecology. This policy extends to product development, the green architecture of its main factory in Belgium and the international distribution network through twelve thousand small health food shops, still accounting for 45% of turnover, as well as the supermarket giants. Company policy dictates that products must originate from a natural source with a low level of toxicity to minimize their burden on the environment and they must be equally efficient as conventional, more polluting products. Ecover products are not permitted to include petrochemical detergents/perfumes/ solvents/acids, polycarboxylates, phosphonates, animal soaps, perborates, sulphates, colourings, phosphates, EDTA/NTA, optical brighteners and chlorine-based bleaches. Animal testing is also banned. Ecover also have an integrated packaging policy and encourage consumers to refill 1-litre (1³⁄₄-gal) containers of washing-up liquid at shops stocking 25-litre (44-gal) bulk containers, which are themselves refilled at the factory. These polyethylene bottles (with polypropylene tops) have extra-wide necks, a level indicator, plastic labels (also recyclable) and a life expectancy of twenty refills before recycling, saving on waste and landfill space. Ecover's product range includes washing powder, bleach, water softener, liquid wool wash and fabric conditioner.

✎	Ecover Products, Belgium	314
⚙	Ecover Products, Belgium	314
▤	Various cleaning agents, plastic containers	
♻	• Reduction of water-borne toxins and pollutants	326
	• Reuse and recycling of containers	327
	• Retailing system geared to small and large outlets	

Bon-Bon

Two billion greetings cards are sent each Christmas in the UK. Most of this veritable forest of paper goes on a one-way trip to from sender to receiver to landfills. Batchelor considered the lifecycle of traditional cards in order to understand where environmental impacts could be reduced. Most cards are made using virgin paper pulp and synthetic printing inks; they are usually a folded card inserted into an envelope. Bon-Bon cards have no envelope and, like a postcard, are unfolded so they use only half the amount of card. This significant reduction in weight (66% over traditional cards) saves on embodied energy of the printed product and transportation energy. Bon-Bon cards are printed using 100% biodegradable soy-based inks from Abba Litho Sales, London, and Metaphor 100% recycled fibre paper (51% post-consumer, 49% production waste) from the Eco range, Curtis Fine Papers, Scotland (p. 291). The printable area per original sheet of Metaphor is designed to minimize wastage. Eco-design principles are also embedded in the cards, ensuring that the lifespan can be extended. On the reverse of the card encouragement is given to the consumer to recycle or reuse it and, on the back of each card simple instructions are given on how to make something, for example, tree decorations from scrap gift wrapping, a recipe for seasonal gingerbread and so on.

✏	Lucy Batchelor, UK	304
⚙	Lucy Batchelor, UK	304
🗐	100% recycled paper, vegetable printing inks	
↻	• Reduction in materials and embodied energy	325
	• Reduction in waste	327
	• Dual-function	327
	• Lifecycle analysis	324

Building Permit (Permis de construire)

A minimalist graphical sofa deconstructs into a veritable playground of castles, dens, huts and other mini-edifices born of children's creativity. A formerly dead object (the sofa) finds a daytime use blurring the boundary between adults' and childrens' needs. When the children finally slumber the adults reconstruct their lives...or play on....

✏	Matali Crasset, France	305
⚙	Domeau & Perés, France	313
🗐	High-resilience foam Cotton fabric covers	
↻	• Multifunctional	327
	• Encourages social interaction	

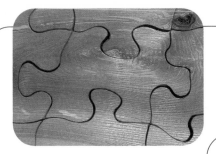

Ash profile jigsaw

Sculpture, game or both? This monumental jigsaw engages the viewer in the history of the ash tree in its knots, growth rings and blemishes. Each piece invites the viewer to pick it up, touch it and examine it. Children and adults alike can delight in this whimsical object, and reflect on its origins.

✎	Julienne Dolphin-Wilding, UK	305
⚙	One-off	
▥	Ash wood	
⏷	• Renewable material	325

Oritapi

Transforming from a two-dimensional carpet into a 3D playspace, Oritapi offers a versatile learning and play environment for young children, a cross between origami and carpet. Constructed entirely of felt, it offers a tactile experience yet structural integrity.

✎	Matali Crasset, France	305
⚙	Domeau & Perés, France	313
▥	Felt	
⏷	• Multifunctional • Renewable single material • Recyclable	327 325 324

Grid™ wall pockets

Grid-iron plans of the city, controlling the flows of people, their work and space, metamorphose into the Grid™. This stitched, die-cut felt organizer is mounted on a wall and offers numerous pockets to reorganize vital artefacts, mementos and notes in your life. The design derives from the idea that space is a scarce urban resource, so the wall in your home liberates vital horizontal space for other objects or activities.

✎	Jamie Salm, MIO, USA	318
⚙	MIO, USA	318
▥	Felt	
⏷	• Renewable and recyclable materials	324/ 325

2.0 Objects for Working

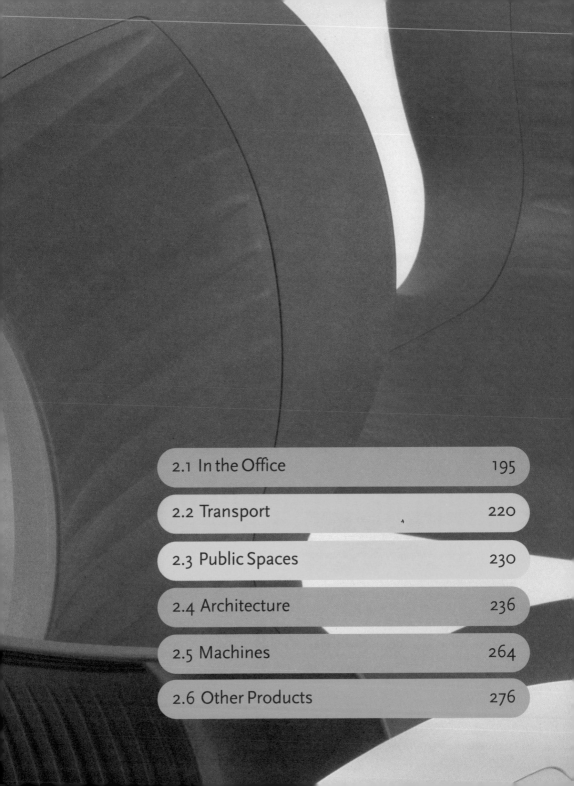

Work: An Evolving Concept

Although there is a tendency to think of the world as one huge post-industrial society, the reality is that there are myriad societies, some still firmly rooted in feudal agrarian systems, others heavily industrialized and still others dominated by service industries. It is therefore untrue, and possibly dangerously misleading, to think that everyone perceives problems of sustainability and work in the same way.

In the developed world 'information' is just as much a raw material as timber, iron, steel and chemicals are in an industrialized society. The main difference is in their environmental impacts. In the information society the worker needs access to a workstation, which may be in the office, at home or (in the case of a laptop computer) somewhere in between. Mobile phone and wireless technology means that workstations no longer have to be connected to physical local area networks (LANs) or fixed telecommunication points. The worker may not need to travel to a physical place of work and thus less transport energy is used. In an industrialized society, however, the worker has to travel to the factory where other workers and physical materials are gathered for the purpose of fabricating a product. But both the information and industrial workers

consume finite resources and energy and produce waste, toxins and hazardous chemicals. All societies must therefore design products, materials and services that reduce their environmental impacts.

Work continues beyond the workplace. Domestic products that have become essential to a way of life, such as washing machines and toilets, need improvements to increase their efficiency of energy and water use. Other appliances such as kettles and cookers must also become more efficient during their lives as such and be capable of disassembly for recycling of the materials at the end of their lives.

Transporting people and distributing goods
Work involves transporting people, distributing goods or both. While electronic networks can reduce the need to move people physically, most work involves some travel. More efficient transport systems are therefore critical. Fuel efficiency needs to be improved for modes of transport that run on internal combustion engines. At the same time, lower-impact fuels and transport products powered with renewable energy need to be developed. Above all, public transport systems have to be coordinated to provide people with flexibility and freedom.

The average supermarket, furniture store or trade outlet, especially in the developed northern hemisphere, will have products from all over the world. Transported over great distances, expending vast quantities of energy, a product's transport energy can sometimes exceed the energy used to make it. Reduction in packaging weight and volume is a perennial challenge to distributors. Even the smallest saving in packaging for each product can represent huge savings in transport energy and waste production for the retailer or middle-man in the distribution chain. One-way-trip packaging can often be replaced by lightweight, reusable packaging systems, and an emphasis on local products sold in local markets could also result in large savings.

Working lightly: a sustainable day
At the office and factory more efficient working practices are aided by well-designed, durable, easily maintained products. It is now possible to have one office machine to serve a network and provide facilities to fax, photocopy, print and scan. Digital files can be shared on local and international networks: the paperless office is a partial reality. Offices can be equipped with durable, modular furniture systems and carpets can be

replaced under a lease-maintenance contract. Office consumables can use recycled content and reused components.

In industrial production facilities designers, in coordination with environmental managers, can reduce inputs of energy and materials and increase efficiencies in production and distribution. Waste streams provide another source of raw material and closed-loop recycling of process chemicals and materials ensures improved eco-efficiency, better profits and improved worker health. Design can help deliver a 'triple bottom line' of reduced impacts on the environment, improved social benefits and profitability.

Work tasks in the home vary from washing the dishes, the clothes or the car to keeping the house in good repair. Future activities might include maintenance to check the efficacy of renewable power appliances or water-conservation systems or removing compost from the waterless toilet.

A sustainable working day in 2025 might involve some of the products that follow on these pages.

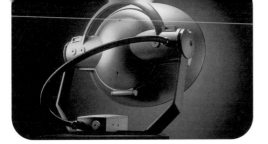

Arena Vision 401

This oval-shaped unit is suitable for exterior and interior lighting of sports facilities. Light output has been improved by 10–15% for the same power input. Disassembly allows separation of the component materials and most parts can be recycled.

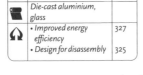

✏	Philips Design, the Netherlands	319
⚙	Philips Electronics, the Netherlands	319
🗏	Die-cast aluminium, glass	
♼	• Improved energy efficiency	327
	• Design for disassembly	325

XK series

Exit signs fitted with incandescent lamps can now be fitted with energy-efficient LED lamps with potential energy savings of up to 90%. A retrofitting kit includes the appropriate screw bases and sockets.

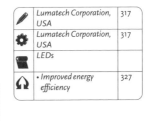

✏	Lumatech Corporation, USA	317
⚙	Lumatech Corporation, USA	317
🗏	LEDs	
♼	• Improved energy efficiency	327

Virtual Daylight™ systems

Many office workers suffer from fatigue and illness as a direct result of the poor lighting in their work environment. Virtual Daylight™ Systems use a combination of polarization, high frequency and full-spectrum technology to reproduce daylight-quality illumination. The systems are designed to use less energy than conventional office lighting and offer a significant boost to those prone to Seasonal Affective Disorder (SAD).

✏	Clearvision Lighting, UK	313
⚙	Clearvision Lighting, UK	313
🗏	Various	
♼	• Improved health and work environment	327

LED 100-TE

The LED 100-TE system offers a range of customized diffusers and reflectors for down-lighting, which can be fitted into two basic housings. There are nine 'Architectural', four 'Deco' and six 'Eco' diffuser/reflector options. Each housing incorporates a tilting light holder for low-energy bulbs and is fitted with electronic control gear, saving 30% over conventional ballasts.

✏	Concord Lighting, UK	313
⚙	Concord Lighting, UK	313
🗏	Die-cast aluminium and polycarbonate	
♼	• Energy-saving (low voltage) customizable lighting system	327

LED lighting

LED technology is driving rapid developments in domestic and industrial illumination products. Maintaining backwards compatibility in the lighting industry is important for competitive, economic and environmental reasons. Bayonet and other industry standard fittings to clusters of LEDs fitted to specialist bulbs produce a low-voltage, low-current, high-energy-efficiency option for everything from task lighting to advertising panel illumination. Since 1983, LEDtronics's research has developed LED technology. LEDs offer power savings of up to 80–90% compared with incandescent lamps, a wide range of lux and lighting spectra, and little or no heat output, so they are suitable where a cool environment needs to be maintained.

✏	LEDtronics, USA	317
⚙	LEDtronics, USA	317
▦	Solid-state technology, optical-grade epoxy, electronic circuitry	
↻	• Low-voltage lighting • Reduction in materials used	327 325

QL Induction Lighting

Unlike incandescent lamps, induction lighting uses an induction coil rather than filaments to excite the gas within the bulb. A high-frequency generator creates an electromagnetic field in the power coupler (induction coil) that excites the mercury atoms in the gas, causing them to emit UV radiation. This radiation hits the inside of the phosphor-coated bulb, causing it to fluoresce and emit light. Since there is no filament to burn out, induction lamps can last for up to fifteen years based on 4,000 burning hours per year. Induction lamps have a number of other advantages including low energy consumption, constant illuminance independent of voltage fluctuations, flicker-free start and an automatic stop circuit in the case of lamp failure. These 'electrodeless' lamps are available in 55W, 85W and 165W ratings so are suitable for general indoor lighting, outdoor lighting with or without infrared sensing and public space applications. Being a source of electromagnetic fields, installation must comply with Electro Magnetic Compatibility (EMC) standards and luminaires must have adequate cooling systems.

✏	Philips Lighting, the Netherlands	319
⚙	Philips Lighting, the Netherlands	319
▦	Glass, various metals, insulators, mercury	
↻	• Low-energy lighting system • Low maintenance requirement	327 327

Zeno

Seasonal Affective Disorder, otherwise known by the appropriate acronym SAD, is a condition resulting from hormonal imbalance caused by inadequate exposure to sunlight. It is thought that many people suffer from SAD in temperate climates during the winter months. Office workers, who predominantly work in artificially lit environments, are a particularly susceptible group. Zeno addresses the issue of light quality by creating luminance using natural sunlight and a variety of artificial lighting spectra. Sunlight is captured using an external array and fed down optic fibres to a complex reflecting disc whose surface has a high reflection co-efficient. Natural sunlight can be blended with light from compact fluorescent bulbs (100W halogen or 70W iodine) to vary the luminosity and quality, and hence the atmosphere created by the lighting.

✐	Diego Rossi and Raffaele Tedesco, Italy	309
⚙	Luceplan, Italy	317
▤	Various technosphere materials	
♺	• Reduced energy consumption	327
	• Improved well-being	327

Duck Light

Building on the success of their 'e-light', Artemide have created another versatile light suitable for the office or home environment. The Zamak diffuser gives an even light distribution, the strong polycarbonate base and shade ensure a long life and all components are easily disassembled for repair.

✐	Ernesto Gismondi, Artemide Design, Italy	311
⚙	Artemide SpA, Italy	311
▤	Zamak, polycarbonate, low-energy bulb	
♺	• Reduced energy consumption	327

Biomorph multidesk

Correct posture while working at a computer is essential for good health. This desk permits adjustment of the height of the platforms holding the computer monitor and the keyboard and features safe, rounded edges to all components.

✏	Stephen Barlow-Lawson, USA	304
⚙	Ground Support Equipment (US) Ltd, USA	316
📇	Painted fibreboard, steel	
♻	• Improved health and safety	327

Viper

Elliptical cross-section cardboard tubes made from recycled paper are connected to each other at top and bottom by a specially moulded plastic capping. Extensive articulation between adjacent tubes permits the screen to be rolled up when not in use.

✏	Hans Sandgren Jakobsen, Denmark	307
⚙	Fritz Hansen, Denmark	315
📇	Cardboard, plastic	
♻	• Recycled material	324

Cloud

Cloud was created as part of an on-going project called Creative Meeting Places by OFFECCT. It is a portable, inflatable meeting room with an integral floor that is contained in a transport bag and comes complete with an integral fan to inflate the space. Made of white ripstop nylon, it is lined with an acoustically absorbent textile surface. Cloud aims to forge a relaxed, unpretentious atmosphere for fostering a different environment for communication. Inside this ethereal form, normal barriers and traditional 'roles' evaporate as people are cocooned in the whiteness of the interior. Although aimed at the corporate market, this sociable space has considerable potential to serve a wide range of communities.

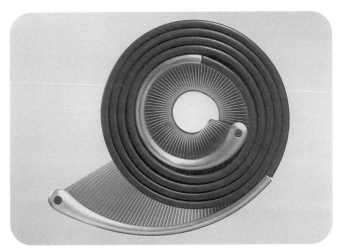

Cartoons

Cartoons is a flexible, free-standing screen suitable for partitioning in domestic and office spaces. Corrugated paper board extracted from pure cellulose is stiffened at the edges with a closure of cold-processed, CFC-free polyurethane and at the ends with die-cast aluminium. This configuration allows the screen to be positioned in a sinuous style to suit the user and to be rolled up when not in use.

✎	Monica Förster, Sweden	306
⚙	OFFECCT, Sweden	319
📰	Ripstop nylon, acoustic textile, fan	
🎧	• Lightweight, portable, multifunctional building • New social space	325/ 327

✎	Luigi Baroli, Italy	304
⚙	Baleri Italia SpA, Italy	312
📰	Corrugated paper board, aluminium, CFC-free technopolymer	
🎧	• Renewable and recyclable materials • Clean production	324/ 325 325

Jump Stuff, Jump Stuff II

Everybody customizes their domestic space, so why not the work space too? Through an extended series of projects and development of conceptual prototypes in the mid- to late 1990s, such as the Flo & Eddy workstation, Haworth examined the cognitive ergonomics of the desk area. The outcome is the Jump Stuff system, which allows individuals to select the components they require to maximize the functionality and comfort of their own desks. The spine of the system is a free-standing or panel-/wall-mounted rail to which the modular components can be attached. Whatever your regular tasks, you can attach and orient the appropriate accessory to the mounting rail. Different types of task lights can be attached to the rail and all the accessories can be easily adjusted for a 'hot desking' role. Although there are four basic variations to the system it is also possible to purchase each module independently so you can 'grow' the system to suit your needs.

✏	Haworth, Inc., USA	316
⚙	Haworth, Inc., USA	316
📜	*Various metals and polymers*	
🎧	• *Multifunctional, modular system*	327
	• *Design for need*	326

Unitable

Those who work from home or have occasional need for an office at home will appreciate the versatility of the Unitable. Height, length and angle of inclination of the work surface are all adjustable. Table sections are modular and hollow or solid depending on the user's requirements. Natural beech and birch, birch linoleum and laminates are available together with a birch ply stained black. All tops fix to a metal frame that accommodates table tops of 1.5–4 m (5–13 ft) length and 750–900 mm (30–35 in) width. For the tilting version of the table, the Unidrawing table, a lightweight multiplex webbed panel is formed of corrugated aluminium sandwiched between ply or laminate. And yet another variant, the Unifoldingtable, can be folded for use as a display board or conference table.

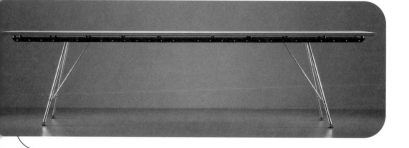

✏	Atelier Alinea, Switzerland	311
⚙	Atelier Alinea, Switzerland	311
📜	*Various woods, plywoods and wood laminates, corrugated aluminium, steel*	
🎧	• *Multifunctional*	327
	• *Reduction in materials used*	325

System 180

This is a modular system of cube or rectilinear shapes for constructing multipurpose furniture. Its flexibility derives from the tubular steel frame elements with flattened ends that can be bolted together. This 'meccano' approach allows users to continuously construct, deconstruct and reconstruct as needs demand. So it is possible to build everything from desk- and work-spaces to shelving and even tea trolleys. The designer–manufacturer offers a closed-loop system with return of system elements if they are excess to requirements.

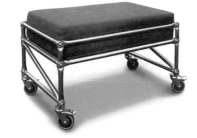

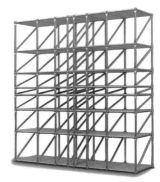

✏	*Futureproof/ed, USA*	315
⚙	*Futureproof/ed, USA*	315
📑	Steel, various materials for tops/shelves	
🎧	• Modular system	327
	• Durable and recyclable materials	324
	• Product take-back	328

✏	Bernard Vuarnesson, Sculptures-Jeux, France	310
⚙	Bernard Vuarnesson, Sculptures-Jeux, France	310
📑	Recycled and recyclable fibreboard, beech wood press studs	
🎧	• Recycled, recyclable and reusable materials	324
	• Self-assembly	325
	• Low embodied energy	325

CLARK collection

It takes ten-to-fifteen minutes to assemble this flat-pack storage unit for home and office, made of 100% recycled and recyclable fibreboard and varnished solid beech wood. Drawer dimensions vary, so a number of variations are possible. Wheels facilitate moving the chest of drawers for cleaning or moving furniture.

Eco-Aspirator and The Office Oasis 'Design Organism'

Biomimetics, looking to nature for design inspiration, suggests a different design process where nature's ability to evolve, integrate and create synergistic sustainable solutions is mirrored in human lives. The Eco-Aspirator or 'Living sculpture for enhancing indoor air quality' is part of an on-going experiment to create multifunctional designs that are actually mini-ecosystems or microcosms. A toughened glass column provides a habitat for species across the different kingdoms of life from a rainforest environment: animals (fish, insects), mosses, ferns, orchids, fungi, algae and micro-organisms. These co-evolve to create their own communities where gaseous exchange, nutrient balance and growth follow cycles dictated by macro inputs (sunlight), the interior climate and the living organisms (including the human inhabitants). These Eco-Aspirators are 'positive impact' solutions, they regenerate their surroundings and are ideal for warding off symptoms of 'sick building' syndrome where indoor air quality can be a factor poorer than that outside. Eco-Aspirators offer educational opportunities and community involvement. Being self-healing systems they are easier to look after than a fish-tank since the water is cleaned by the system.

On a larger scale biomimetics can create a micro-habitat in the office complex where the senses are awakened to a world beyond the workstation. This eco-engineered living system, designed by John-Paul Frazer and Jacques Abelman, would deliver fresh oxygen as the plants process spent carbon dioxide from human respiration and reconnects, refreshes and regenerates the environment.

✎	John-Paul Frazer, Bioinspiration, UK	306
⚙	One-offs and conceptual prototype	
▰	Glass, plants, water and various materials	
↻	• Improved health and social well-being	327

Leap™ seating

Building on the lessons learnt from the design of the Protégé Chair in 1991, Leap™ Seating is one of Steelcase's leading products with respect to recycling, waste reduction and low-impact manufacturing. The basic design is very durable but parts can be easily removed for repair or upgrading if required. At least 92% of the chair's parts are recyclable and the cushioning used in the upholstery is made of 50%-recycled PET. During manufacture adhesives and paints with no or limited volatile organic compounds (VOCs) and water-based metal-plating processes considerably reduce aquatic and aerial emissions. Leap is Greenguard® Indoor Air Quality certified. Employee working conditions at the Grand Rapids, Michigan, factory have also been re-engineered to provide a more healthy environment. What will happen to the chair at the end of its life has not yet been defined, but leasing and take-back are options all responsible manufacturers will have to consider in the near future.

✐	Steelcase, Inc., USA	321
⚙	Steelcase, Inc., USA	321
📃	Recycled PET, plated steel, polyester	
♻	• Design for disassembly and ease of repair	325/ 327
	• Recycled and recyclable materials	324
	• Low-impact manufacturing	

X & Y chairs

Available in X or Y configurations, these versatile stacking office or conference chairs are robustly constructed. Improved comfort is afforded by the moveable backrest that is on spring bearings. The range comes with or without upholstery depending on user requirements. Both X and Y versions can be assembled into benches of two, three or four seats by the inclusion of an additional beam.

✐	Emilio Ambasz & Assoc., USA	306
⚙	Interstuhl Büromöbel GmbH, Germany	317
📃	Metal, plastics, textiles	
♻	• Durable	327
	• Multifunctional	327
✪	iF Design Award, 2003	330

X-In Balance workplace screen

Economic use of materials is of direct benefit to the environment, yet achieving this aim is often a daunting task in which lightness has to be balanced against the need to fulfil functional requirements. X-In Balance achieves this goal and more.

✎	Gerald Wurz, Austria	310
⚙	Nova Form/Kautzky Mechanik, Austria	319
🗋	Balloon silk, steel	
⚓	• Reduction in materials used	325
	• Lightweight recyclable materials	324
	• Low energy manufacturing	325
	• Reduction in transport energy	326

Picto, FS line

Wilkhahn initiated a project in 1992 entitled 'Environmental Control' with the support of the Ministry for Environmental and Economic Affairs for the state of Lower Saxony. Following an audit of their corporate eco-balance of inputs and outputs, teams were set up to reduce environmental impacts in production and to select materials within an integrated IT framework. The Wilkhahn range of office seating is designed to minimize polluting processes during production. Chrome plating of metals is avoided and upholstery is made from durable, wear-resistant wool and polyester fabrics without gluing or welding. All furniture is easily assembled, disassembled and maintained, and individual components can be recovered for recycling upon disassembly.

✎	Produkt Entwicklung Roericht, Germany	309
⚙	Wilkhahn, Germany	323
🗋	Pure-grade metals, thermoplastics	
⚓	• Clean production	325
	• Design for disassembly and recycling	325

Mirra™ work chair

Following on from the success of the up-market Aeron® office chair in 1994, the Mirra™ aims to provide levels of comfort, performance and recyclability to the mid-price work chairs. Herman Miller's participation in an extensive study of body shape and size conducted by CAESAR (Civilian American–European Surface Anthropometric Resource) provided background research for the Mirra™. Laser-scanning the seating positions of hundreds of people in different postures provided raw data to assist with the its design. In particular, the differences between the back shape of men and women in different postures helped develop the one-piece TriFlex™ backrest with built-in flex zones and FlexFront™, a flexible, adjustable front edge that permits individual control of the depth of the seat. Further features that allow the user to control configuration include the Harmonic™ tilt mechanism. Herman Miller claims that Mirra™ is designed to suit 95% of the world's population. In accordance with the manufacturer's stringent Design for the Environment (DfE) protocols the chair's lifecycle has been examined to assure that its material chemistry, recyclability, manufacturing, packaging and ease of disassembly minimize environmental impacts. At end of life 96% of the materials can be recycled.

✎	Studio 7.5, Germany	309
⚙	Herman Miller, Inc., USA	316
▤	Various recyclable materials	
♺	• Recyclable materials	324
	• Lifecycle analysis	324

Torsio

Comprising just two separately moulded wooden parts, the Torsio range is designed to be stackable and can be linked in rows for conference seating. The twisted back section tapers from 22 mm (1 in) at the feet to 7 mm (¼ in) for the backrest, giving it a subtle flexing for improved comfort. Surface finishes include maple, walnut or black.

✎	Hanspeter Steiger Designstudio, Switzerland	306
⚙	Röthlisberger, Switzerland	320
▤	Wood laminates	
♺	• Reduction in materials used	325
	• Renewable materials	324
	• Durable	327
⬤	iF Design Award, 2003 Swiss Design Prize, 2003	330

Sundeala medium board screen

The original Sundeala company began manufacturing fibreboard from waste cellulose in 1898, and for the last seventy years Sundeala boards have utilized recycled newsprint as the primary material. 'K' quality unbleached natural board is for interior use, while 'A' quality with natural binders and colouring to reduce moisture penetration is suitable for sheltered exterior use.

✎	Celotex, UK	313
⚙	Celotex, UK	313
▤	Recycled newsprint, natural binders	
♺	• Waste materials used	325
	• Recycled and recyclable material	324

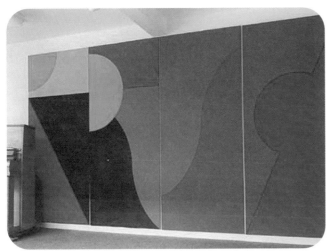

Aeron

The Aeron chair represents a step change in the way office chairs are designed. It is manufactured in three sizes to accommodate diversity of the human form and weight, making it suitable for users up to 136 kg (200 lb) in weight and from the first percentile female to the ninety-ninth percentile male. It has very advanced ergonomics. Pneumatic height adjustment, a sophisticated Kinemat tilt system and the Pellicle, a synthetic, breathable, membrane, are components of the seat pan, which adjusts to individual body shapes. The manufacturing process uses less energy than conventional foam construction and the use of discrete components of synthetic and recycled

materials facilitates disassembly and ease of repair for worn components (which are subsequently recycled). Such design improves the longevity of the product. Components are made of one material rather than a mixture of materials to facilitate future reuse and recycling.

✏️	Bill Stumpf and Don Chadwick, Herman Miller Inc, USA	30
⚙️	Herman Miller, Inc., USA	316
🗔	Plastic (PET, ABS, nylon and glass-filled nylons), steel, aluminium and foam/fabric	
♻️	• Improved ergonomics • Design for disassembly, recycling and remanufacturing • Single-material components	327 325 326

Caper chair

Office workspace is a precious commodity, so hot-desking and multifunctional spaces are becoming more evident. In principle these ideas for improving the economic and eco-efficiency of spaces are sound, but human nature demands that such practices also meet real needs. The key area of design focus for the Caper was to produce a chair that was easy to move around and personally re-configure. Lightweight stacking and multi-task models help deliver this flexibility to multiple users. A contoured and perforated polypropylene seat and back minimize pressure points while providing

aeration. An optional FLEXNET breathable suspension material can further enhance comfort and weight distribution. Caper also embodies the company's full Design for the Environment (DfE) protocols, being assembled with a high percentage of post-consumer content materials and being 100% recyclable.

✏️	Jeff Weber, Stumpf/Weber + Associates, USA	310
⚙️	Herman Miller, Inc., USA	316
🗔	Various post-consumer and recyclable materials	
♻️	• Recycled and recyclable materials • Lifecycle analysis	324 324

Remarkable recycled pencil

Used polystyrene cups from vending machines are shredded and re-processed into a new 'plastic alloy', in which graphite and other materials are mixed with polystyrene and extruded in a special die to create a new type of pencil. It performs as well as traditional 'lead' pencils and helps reduce consumption of the timber that traditionally encases the lead.

✏️	Edward Douglas-Miller, Remarkable Pencils, UK	320
⚙️	Remarkable Pencils Ltd, UK	320
📜	Recycled polystyrene, graphite, additives	
♺	• Recycled materials • Reduced resource consumption	324 325

Karisma

Sanford UK is part of the Sanford Corporation, which is the world's largest manufacturer of pencils based upon waste wood products, a mixture of wood flour and polymers. All wood-cased pencils manufactured by Sanford UK use wood from managed forests and, where possible, pencils are protected by water-based varnishes, which are hardened by ultraviolet light, rather than using solvent-based inks. Packaging and plastic waste are recycled at the production plant.

✏️	Sanford UK Ltd, UK	320
⚙️	Sanford UK Ltd, UK	320
📜	Wood, water-based varnishes	
♺	• Recycled materials • Clean production • Supply-chain management	324 325 325

Epistola

Using a pencil as the fulcrum, this exquisitely simple set of scales allows letters to be graded for correct stamping. It also serves as a letter opener and involves minimal use of materials and energy during manufacturing.

✏️	Teo Enlund, Sweden	306
⚙️	Simplicitas, Sweden	321
📜	Metal	
♺	• Reduction in materials used • Low-energy manufacturing	325 325

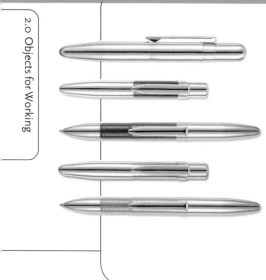

Millennium Pens

Since 1948 the Fisher Space Pen Company has manufactured high-quality pens using their unique pressurized brass-cased ink refills. The company claims that the seals in the refill are so good that the thixotropic ink will not dry out for over a hundred years. Pressurized with nitrogen gas to 50psi, the ink is delivered to an ultra-hard tungsten carbide ball held in a stainless steel collar and is capable of flowing at temperatures of −45 to +121° Celsius. Released for the millennium 2000 celebrations, the Millennium pens are made of extremely durable titanium nitride and chromed steel in a black or gold finish. These are luxury pens that should last a lifetime.

✐	Fisher Space Pen Company, USA	314
⚙	Fisher Space Pen Company, USA	314
📜	Titanium nitride, chrome, brass, tungsten carbide, stainless steel, ink	
♻	• Durable materials and product	324/ 327
	• Repairable and refillable	327

Remarkable Medium Collection

Following on from the success of the Remarkable recycled pencils the company has diversified to produce a range of rulers, writing pads, pencil cases, pens and colouring pencils using recycled or certified materials. Rulers are made from seven recycled vending cups; pads from recycled paper and board; pencil cases from recycled car tyres and colouring pencils from FSC certified sustainable timber. As European legislation comes into effect regarding the disposal and recycling of electrical and electronic equipment (the WEEE Directive) Remarkable have produced pens made from recycling plastic components from computer printers. Similarly old fridges are being tested for recycling materials to produce new fridge magnets. Significantly Remarkable struck a deal with two leading supermarket companies in the UK – Tesco and Sainsbury's – to sell recycled branded products, showing that retailer and consumer perceptions are shifting in favour of certain types of recycled goods.

Paperfile

Unused billboard posters find a new life as abstract graphical folders for storing precious notes, documents and paperwork. These tough everyday files have out-lived the semantics of the advertising hoarding, their corporate messages deconstructed and abandoned.

✐	Jan Neggers, the Netherlands	308
⚙	Goods, the Netherlands	315
📜	Paper, printing inks	
♻	• Reused materials	326

✐	Remarkable Pencils Ltd, UK	320
⚙	Remarkable Pencils Ltd, UK	320
📜	Various recycled materials	
♻	• Recycled materials	324

Save A Cup

Drinks vending machines daily consume vast quantities of standard 80 mm (3 in) polystyrene cups to satisfy the thirst of office workers and users of public spaces. All those spent cups – what a waste! Save A Cup has organized direct or third-party collection of used cups in all the major UK cities, using specially designed bins and machines to shred the cups. Companies registered with the UK's Environment Agency can obtain a Packaging Recovery Note (PRN) for the tonnage recycled to comply with the UK Packaging Waste Regulations. The feedstock recyclate is suitable for low-grade use such as pens, rulers and key rings.

🖊	Save A Cup Recycling Company, UK	320
⚙	Save A Cup Recycling Company, UK	320
🗒	Polystyrene	
♺	• Recycled materials	324

Sensa™ pen

Gripping a pen for an extended period can cause discomfort. A soft, non-toxic gel around the grip area moulds itself to the user's fingers as it warms up and consequently improves comfort. Once the gel cools it returns to its original shape.

🖊	Boyd Willat, USA	310
⚙	Willat Writing Instruments, USA	323
🗒	Metal, gel	
♺	• Improved ergonomics	327

GreenDisk

Ex-Microsoft program disks are recycled by deleting the data and triple-testing so as to guarantee that they are virus- and error-free. A new label is added and the disks are boxed in packs of ten or twenty-five using recycled brown cardboard.

🖊	GreenDisk, USA	315
⚙	GreenDisk, USA	315
🗒	Recycled floppy disks	
♺	• Reusing end-of-life components	326

IBM ThinkPad TransNote

It could be argued that desktop and laptop computers have impacted on our ability to communicate by other means. Our drawing, drafting and hand-writing skills have atrophied, and yet there is often a longing to simply turn off the computer and put pen or pencil to paper. IBM are aware of these shifts in behaviour and have originated the ThinkPad TransNote, which combines a laptop with a touch-screen input, scanner and an accompanying pen and pad of paper, so choice is restored. There is no need to work out how to create a Venn diagram in a complicated graphics program, you can simply sketch on paper, scan and archive or email the result, while keeping your original hard copy. While this design might seem a little contrived it meets a genuine need for regular computer users to choose the way in which they work. This is important for our sense of well-being and permits creativity to wander beyond the boundaries of software interfaces designed by others.

✏	IBM Corporation, USA	316
⚙	IBM Corporation, USA	316
📜	Various materials and electronics	
🎧	• Multifunctional	327
🏆	iF Design Awards, 2002	330

Multimedia LCD projector VPL-CX5/C-55

Key function buttons are ready to hand on this slimline LCD projector, while less frequently used connections and controls are placed behind a lift-up panel. On powering-up the lens pops open automatically, so it is well protected in transit. Sony have extended the capability of this projector by including a Memory Stick slot on the VPL-CX5 for enabling presentations even if the user doesn't have a laptop or desktop PC. Both models are USB compatible. This is a robust projector for everyday use in the office or out on the road. It is built to the quality and environmental standards expected from Sony, a company that is a member of the World Business Council for Sustainable Development and the Dow Jones Sustainability Indexes.

✏	Sony Design Center, Japan	321
⚙	Sony Corporation, Japan	321
📜	Various materials and electronics	
🎧	• Lightweight, compact design	325
	• Improved functionality	327
🏆	iF Design Award, 2003	330

Portégé 2000

In the quest of the electronics industry to miniaturize, streamline and reduce, the Portégé 2000 represents something of a milestone. Weighing just 1.19 kg (2 lb 10 oz), this ultra-portable wireless Notebook is a complete mobile office product. An advanced lithium ion polymer battery and a 750MHz Intel® Pentium™ - M Ultra Low Voltage (ULV) CPU ensure fast, reliable working in battery mode. An 8 mm-thick (⅓ in) 20Gb hard drive provides plenty of storage and slips neatly within the 14.9 mm (½ in) external Notebook casing. A recent report by the UN university in Tokyo noted that the average desktop computer consumes an incredible 240 kg (528 lb) of fossil fuels, 22 kg (48 lb) of chemicals and 1,500 kg (3,300 lb) of water during manufacturing. So the reductionism applied to the material content of the Portégé 2000 is commendable. Let's hope that such reductionism is not at the expense of durability. If this Notebook can provide a reliable service to its owners for five years that is two more than the average life of a PC. Time will tell.

✏	Toshiba Europe, Germany	322
⚙	Toshiba Europe, Germany	322
📜	Various materials and electronics	
🎧	• Reduction in materials used	325
🏆	iF Design shortlist 2003	330

Multisync® LT 140

This lightweight, compact data projector has a PCMIA-card drive, which permits the unit to be used without linking up to a PC. The total number of components has been kept to a minimum and upon disassembly materials are easily separated into pure-grades. A slot is provided to store the remote control and a peripheral mouse enables annotations to be made on the projected data.

✏	IDEO Product Development, Japan	307
⚙	NEC Deutschland GmbH, Germany	318
📖	Various	
🎧	• Reduction in materials used	325
	• Design for disassembly	325
❓	iF Ecology Design Award, 2000	330

Single-handed keyboard

'Access for all' is the clarion cry of the proponents of the Information Age but conventional keyboard design denies access to individuals with disabilities. Maltron's single-handed and head/mouth stick keyboards are tools to help them overcome this hurdle and to enjoy what others take for granted. The Etype keyboard is also a tool for those suffering from repetitive strain injury (RSI), that most modern of ailments. A curved keyboard and palm resting pads ensure less tiring movements.

✏	PCD Maltron Ltd, UK	319
⚙	PCD Maltron Ltd, UK	319
📖	Various plastics, electronics	
🎧	• Improved user health	327
	• Improved access to information for those with disabilities	326

The Compostable Keyboard

New product take-back legislation is forcing computer manufacturers to examine systematic improvements in their capability to recycle components and materials at end-of-life. For example, the European Directive on Waste from Electrical and Electronic Equipment (WEEE), enforceable from 2004 onwards, means that manufacturers of computers, white goods and household appliances have the legal responsibility to ensure their products are taken back and recycled. The cost implications of such legislation are significant. Design for disassembly is essential to separate waste streams, but markets have to be found for the reused components or recycled materials if their use in remanufacturing is not feasible. Biodegradable plastics and composites offer an easier option since, once separated, they can be composted at a municipal waste facility. While enthusiasm for biodegradable plastics seems muted among the North American and European plastics manufacturers, it seems an ideal opportunity to experiment with more prosaic solutions. The Compostable Keyboard uses fibrous pulp from carrots, spinach and celery bound with cornstarch. The pulp can be injection moulded into a variety of forms and, once protected by a clear water-based finish, is moisture resistant and durable. Synthetic colourings are avoided as each plant has its unique colour signature. This tongue-in-cheek approach offers many advantages, although it is difficult to see how it would pass muster with the corporate marketing department and the need to use a brand colour palette.

✏	Jason Iverson and Shayan Rafie, University of Washington, Industrial Design Dept, USA	307
📖	Prototype	
⚙	Plant fibre, corn starch	
🎧	• Biodegradable, compostable materials	324
❓	IDRA award, 2000–2001	330

Digital iR series

Twenty-nine of Canon's copiers are certified to the German Blue Angel eco-label, including seven copiers in the Digital iR series, all-in-one networked devices capable of fax, scan, email and scan-to-database operations. Even the smallest in the range, the iR1600, offers a very capable specification including 16ppm output, network printing, digital copying, A3 paper handling and a relatively small machine footprint. Digital scanning at 1,200 x 600 dpi, or 2,400 x 600 dpi with AIR smoothing technology, ensures high-quality outputs. Automatic document feeds, two-way copy/print tray, and 350-page input trays ensure high throughput when required. Efficiency can be improved by adding a 10Gb hard disk for more storage and faster spooling. Aside from providing a range of new copiers to meet every office requirement, Canon have copier remanufacturing plants in the USA and Germany, and re-conditioning plants in Japan and China, so reliable used copiers are available too. Take-back of old copiers is most advanced in Germany where regulations, in the form of the Electronic Scrap Ordinance, ensure appliance manufacturers are responsible for the collection and recycling of electronic products no longer required by consumers. Copier consumables have also been targeted to reduce waste. Canon Bretagne in France is one of three global plants that act as regional collection centres for old cartridges. Canon shows commitment to the environmental and corporate social responsibility agenda, as would be expected from a company that is listed on the Dow Jones Sustainability Indexes.

✎	Canon, Europe	312
⚙	Canon, Europe	312
▤	Various materials and electronics	
⌁	• Improved functionality	327
	• Reduced materials and consumables	325/327
	• Blue Angel Eco-label	328

Xerox® Document Centre 470 and 460® series

The Xerox Corporation was an early adopter of greener manufacturing, environmental management systems and product take-back. This has meant designing standardized parts that are interchangeable across a range of products and are clearly identified to assist with reuse and recycling. Products are also designed for ease of disassembly to facilitate reuse, recycling or materials recovery. A reduction in the total number of parts in each machine is also central to the company's design philosophy. The Xerox® Document Centre 470 and 460® series are typical of networked, multifunctional machines capable of copying, printing, faxing and scanning. The series is Energy Star compliant, the sleep mode requiring just 65 watts, compared with 1,425 watts in operational and 260 watts in low power mode.

✎	Xerox Corporation, USA	323
⚙	Xerox Corporation, USA	323
▤	Various	
⌁	• Multifunctional	327
	• Design for disassembly	325
	• Reduction of production waste	327
	• Product take-back	324

ECOSYS

The current range of nine ECOSYS laser printers manufactured by Kyocera constitutes a modular design system that permits upgrading from a simple desktop personal printer, the FS-680, capable of 8ppm, to the free-standing FS-9000, which has an output of up to 36ppm. These printers use specially developed long-life drum technology rather than cartridges, so only the toner has to be replenished. Consequently maintenance requirements and operating costs are low.

✎	Porsche with Kyocera, Japan	309 317
⚙	Kyocera, Japan	317
▤	Various	
⌁	• Reduction in consumables	327
	• Modular upgrade path	327

USB FlashDrives

Slotting into the USB drives of PCs and laptops, these lightweight FlashDrives have revolutionized information transfer and back-up. Weighing just a few grammes, USB FlashDrives are cross-platform (PC and Mac) devices that store from 28Mb up to 256Mb of digital data. They can be attached to your key ring, sit in your pocket with loose change or hang around your neck. These versatile devices download or upload data rapidly, utilizing the speed of the Universal Serial Bus (USB) in the computer. They are ideal for backing up PowerPoint presentations, digital images or other important data and can be locked with a security ID and password. No more messing around burning a whole CD, DVD or ZIP disc just to store a couple of files. USB FlashDrives could make the humble floppy disk redundant for the next generation of computer users.

✏	Various designers, worldwide	
⚙	Various manufacturers, worldwide	
🔋	Polymer casings, various metals, electronic circuitry	
♺	• Reduction in materials used	325
	• Multifunctional	327

CanoScan LiDE 50

Since the 1960s, when the transistor made minaturization a reality, electronic goods manufacturers have jostled for accolades over their latest miniature gizmo. Canon have an enviable reputation in this field (examples include the Canon IXUS range of cameras; p. 165), and it is a company with strong environmental and social corporate responsibility strategies. The CanoScan LiDE 50 is an exceptionally compact high-resolution USB 2.0 flatbed scanner capable of 48-bit colour depth at 1,200 x 2,400 dpi. Scanner functions are activated by a series of 'EZ Scanbuttons' and it is possible to scan up to ten sheets in automatic mode. All this functionality is delivered in a very small-footprint device.

✏	Canon Design Center, Japan	312
⚙	Canon, Japan	312
🔋	Various polymers, metals and electronics	
♺	• Reduction in materials used	325
✹	iF Design Award, 2003	330

Belkin USB Media Reader/Writer

A perennial problem with the IT and computer industries is the delicate issues of protectionism, proprietary rather than open source standards, and lack of backwards compatibility. So any device that facilitates communication between and within the PC world is to be lauded. Such devices help slow down or prevent equipment becoming obsolete and generally improve our working lives. One such device is this media reader/writer that is compatible with SmartMedia, Compact Flash, Multimedia, Secure Digital and Memory Stick media cards. Connecting via a USB port to the PC allows rapid data transfers. With a small footprint, the device is suitable for vertical or horizontal standing, and keeps consumption of raw materials to a minimum.

✏	Belkin Corporation, USA	312
⚙	Belkin Corporation, USA	312
🔋	Various polymers, metals and electronics	
♺	• Reduction in materials used	325
	• Cross-media compatibility	
✹	iF Design Award, 2003	330

IFCO returnable transit packaging

Eleven standard sizes of flat-pack, reusable plastic containers with ventilated sides are meant for transit packaging for all types of fresh produce. The IFCO system is used in over thirty countries and an estimated seventy million packaging units are in circulation. Compatible with loading on Euro and ISO pallets, the units are of constant tare, are easily cleaned and when folded reduce storage space requirements by 80%. Unit weight-to-volume ratios are economical: tare weights vary from 0.65 kg (1 lb 7 oz) to 1.75 kg (3 lb 14 oz), giving respective storage volumes of between 0.01 sq. m (0.11 sq. ft) and 0.05 sq. m (0.54 sq. ft). At the end of their useful

working lives the polypropylene is recycled. An ecobalance study by Ecobalance Applied Research GmbH revealed significantly less environmental impact from the IFCO system than from conventional one-way corrugated cardboard boxes.

✎	International Food Container Organization (IFCO), Germany	317
⚙	International Food Container Organization (IFCO), Germany	317
📜	Polypropylene	
⌂	• Recyclable single material	324/325
	• Closed-loop system	325
	• Reductions in transport energy	326
	• Blue Angel eco-label	328

consignments. It can be stored into one-third of its original size by folding the sides, thus saving valuable cargo space. Meeting EU standards and with an expected service life of ten years, the Pallecon 3 Autoflow can be entirely recycled at the end of its useful life.

Pallecon 3 Autoflow

Made of sheet and solid steel, this container is suitable for transporting a wide range of industrial liquids from pharmaceutical products to foodstuffs. It is emptied via a sump through valves, which are recyclable, and is easily cleaned between

✎	LSK Industries Pty Ltd, Australia	317
⚙	LSK Industries Pty Ltd, Australia	317
📜	Steel	
⌂	• Single material	325
	• Reusable product	324
	• Recyclable	324
	• Potential reduction of transport energy	326
🔍	iF Design Award, 2000	330

VarioPac® / Ejector®

The struggle to extract CDs from their protective covers is consigned to the past thanks to this well-conceived and -manufactured product. Simply press the lever in the corner to eject the CD. An assessment by FH Lippe of the VarioPac Rover conventional cases revealed that the tactile, translucent and shatter-proof Metocene (metallocene polypropylene) requires 46% less energy during manufacturing by injection moulding and reduces transport volume by 33% – ample proof that this redesign reduces environmental impacts.

✏	VarioPac Disc Systems GmbH, Germany	310
⚙	Ejector GmbH, Germany	314
📜	Metocene X 50081	
🎧	• Reduction in materials used	325
	• Reduction of embodied and transport energy	325/326
✪	iF Ecology Design Award, 2000	330

Presswood pallet

Unlike traditional timber pallets, the 'Inka' pallets don't need to be fixed with staples or nails since they are manufactured from recycled timber waste bonded with water-resistant synthetic resins. Other advantages over conventional pallets include more compact stacking and lower tare weight. Standard pallet sizes meet current European regulations and are recyclable. As of January 2000, wood is included within the EU packaging waste recovery and recycling regulations, so the reduced wood content of these pallets lowers the costs associated with these obligations.

✏	INKA Paletten GmbH, Germany	316
⚙	INKA Paletten GmbH, Germany	316
📜	Waste timber, resins	
🎧	• Recycled materials	324
	• Reduction in transport energy	326

Disk Lev

There are numerous synthetic plastic CD 'jewel' cases on the market but most need secondary packaging in transit, which is often discarded by the user and inherently high in embodied energy. Disk Lev is a lightweight cardboard design that can be mailed but also provides protection for archiving and storage. When folded, a central tab fixes the CD in position and on opening, the CD is automatically raised from the tab, without applying any pressure, for easy removal.

✏	Philipp Prause, Austria	309
⚙	Ernst Schausberger & Co., Austria	314
📜	Cardboard	
🎧	• Reduction in materials used	325
	• Recyclable or compostable	324
✪	iF Design Award, 2003	330

Jiffy® padded bag

Jiffy produce a range
of bags to protect goods
transported via postal
and courier systems.
The padding is made of
72–75%-recycled, shredded
newsprint and the exterior
is a tough brown paper,
which permits the bags
to be reused.

✏	Pactiv Corporation, USA	319
⚙	Pactiv Corporation, USA	319
▤	Recycled newsprint	
♻	• Recycled materials • Reusable product	324 324

PataTotal

In the same way that we
eat the ice-cream cone,
Ela suggests we should
eat our container of potato
chips, enabling a zero-
waste strategy. This edible
container can be flavoured
with your favourite spices
and herbs, their aromas
mingling with that of the
chips. As national cuisines
go global it is a surprise
that the taco shell hasn't
morphed into the potato
shell. Seaweed-covered
sushi, rice-paper-coated
tempura and other
delicacies reveal a lexicon
of possibilities for edible
packaging.

✏	Adital Ela, graduate, Design Academy Eindhoven, the Netherlands	305
⚙	Prototype	
▤	Potatoes	
♻	• Renewable materials • Zero waste	325 325

Schäfer Eco Keg

Die-cast, injection-
moulded, thermoplastic
base and top clip on to
the stainless steel body
of this beverage container,
avoiding the need to
glue in place rubber or
polyurethane sealing
rings. All parts can be
disassembled for repair,
replacement and pure-
grade recycling. The
container is suitable for
all automated KEG plants
and can be stacked
more easily than
conventional containers,
making for space savings
and improved transport
efficiency.

✏	Schäfer Werke GmbH, Germany	320
⚙	Schäfer Werke GmbH, Germany	320
▤	Stainless steel, thermoplastic	
♻	• Design for durability, disassembly and recyclability	325
	• Reduction in transport energy	326
✹	iF Ecology Design Award, 2000	330

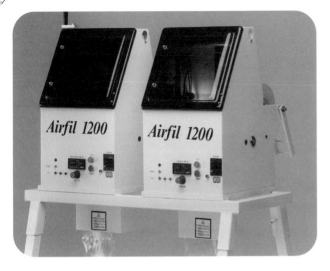

Airfil

This lightweight packaging made of recyclable PE uses air as the shock-absorbing material to protect goods in transit. Airfil air bags are produced in standard and bespoke sizes, providing a viable, less expensive alternative to polystyrene 'chips' and 'bubble wrap'. The Airfil system significantly reduces storage space requirements and allows a reduction in the thickness of the outer packaging material. It is also reusable, clean and free of dust.

✏	Amasec Airfil, UK	311
✿	Amasec Airfil, UK	311
▤	Polyethylene	
♺	• Reduction in materials used	325
	• Recyclable materials (plastic)	324

3M™ recycling compatible label material 8000

Contamination of recycled plastic feedstock with unknown types of plastic can render recyclate unusable and damage the production plant. It is not always possible to create labels by embossing the information on components or products, so 3M have produced a stick-on label that can be used when recycling ABS and polycarbonate, ABS/polycarbonate mixes and polystyrene. Such plastics are common in the electronics industry where identification of materials at disassembly will become more critical as the EU WEEE Directive on recycling of electronic equipment is applied over the next few years.

✏	Hiep Nguyen, Gerald Schniedermeier, Yolanda Grievenow, Den Suoss, 3M Deutschland, Germany	311
✿	3M Deutschland, Germany	311
▤	Plastics	
♺	• Materials labelling	325
	• Encourages recycling	326
✪	iF Ecology Design Award, 1999	330

Earthshell Packaging®

Earthshell Packaging® is a heat-moulded foam laminate derived from plant starches from annual agricultural crops, such as potatoes, corn, wheat, rice and tapioca. These starches are mixed with limestone, fibre and water into a batter-like consistency. The resultant 'batter' is injected into moulds, which are heated and pressurized releasing the water as steam and allowing the foam laminate to take its final rigid shape. Plates, bowls, hinged-lid sandwich and salad containers are grease- and stain-resistant, and are suitable for cold and hot food and microwave cooking. Life Cycle Inventory (LCI) analysis was applied by leading US consultants Franklin Associates, a pioneer in life cycle analysis (LCA), in order to minimize the environmental footprint of the packaging. Throughout the lifecycle Earthshell Packaging® shows a smaller footprint than traditional synthetic polymer foam packaging as it consumes less energy, uses less fossil fuel, produces lower greenhouse/air/water emissions and reduces its landfill volume requirement.

✏	Earthshell Corporation, USA	314
⚙	Earthshell Corporation, USA	314
▦	Starch, fibre, limestone	
♻	• Renewable and geosphere materials	324/325
	• Biodegradable	324
	• Reduced energy of manufacture	325
	• Reduced emissions	326

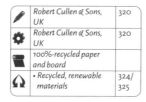

Veltins Steinie

Just fifty years ago most beer bottles were reused by the (local) brewery that produced them, but as beer manufacturing went global the distribution and retailing chains changed beyond recognition, resulting in many bottles making a one-way trip to the landfill. Although 'bottle banks' placed strategically at supermarkets and local sites have encouraged consumers to recycle their empties, and many breweries operate return systems for their pubs and bars, vast amounts of virgin glass is still thrown away. Veltins saw the challenge of re-designing their bottles to improve reuse efficiency as a means to re-brand their beer. This reusable bottle provides a distinctive ergonomic profile and doesn't need a conventional paper label as information is embossed on the base of the bottle.

✏	Formgestaltung Schnell-Waltenberger, Germany	306
⚙	Brauerei C & A Veltins, Germany	
▦	Glass	325
♻	• Reduction in materials used	324
	• Reusable product	
◉	iF Design Award, 2003	330

Cull-Un Pack

This is a UN-certified packaging design for the transport of hazardous chemicals in glass containers. A strong moulded pulp base and top, made from used cardboard boxes, protects the containers, which are enclosed in a corrugated cardboard outer. It meets stringent safety tests, including a 1.9 m (6 ft) drop, which had previously been met only by using expanded polystyrene packaging. All the materials are from recycled sources and the packs can be delivered flat, saving valuable delivery space.

✏	Robert Cullen & Sons, UK	320
⚙	Robert Cullen & Sons, UK	320
▦	100%-recycled paper and board	
♻	• Recycled, renewable materials	324/325

Transfert

Take a close look at the equipment you use in your daily working life, then ask whether it suffers from innovation lethargy because it has become a default industry standard that no-one questions despite its obvious drawbacks. This is the process that led to the development of Transfert, a new modular stainless steel trolley system for restaurants and corporate or municipal kitchens. Existing trollies conform to dimensions set by the GastroNormes international standard containers, which vary in depth from 20–65 mm (³⁄₄–2¹⁄₂ in). These containers slide on L-shaped runners set in a fixed frame that usually accommodates ten trays, but the trolley structure is difficult to access to clean properly. Transfert accepts the GastroNormes containers but makes a radical departure from the usual trolley. It is made of a series of individual stacking modules made of a continuous single 10 mm-(²⁄₅ in) diameter stainless steel wire bent in three dimensions without any welding. This modular solution and the unique geometry make Transfert more easily hand-washable and also the first machine-washable trolley. It can be stacked in a stationary or mobile unit to provide flexible work practices and minimize workers' stress and strains. Transfert is the first piece of professional kitchen equipment to be awarded a prestigious Etoile du Design award in 2004.

✏	François Tesnière & Anne-Charlotte Goût, 3bornes Architectes, France	310
⚙	TSA Inox, France	322
📄	Stainless steel 204Cu	
♻	• Durable	327
	• Modular design	327
	• Improved health & safety	327
	• Improved ergonomics	327

Potatopak

Large-scale commercial manufacturing of polystyrene began by Dow Chemicals in the USA and I G Farben in Germany in the 1930s but it wasn't until the rise of global fast food corporations and supermarkets that polystyrene began getting some negative publicity. Polystyrene containers tend to have short lives, ending up in landfill sites where there they will remain for many generations. Potatopak products offer a biodegradable alternative. Using starch obtained from low-grade potatoes, the plant can produce a wide variety of food containers with similar handling and rigidity characteristics of synthetic plastics. Labels are not required as the customer brand identity and other product information can be embossed on the bottom of the container. Although not intended to be consumed, the containers are harmless if eaten by mistake.

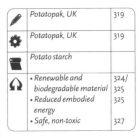

✏	Potatopak, UK	319
⚙	Potatopak, UK	319
📄	Potato starch	
♻	• Renewable and biodegradable material	324/ 325
	• Reduced embodied energy	325
	• Safe, non-toxic	327

Allison bus

An improvement in fuel economy of 50% over conventional powered units is achieved by using the Allison Electric Drives hybrid system developed by Allison Transmission, a subsidiary of General Motors. The system also gives a significant reduction in emissions of carbon monoxide, hydrocarbons, particulate matter and nitrous oxide. The Allison Bus represents an opportunity to create cleaner public transport systems. A well-peopled bus equipped with such clean technology is clearly more sustainable than lots of single-occupancy cars.

✏	Allison Transmission / General Motors, USA	311/ 315
⚙	Allison Transmission / General Motors, USA	311/ 315
📜	Various	
⬇	• Fuel economy • Reduced emissions	327 326

Metropolitan express train

In contrast to most modern trains, the interior of this new train system is made of entirely natural or recyclable materials. With moulded laminated wood shells and leather upholstery, this design makes significant reference to the original 1956 Herman Miller Model No. 670 and 671 lounge chairs designed by Charles and Ray Eames. For those who travel regularly by rail the little touches, such as a padded head pillow, will be much appreciated.

✏	gmp - Architekten, Germany	306
⚙	Deutsche Bahn AG and Metropolitan Express Train GmbH, Germany	
📜	Leather, plywood, stainless steel	
⬇	• Renewable and recyclable materials	324/ 325
🏆	iF Design Award, 2000	330

Checkpoint and Checktag

Loose wheel nuts can lead to accidents, with loss of life and possible spillage of pollutants and toxins into the environment. Checkpoint and Checktag are two types of plastic cap that are pushed over a nut once it has been tightened to the correct torque. The arrows on each cap should be aligned unless the nuts have worked themselves loose. A quick visual check is all that is needed to identify a rogue nut.

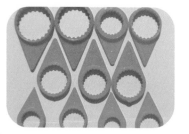

✏	Mike Marczynski, UK	308
⚙	Business Lines, UK	312
📜	Plastic	
⬇	• Improved road safety • Reduced pollution	327 326

Solo

Solo takes midi-bus design to new lows – that is, it provides a low-level platform to enable wheelchair users to mount from pavement to bus by an extendible automatic ramp. A low centre of gravity also produces less roll and a more comfortable ride for all. And in the absence of any steps, buses can pick up and drop off passengers more quickly, enabling them to keep accurately to their specified timetables and reduce emissions while idling.

✏️	Optare Group Ltd, UK	319
⚙️	Optare Group Ltd, UK	319
📜	Various	
🎧	• Alternative transport for improved choice of mobility	326
	• Improved functionality for passenger loading/unloading	327

Flexitec

Traffic-calming systems installed using conventional techniques require considerable manpower and cause disruption to traffic during installation. Flexitec, a hard-wearing modular system of kerbs, blocks and ramps, manufactured from recycled rubber, is installed by bolting each module to the existing road surface. It reduces road congestion during installation and can be used for permanent or temporary traffic calming.

✏️	Prismo Travel Products, UK	320
⚙️	Prismo Travel Products, UK	320
📜	Recycled rubber	
🎧	• Recycled materials	324
	• Reusable product	324
	• Reduced energy of installation	

DRIVE ON

Energy reclamation could be big business in the future as fossil fuel resources diminish. Normally energy is reclaimed from each car in systems like Toyota's New Prius where braking is converted back into electrical energy stored in a battery. This prototype looks at reclaiming energy from the public domain network of roads. As vehicles travel over the pad, which is just proud of the road surface, the piston moves up and down, generating a small electrical current. Those managing traffic networks can examine their data to get a best fit between vehicle flows and speed requirements in order to calculate the points to install these mini generators on the road system.

✏️	Andreas Unterschuetz, USA	310
⚙️	Conceptual prototype	
📜	Various materials	
🎧	• Energy generation and conservation	327
🏆	IDRA award, 2002–2003	330

Buses

Since the mid-1990s the German commercial vehicle manufacturers MAN Nutzfahrzeuge have been testing working prototypes using natural gas (CNG), liquefied petroleum gas (LPG), hydrogen fuel cells and a biofuel called rapeseed oil methylester (RME) as an alternative to diesel fuel. An articulated bus powered by CNG develops 310 bhp but, in combination with closed-loop catalytic converters, conforms to Euro 3 emissions levels proposed by Germany, which are less than or equal to 2 g/kWh carbon monoxide, 0.6 g/kWh hydrocarbons, 5 g/kWh nitrous oxides and 0.1 g/kWh of particulate matter (PM10s). These levels show a reduction factor of between 3 and 5 of the Euro 1, 1990, exhaust legislation. A further benefit is a reduction in noise to almost half the normal level of a diesel-powered bus. MAN's commitment to reducing environmental impacts is reflected in their accreditation to environmental management standards including EMAS and, at the Steyr factory, ISO 14001.

✏️	MAN Nutzfahrzeuge, Germany	318
⚙️	MAN Nutzfahrzeuge, Germany	318
📜	Various	
♻️	• Reduced emissions • Reduced noise	326 326

Liquefied natural gas-powered vehicle

Chilled food is delivered daily to each store in the Marks & Spencer retail chain using articulated lorries with refrigeration units. Following a review of their distribution system with their lead contractor, BOC (now called Gist), the company examined ways of reducing vehicular emissions and noise pollution. This culminated in the development of a new fleet of natural-gas-powered vehicles equipped with quiet, non-polluting cryo-eutectic refrigeration units. Compared with the original diesel-engined vehicles, emissions from the natural-gas vehicles produce 89% fewer particulates, 69% fewer nitrous oxides and approximately 10–20 % less carbon dioxide.

✏️	Marks & Spencer with Varity Perkins, ERF, Gray and Adams and Gist (formerly BOC Distribution Services), with the Energy Savings Trust, UK	315
⚙️	Joint venture with Gist, UK	315
📜	Various	
♻️	• Reduced emissions • Reduced noise	326 326

caring for the environment

FRESH St Michael FOODS FOR

MARKS & SPENCER

This vehicle is powered by natural gas.

QUALITY · VALUE · FRESHNESS

Citadis & Flexibility outlook T3000

Enthoven Associates see accessibility and liveability as the key drivers in addressing sustainable mobility issues. The city environment and its infrastructure are threatened by aerial and water-based pollution caused by current transport modes. Furthermore, the erosion of access to public spaces, gentrification collective space to compensate for public space eroded by privatization and building for commercial gain.

Transport systems have a long history of driving urban development. Lately, many cities in Europe, recalling the efficiencies of nineteenth-century systems, have reintroduced trams. The modern tram is, safety features. Citadis is a tram for Rotterdam (2002). This concept is about to be released in Jerusalem. Flexibility outlook T3000 for Bombardier Transportation is for Brussels. The rich Art-Deco heritage of the city is

reflected in the interior with a rich palette of colours and the use of luxurious seating materials such as leather. Dare we use the words 'travel in style' when we mean public transport? The answer looks set to be yes.

and loss of urban diversity challenge socio-cultural functions. Private transport, i.e. the car, shows a growing incompatibility with accessibility and liveability. Low-emission vehicles will not deliver a sustainable solution. Public transport systems have to embrace strategic and spatial functionality, and so address a wider range of sustainability issues, offering a de facto however, a long way from its noisy, clattering ancestors. It must be different in order to lure modern commuters out of their cars. Enthoven Associates have been involved in designing tram systems for Amsterdam, Brussels and Jerusalem. Their designs reveal a passion to create high-quality interiors, universal access for special-needs commuters and numerous

✎	Enthoven Associates Design Consultants, Belgium	306
⚙	Various clients including Bombardier Transportation, Brussels and Alstom, Rotterdam	
📃	Various materials	
🎧	• Alternative, sustainable, transport	326
	• Eco-efficient mass product-service-system (PSS)	324

Citaro Fuel Cell Bus

Following the successful trialing of NEBUS (New Electric Bus), a prototype hydrogen fuel-cell-powered Mercedes Benz O 405 bus in 1997, the Mercedes-Benz Citaro is a new bus being tested in Amsterdam, Barcelona, Hamburg, London, Luxembourg, Madrid, Porto, Reykjavik, Stockholm and Stuttgart. The 12 m- (39 ft-) long Citaro has a top speed of 80 km/h (50 mph) a range of 200 km (125 miles) and can accommodate 70 passengers. The electric motor and automatic transmission are at the rear, while the fuel cell, with an output of over 200 kilowatt, and 350 bar (5,000psi) compressed hydrogen cylinders are in the raised roof. Citaro is supported by the CUTE (Clean Urban Transport for Europe) and ECTOS (Ecological City Transport System) projects in order to monitor passenger feedback. Low noise levels, high comfort specifications and an emission-free exhaust should impress commuters. The wider implications of hydrogen economy systems remain to be seen but the Citaro will definitely show a reduction in local environmental impact.

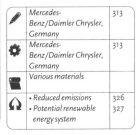

✎	Mercedes-Benz/Daimler Chrysler, Germany	313
⚙	Mercedes-Benz/Daimler Chrysler, Germany	313
📃	Various materials	
🎧	• Reduced emissions	326
	• Potential renewable energy system	327

CIVIA 501M

In order to reduce the substantial environmental burdens of millions of urban commuters going to work in their cars, there have to be significant improvements in public transport systems. A consortium, including CAF, Siemens and Alstom Bombardier, worked with their client Renfe to produce the Suburban Train, CIVIA. Bright, light, airy interiors, generous gangways and wide opening doors all contribute to a sense of internal space. Doorway ramps enable access for wheelchairs and those with mobility difficulties. Six TFT screens in each carriage inform passengers as to the destinations and timetabling and a CCTV system provides improved passenger security.

Colourful, robust synthetic resins and alloys cope with everyday wear and tear while offering comfort for short-haul suburban journeys. The CIVIA can operate with one to four carriages, providing flexible operation. A range of modular carriage designs (A1, A2, A3) and two types of motor-driven bogies enable operators to mix and match a train to suit their particular passenger and route needs.

Electronic systems monitor energy consumption for internal climate control and obtain the best eco-efficient performance for the passenger load and track conditions.

✎	A consortium for Renfe, Spain	320
⚙	Renfe, Spain	320
▬	Various materials	
↻	• High-quality public transport product	326

Passenger information system

Saving up to 60% of the energy consumption of other pulse technology displays, this modular aluminium-framed passenger information system uses hundreds of LEDs controlled with a patented system. With a legible display and clean lines, this system conveys information with maximum efficiency and minimum fuss. It is also easily maintained by one person.

✎	Interform Design, Germany	307
⚙	LUMINO Licht Elektronik GmbH, Germany	318
▬	Ceramic-coated dispersion glass, aluminium, LEDs	
↻	• Energy efficiency • Modular design • Ease of maintenance	327 327 327
✪	iF Ecology Design Award, 2000	330

FanWing

High speed, noise and huge fuel bills are the hallmarks of today's fixed-wing aircraft. The FanWing, currently tested as a working model prototype, is an aircraft with near-vertical take-off capabilities that serves as a quiet, slow but fuel-efficient, load-carrying transporter. In an

Moreover, it is simple and inexpensive to construct and therefore offers an economical air-transport system for everything from disaster relief work to fire-fighting and reconnaissance or traffic monitoring. Preliminary specifications for a three-passenger version show

intriguing innovation, the designers have introduced a large rotor along the entire leading edge of the wing. The engine directly powers the rotor, which is capable of producing both lift and thrust as the cross-flow fan pulls air in at the front and accelerates it over the trailing edge of the wing. Wind-tunnel testing reveals 15 kg (33 lb) of lift per horsepower, equivalent to a payload capacity of 1 to 1.5 tonnes for a 100-horsepower power unit.

it weighing in at just 350 kg (770 lb) empty and having a top speed of 60 km/h (37 mph), a wingspan of 10 m (33 ft) and a flying time of ten hours.

✏	FanWing, UK/Italy	306
✹	Model prototype	
▤	Various	
⌂	• Fuel economy • Simple, low-cost construction	327 325

Helios

The Helios is an enlarged version of the Centurion 'flying wing'. It has a wingspan of 75 m (247 ft), which is two-and-a-half times that of the Pathfinder flying wing and longer than that of a Boeing 747 jet. AeroVironment's ambition is to enable Helios to fly at 30,500 m (100,000 ft) continuously for twenty-four hours and at 15,259 m (50,000 ft) for four days, all under solar power. This aircraft is known as an uninhabited aerial vehicle (UAV) and is suitable for remote sensing and reconnaissance with

a multiplicity of applications for recording the weather, changes in vegetation cover and military operations.

✏	NASA, Dryden Research Center, USA, with AeroVironment, Inc., USA	308/311
✹	NASA Dryden Flight Research Center, USA, with AeroVironment, Inc., USA	308/311
▤	Photovoltaic modules, lightweight metals and composites	
⌂	• Zero emissions • Renewable energy	326 327

Trent turbofan engines

Rolls Royce, General Electric and Pratt and Whitney dominate engine manufacturing for aeroplanes. Rolls Royce have taken the lead in producing a fuel-efficient, lightweight and low-cost jet engine. Weight reduction was achieved with titanium fan blades comprising three sheets in close proximity, which were subjected to heat, causing flow of material and bridging between the layers to form a honeycomb structure. Other innovations include 'growing' metal by controlled cooling of the molten alloy in the mould to align all the molecules in one direction, forming an extremely strong single crystal. The turbine blades, which are made of this, can operate at almost twice the melting point of normal crystalline metal. A three-shaft design is also more efficient and easier to maintain and upgrade. Aeroplanes with Trent

✏	Rolls Royce, UK	320
⚙	Rolls Royce, UK	320
▤	Various	
♻	• Fuel economy • Reduced air pollution	327 326

engines, such as the Airbus A330, can carry more passengers for the same fuel consumption.

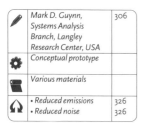

Quiet Green Transport

Revolutionary Aerospace Systems Concepts (RASC) is a branch of NASA concerned with conceptual explorations of new mobility and communication systems for air and space travel. Project FY01 included the concept of Quiet Green Transport, by Langley Research Center, whose dual objectives were to substantially mitigate noise and emissions from commercial aviation. Thirty new configurations for aeroplanes were created and several variants identified for further development. A common feature of these concepts was the use of liquid hydrogen fuel, distributed by electric propulsion through ducted fans over the wing.

✏	Mark D. Guynn, Systems Analysis Branch, Langley Research Center, USA	306
⚙	Conceptual prototype	
▤	Various materials	
♻	• Reduced emissions • Reduced noise	326 326

Boeing 7E7 Dreamliner

There's a new 'lean green' agenda reaching the research and development programmes of all the major aeroplane manufacturers as regulatory pressure increases to curb greenhouse gas emissions and reduce noise levels. Boeing have reached the 'firm concept' stage in developing the 250-seater 7E7 Dreamliner designed to use 20% less fuel than a similar-sized aeroplane. This effectively extends the range of the plane to between 14,500 km and 15,400 km (7,800 and 8,300 nautical miles) – approximately a 15 hour flight – depending on the model. All models will feel airy and light thanks to large 480 mm by 280 mm (19 by 11 in) windows and arched ceilings lit with LEDs whose colour can change to create various 'sky' effects. There's a long way to go as final designs are not expected until 2005, test flights by 2007 and air-worthiness and safety certification by mid-2008. While the concept clearly offers potential reduction of environmental impact and improved passenger comfort, the airline industry is expected to continue rapidly expanding in the next few years. By the time the 7E7 comes along air travel could be operating at significantly higher volumes. Fuel efficiency gains of 20% might be academic in the overall trend of increased greenhouse gas emissions.

✏	Boeing, USA	304
⚙	Conceptual design	
▤	Various materials	
♻	• Fuel economy • Improved passenger comfort	327

✏	Springboard Design Partnership, UK	309
⚙	Conceptual prototype	
🗞	Aluminium, glass, photovoltaics, vertical wind turbine	
🔄	• Renewable energy • Reduced noise	327 326

Dock Shuttle

Springboard Design's native city, Bristol, UK, possesses the remnants of the world-famous dockyard where Isambard Brunel's first iron steamship, the *SS Great Britain*, was built and launched. Building bridges across the old docks is not always feasible, because of shipping traffic, and tends to be prohibitively expensive. The Dock Shuttle provides an imaginative solution. The lightweight aluminium and glass passenger 'pod' is attached to a huge swinging arm pivoted in the centre of the river on a central 'island'. Electric motors in the pod are trickle-charged by photovoltaic cells on the swinging arm and a vertical axis wind turbine. Shipping simply passes by the island on whichever side is open. This design could be applied to passenger movement across any linear water structure.

RA

RA is a zero-emissions, solar-powered boat, which is ideal for freshwater transport where the pollution of conventional diesel or petrol motor boats is damaging to water quality. Built to a high specification using Burmese teak and stainless steel, it contains raw materials that are extremely durable, low-maintenance and 100% recyclable. Greenpeace, the international NGO, assisted in obtaining the construction materials. An added benefit of the solar generation and electric motor system is its quietness of operation, making it a more fitting companion for aquatic wildlife.

✏️	Kopf Solardesign GmbH, Germany	317
⚙️	Kopf Solardesign GmbH, Germany	317
🪑	Stainless steel, teakwood, photovoltaics, batteries	
🎧	• Zero emissions	326
	• Solar power	327
	• Durable	327

Solarshuttle 66 (Helio) and RA82

Kopf are pioneers in developing solar-powered ferries for inland waterways. The Solarshuttle 66, otherwise known as the Helio, is a scaled-up version of a ferry, which has operated between Gaienhofen, Germany, and Steckborn, Switzerland, since 1998. With a maximum speed of 24 km/h (15 mph), the Helio can operate for up to eight hours from the bank of twenty-four batteries without needing a recharge from the photovoltaic panels. The larger RA82 has a capacity of 120 passengers and is in service in Hamburg and Hannover. Low operating costs and negligible environmental impacts could popularize it in urban areas served by waterways and in ecologically sensitive areas.

✐	Dr Herbert Stark, Kopf Solardesign, Germany	309
⚙	Kopf Solardesign GmbH, Germany	317
▬	Stainless steel, teakwood, photovoltaics, batteries	340
⌗	• Zero emissions	326
	• Solar power	327

Furniture

Giulietta

The bench, in a public or private setting, provides a moment's respite, a counter-balance to the rush of modern life. Re-working the classical typology of the English wooden park bench, Rizzatto has created an animated bench in bright, bold polyethylene. Rotationally moulded in a single piece, the Giulietta possesses an air of permanence, toughness and durability. Equally suited to indoor and outdoor use, this bench is weatherproof and rot-proof, so requires no maintenance and is easily recycled.

✏	Paolo Rizzatto, Italy	309
⚙	Serralunga, Italy	321
▤	Polyethylene	
♺	• Single material • Recyclable	325 324

Olympic

This bench, for public spaces and school grounds, builds upon observing human nature, especially how people like to sit on the upright of traditional wooden benches rather than the seat itself. Olympic uses familiar materials, but a new set of fun references, to encourage social gathering and participation.

✏	Thomas Bernstrand, Bernstrand & Co, Sweden	304
⚙	Nola Industrier, Sweden	318
▤	Wood, steel frame	
♺	• Improved social well-being	327

Games-in-Between project

Encouraging people to use public transport requires fresh, imaginative initiatives. Such initiatives are not necessarily concerned with the vehicles themselves but can focus on our perceptions of the public transport experience. Realizing that waiting is an integral part of the experience, Vreni Bazzan designed a temporary campaign to enhance the image of VBZ, the public transport executive of the City of Zurich and promote the organization. Various games were installed in bus and tram shelters. Strong visuals, the green–yellow colour scheme and the iconic symbolism of the individual games gave a strong identity to the whole project. Games included dice, crosswords, labyrinths, drawing, water squirters, xylophones, swings and more. Projects such as Games-in-Between provide valuable feedback on how design can focus on the experience rather than the product or service. Encouraging behavioural change to reduce environmental impacts and enhance individual and community well-being is essential to deliver sustainable development. Creating opportunities to change behaviour requires understanding the experience–object and experience–location interfaces.

✏	Vreni Bazzan, Hochschule für Gestaltung und Kunst, Zurich, Switzerland	304
⚙	One-off project	
▤	Various materials	
♺	• User involvement • Improved well-being	327 327

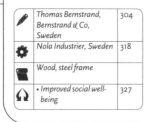

Sussex

With its ranks of straight benches public seating all too often enforces social formality. Robin Day, a British designer who brought a playful modernism to dreary 1950s Britain, continues to show his lightness of touch with this modular seating system. Sussex is based upon a single modular plastic unit attached to a metal frame. It enables subtle curvilinear forms and generates an informal, convivial environment that is much more conducive to social interaction. There is potential for Magis to experiment with the use of a recycled fraction in the plastic and metal components without detriment to the positive social and aesthetic impacts of this system.

✏	Robin Day, UK	305
⚙	Magis SpA, Italy	318
🗐	Plastic, metal	
♻	• Modular • Ease of repair	327 327

Outdoor bench

Baccarne produce a range of outdoor and public seating exclusively from recycled plastics, a mix of polypropylene, polyvinyl chloride and polyethylene obtained from post-production waste streams such as window-frame manufacturing. Planks and sheeting provide basic yet tough functional furniture.

✏	Baccarne Design, Belgium	311
⚙	Baccarne, Belgium	311
🗐	Recycled polypropylene, polyvinyl chloride and polyethylene	
♻	• Recycled materials	324

virtuRail

Any frequent rail traveller is struck by the lack of adequate public seating at train stations. Moreover, the seating provided is often uncomfortable.

Wijffels recognizes passengers as individuals with diverse needs in his expressive landscape bench whose materials and structure reference traditional park benches. Here a continuum of seating options is provided for short or long waiting periods. This design points the way for extending the functionality and humanism in street and public space furniture.

✏	Manuel Wijffels, graduate student 2002, Design Academy Eindhoven, The Netherlands	305
⚙	Conceptual prototype	
▭	Wood, steel	
🎧	• Renewable and recyclable materials	324/325
	• Design for ease of maintenance	327
	• Improved ergonomics	327
	• Communal facility	

Shade On

There's a precision in Western European time-keeping that seems to link back to its role as cradle of the Industrial Revolution, when time became commodified. Shade On is a whimsical sundial that is beautifully imprecise. As the sun revolves around this sculptural totem it reveals the time twice a day. 'Nearly eleven' is actually 'half past two' and 'some time in the afternoon' is 'around four'. This ambivalent clock reminds us that the 24-hour clock is a convenient, but not the only, way to mark time. Does this work set a challenge for designers to re-examine their relationship with time? New thoughts are emerging from diverse parts of the design diaspora about the concept of 'slow design'. A re-think about the fast industrial design paradigm is overdue. Shade On hints at possibilities.

✏	Maarten Baas, graduate 2002, Design Academy Eindhoven, the Netherlands	305
⚙	One-off	
▭	Bronze	
🎧	• Durable materials	324
	• Socially conscious product	

Bus and tram shelter system

Modularity is the key ingredient of this system of bus and tram shelters for X-City Marketing Hannover and Ströer Out-of-Home Media. Local geography and passenger mobility data at individual locations influence the exact choice of module panels, seats, litter bins, information displays and advertising panels, but a customized solution is always possible. Individual components are easily re-configured or repaired so the shelter system can be more efficiently managed and maintained. Transparency and configuration of the panels with bright lighting ensures confidence in public safety can be high.

✏	James Irvine and Steffan Kaz, Milan, Italy	307
⚙	X-City Marketing Hannover, Germany	323
▭	Metals, glass, polymers and electronics	
🎧	• Modular	327
	• Ease of repair and maintenance	327
🏆	iF Design Award, 2002	330

Un cri jeté

Constructed of gabions back-filled with pebbles from the beach, this U-shaped construction invites exploration. Once cocooned in its walls you experience a different world as the tide enters and retreats. Here you can shout to the wind, laugh and dance as the elements dictate your mood. This design encourages contemplative time by decommodifying design, freeing it from its commercial burdens. In doing so design rediscovers its joy and power.

Sofanco

Stone is an extremely durable natural material. Oscar Tusquets Blanca has captured the strength of this material but rendered it in a fluid, organic form to create a design of great potential longevity, albeit requiring moderate energy input during manufacturing.

✏	Oscar Tusquets Blanca, Spain	304
⚙	Escofet, Spain	314
🗋	Stainless steel, reinforced cast stone	
↻	• Abundant geosphere materials	324

✏	Sonja Kuypers, graduate student 2003, Design Academy Eindhoven, the Netherlands	305
⚙	One-off	
🗋	Stone, steel	
↻	• Multifunctional • Design without consumption	327

SINE seat

Extruded plastic lumber provides the catalyst for this innovative public seating, which can be fabricated to bespoke lengths and curvatures depending on the client's requirements. Two styles of cast aluminium frame, one with a back rest, permit further customization. Achieving similar results in hardwood would prove more costly. Utilizing the recycled plastic also means that expensive resources are released for more valued activities.

✏	V K α C Partnership, UK	310
⚙	V K α C Partnership, UK	310
🗋	Recycled plastics, aluminium	
↻	• Recycled and recyclable materials	324

Street Light F

Power generation using solar cells offers an opportunity to re-examine the design parameters for familiar objects. In this case the rectilinear shape of the 35-cell solar array was the prime component in configuring the polycarbonate diffuser to create the clean lines of this new generation of street lighting. Batteries in the base of the upright are capable of lighting the 18-watt fluorescent lamp for seven days without a recharge.

✏	Roy Fleetwood, UK	306
⚙	YKK Architectural Products, Inc., Japan	323
📇	Aluminium polycarbonate, solar cells	
♻	• Solar power	327

M400 CobraHead-Styled LED Streetlight

Satellite photos reveal the earth at night to be brightly lit in the major urban and mega-metropolitan centres. It is increasingly difficult to escape this 'light pollution' and clearly see the stars. In the USA it is estimated that almost 30% of outdoor lighting doesn't light its target but is lost into outer space. In line with the ambitions of the Dark Skies initiative LEDtronics have introduced an LED streetlight head to replace traditional incandescent streetlights. The M400 CobraHead-Styled LED Streetlight contains 400 warm incandescent-white (3,200K) LEDs arranged to optimize directional light to the target. With only a 19W power drawer and a lifetime of over 100,000 hours for each LED, this represents a massive energy saving against traditional incandescents and means that street lighting is a more feasible option for local renewable energy networks. Old 'Cobra'–style lamps can be retrofitted with the LED units or complete new die-cast aluminium M400 luminaires can be fitted as replacements. Standard voltage is 120VAC but 12V, 24V, 28V and 240V versions are manufactured.

✏	LEDtronics, USA	317
⚙	LEDtronics, USA	317
📇	Die-cast aluminium, LEDs, electonic gear	
♻	• Energy conservation	327

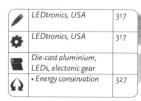

Metronomis

A range of street lighting by Philips is specially designed for energy-saving lamps and low maintenance. Modular components are durable and vandal-proof and permit different design permutations according to customers' preferences.

✒	Philips Design, the Netherlands	319
✿	Philips Electronics, the Netherlands	319
▤	Metal, glass, lamp	
♻	• Energy efficiency	327
	• Modular design	327
	• Design for disassembly	325

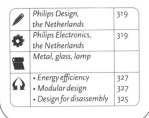

Thylia

Street lighting design is often conservative and lacks flexibility in meeting local lighting requirements. Thylia can provide almost bespoke lighting solution for any location with its wide variety of modular components fitted to the 2.75 m- (9 ft-) high base cast or welded steel base unit, which has a number of decorative options. There are six mast types offering a combination of two curves and three height options; each mast is made from 60 mm (2⅓ in) diameter steel tubing. The IP66 light heads are one half to one third of the size of conventional systems as they deploy a high luminance micro-reflector µR® unit. Easy installation, adjustment and maintenance are integral to the system.

✒	Tortel Design, France	310
✿	Comatelec Schréder Group GiE, France	313
▤	Steel, various electronic components	
♻	• Reduced materials used	325
	• Modular, multifunctional design	327
	• Improved public street illumination	
✪	iF Design Award, 2003	330

Boase

The return of the tree house is imminent according to a group of young Danish architects (Force4) who won The Home of the Future competition in 2001, organized by the Danish Building Research Institute and RealDania. They aim to provide affordable housing on stilts surrounded by planted dense vegetation, which helps clean up polluted sites. The Boase project (from the Danish *bo*, to live, and *oase*, oasis) aims for sustainablility and accessibility. A survey of Denmark found 14,000 post-industrial urban sites that developers and local authorities would not consider for housing. Normally, developers transport polluted soil from a site. Now, thanks to phytoremediation, using plant roots and micro-

organisms to render the pollutants safe, and the suspending of a development above ground, the soil can remain in situ, stilt houses can be erected and fast-growing trees can be planted. The City of Copenhagen has offered the blighted area Nørrebro as an experimental site for the first Boase. Suspended access paths made of technical textile that insulates and collects solar energy and sectional supports/bridges made of strong fibreglass-reinforced plastic link the apartments and house units, which are made of the same material, in a minimum-waste production line.

The modules can be rapidly erected on site. Governments try to encourage developers to use 'brownfield' sites but the cost of decontaminating soils makes such projects uneconomic. In light of the UN World Research Institute's prediction that human alteration of the earth's surface will increase from one-third today to two-thirds by 2100, Boase offers a way of reducing use of bio-productive land for urban developments.

✎	Force4 and KHRAS, Denmark	306
⚙	Conceptual prototype	
▦	Fibreglass-reinforced plastic, trees, technosphere materials and photovoltaics	
♺	• Conservation of land resources	328
	• Affordable city homes	326
	• Cleans contaminated land	
	• Solar power	327
	• Modular, minimum-waste construction	325

Island Wood Center

A building reflects the philosophical diet of an organization, in the same way that 'we are what we eat'. None more so than the Island Wood Center, which is part of a range of sustainable design facilities on a 103-hectare (255-acre) campus. The organization brings a holistic approach to education, integrating scientific inquiry, technology, the arts, energy conservation and community living in order

✎	Mithun Architects, USA	308
⚙	Mithun Architects, USA	308
▦	Hardwoods and softwoods	
♺	• Renewable materials	325
	• Energy conservation	326
	• Educational resource	326
★	IDRA award, 2002–2003	330

to encourage individual and community stewardship and a deeper understanding of the link between biological and cultural diversity. A reclaimed beam supports the roof trusses in a marriage of old and new, bringing attention to the superb engineering and aesthetic qualities of wood.

BRE, New Environmental Office

Designed to use 30% less energy than current 'best practice' in the UK, this building can accommodate over a hundred people. Cooling is achieved by natural automatic ventilation at night combined with ground water pumped through the concrete floors and ceilings, which has an efficiency of 1kWh output for pumping to an equivalent 12–16 kWh cooling energy input. Timber and steel are the primary materials for the structure and originate predominantly from recycled sources. Thanks to a combination of the thermal mass of the building, natural cooling and automated monitoring systems, the building regulates its own climate.

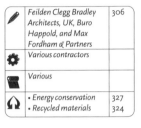

✎	Feilden Clegg Bradley Architects, UK, Buro Happold, and Max Fordham & Partners	306
⚙	Various contractors	
📜	Various	
♻	• Energy conservation • Recycled materials	327 324

Halfcostapartments

Hungarian architect Elemér Zalotay is an influential maverick whose own well-documented, modular house of recycled and found materials in Ziegelried, Switzerland, is a cacophony of eccentricity and serious ecological thinking – a blueprint for lightweight, multi-storey towers. His drawings, annotated engineering sketches and photo collages provide visual relief from the usual CAD plans and sections. Halfcostapartments is a vision of high-rise living in a vertical park, an architectural framework for 4,000–8,000 apartments. With 60% of the world's population living in cities there is a need to re-nurture people's relationship with natural living systems. The vision for Halfcostapartments is presented in a series of self-published booklets. Zalotay argues that architects' concepts generate a given engineering that produces expensive apartments. By contrast his vertiginous tower of 170 storeys demands an economical engineered solution that creates a frame, a series of 'air roots around a tree in a jungle'. The interaction between the frame and vertical park delivers the architecture. Water, steel and reinforced concrete provide structural integrity. A 600 m- (1,970 ft-) high central multi-core tower of parallel tubes, each with earthquake tremor pads, provides the attachment point for all horizontal elements. A contra-weight system of lifts using the dead-weight of the people and cars plus a little supplementary power from dynamo motors gives access to the basement parking for 4,000–7,000 cars in the 200–400 m- (310–650 ft-) deep foundation. Horizontal chromed sheet steel is deployed for a series of sectional modular room units on a horizontal truss that extend outwards from the multi-core tower. Walls and floors are fabricated from 0.5 mm sheet-steel sections providing a water 'sleeve'. Water provides building strength and mass, circulation for heating and waste disposal, fire-safety provision and irrigation for all the planted vegetation (0.5 trees per apartment). This vision challenges conventional tower builds and provides an exciting platform for theoretical discourse.

✎	Elemér Zalotay, Switzerland	310
⚙	Conceptual design	
📜	Steel, water, reinforced concrete, plants	
♻	• Modular suspended apartments • Low energy lift system • Water-based structural and heating system • Trees for cooling and shading	327 327

Microflat

Spiralling property prices in European cities like London have forced key workers such as nurses, teachers and postal staff to the outer suburbs where they commute huge distances to their jobs. The Microflat is conceived as an affordable but compact apartment just 32.5 sq. m (350 sq. ft), two-thirds of a typical one-bedroomed flat. The spatial arrangement looks more like that of a boat interior, with sliding panels and oblique partitions. The angled balcony creates a sense of privacy for the residents but also floods the living area with light. Unusually, the architects-cum-property developers suggest that prospective owners will be vetted and those with an income over £30,000 per annum will not be eligible. Only first-time buyers need apply. Another caveat, which has been successfully tried in Germany, is that properties can be only sold on to other key workers. As architectural circles revive ideas around the compact diverse city, the Microflats offer an option to reclaim gap urban sites and bring life back to blighted areas. Fabrication and construction can be fast-tracked, and exterior materials matched to the locality, but is the concept ahead of the planning authorities? With demand in the next few years reaching nearly 250,000 houses in south-east England, maybe the Microflat's time has come.

	Piercy Connor Architects, UK	308
⚙	Conceptual design, 2002, for London and Manchester, UK	
	Various materials during pre-fabrication	
⌂	• Spatial economy	328
	• Reduced resource use	325
	• Affordable and key worker housing	326

Cardboard Building, Westborough Primary School, UK, 1999–2002

Origami and the intrinsic strengths of folded paper structures inspired the design for this intriguing and imaginative school building whose aim was to use 90%-recycled and recyclable materials in its construction. This was a research and collaborative project between the architects and other professionals, children and staff at Westborough Primary School and the research partners, The Cory Environmental Trust and the Department of Environment, Transport and Roads. The project team included engineering consultants Buro Happold, three cardboard manufacturers and contractor, CG Franklin. All panels were pre-fabricated off-site. The walls and roof are 166 mm (6½ in) thick timber-edged insulated panels made of layers of laminated cardboard sheet (Paper Marc) and cardboard honeycomb (Quinton & Kaines). Recycled interior products included rubber floor tiles, wastepaper structural board, polyurethane core board, pinboard, and Tetrapak board. At the end of its estimated twenty-year life, most of the materials can be recycled. Many of the structural materials have good acoustical and insulation values.

	Cottrell & Vermeulen, London, UK	305
⚙	Various contractors	
	Recycled cardboard, paper, plastics and rubber	
⌂	• Recycled and recyclable materials	324
	• Low-embodied energy materials and construction	324/ 325
	• Design for disassembly	325

EDITT Tower

Singapore is a small island state whose sustainable future may depend on shifting from reliance on imports to increasing its own bio-productivity. Many Singaporeans live in high-rise housing or work in skyscraper offices, and building tall saves precious land for biomass production. A competition for a multi-use exposition building, sponsored by the Urban Redevelopment Authority in Singapore, EDITT (Ecological Design in the Tropics) and the National University of Singapore, generated a conceptual proposal from Dr Ken Yeang, renowned for his bioclimatic skyscrapers such as the IBM HQ in Kuala Lumpur. The EDITT Tower is a 26-storey habitat for humans, plants and birds, designed to improve a degraded ecosystem at a busy city road junction. The key biological aspect is the creation of planted areas of 3,800 sq. m (41,000 sq. ft) compared to usable ground-floor areas of 6,000 sq. m (64,000 sq. ft). Crucially, the lowest six floors will be wide landscaped ramps filled with cafés, shops, viewing decks and performance spaces, integrating the tower with surrounding street life. The building should be 'loose-fit' to facilitate future, unknown uses. This is achieved by including removable partitions and floors, and mechanical jointing. Unlike most skyscrapers, which avariciously devour resources, the EDITT Tower is designed to collect rainwater in a 'roof-catchment-pan' and recycle grey-water to satisfy 55% of its water needs; recover and recycle 15,200 cu. m (537,000 cu. ft) of sewage sludge per annum; generate 40% of its electricity from photovoltaic facades; separate consumable materials at source for recycling; and maximize bioclimatic control of ventilation and humidity by using wind walls to deflect cool air and mechanical fans with de-misters for comfort cooling. This vegetative tower is a realistic prospect in the high humidity of tropical Singapore, but it needs a developer who will take the long view.

✒	Ken Yeang, T R Hamzah & Yeang, Malaysia	310
⚙	Conceptual design, Singapore, 1998	
🗋	Steel, concrete, timber, bricks, aluminium cladding, photovoltaics, vegetation	
🎧	• Low ecological footprint	328
	• Solar power	327
	• Natural cooling system	
	• Modular, multi-use space	327

Pod hotels

Eco-tourism emerged as a niche market in the mid-1990s but new estimates indicate that 'sustainable tourism', a more holistic and demanding brand of eco-tourism, is due to become a £1bn industry by 2025. Ironically even eco-tourists are putting a severe strain on wilderness environments, so M3 Architects is proposing a portable hotel pod. Each pre-fabricated pod would be transported, assembled in-situ and include self-supporting systems. After a predetermined lifespan of up to fifteen years pods would be disassembled and any residual waste secured in the base of the pod would be treated. Pods can be repaired and re-transported to a new 'virgin' wilderness site. Buckminster Fuller would approve, but the concept does assume wilderness will still exist for the next generation when the world's human population hits somewhere between 10 and 20 billion.

✒	Ken Hutt and Nadi Jahangiri, M3 Architects, UK	308
⚙	Conceptual design	
🗋	Various materials	
🎧	• Low ecological footprint	328
	• Integrated, self-sustaining energy and waste system	327

Valdemingómez Recycling Plant

Corporations and municipal authorities should increase recycling volumes in the future, but how inherently sustainable are the facilities themselves? Valdemingómez represents a new generation of recycling plants that consider their own environmental impacts, a critical undertaking for a plant in the future southeast Regional Park of Madrid. A cross-section of the 29,000 sq. m (312,000 sq. ft) building reveals a stepped profile with raw waste entering at the highest point and treated waste or waste separated into recyclable mono-materials emerging at the lowest point. This 'gravitational' recycling, in two complementary plants (recycling and composting), is hidden from view by an enormous living structure, a technological hillside of the sedum turf roof (65% of total roof area), which literally blooms in spring. External cladding, comprising recycled

polycarbonate, wraps itself around a bolted lightweight steel frame. Both these elements can be disassembled and recycled at end of life – about twenty years. This inspiring example of municipal architecture reflects key points in the micro-manifesto of the collaborating architects, emphasizing hybrid models and interactions between natural and lightweight, energetically active, sophisticated, artificial materials. Valdemingómez, built for the price of a chicken farm, sets high standards of holistic architecture for future recycling plants.

✏️	Ábalos & Herreros with Ángel Jaramillo, Madrid, Spain	304
⚙️	For Vertresa-RWE Process, Auntamiento de Madrid, Spain	
📜	Recycled polycarbonate, steel, sedum turf roof	
♻️	• Recycled materials	324
	• Design for disassembly	325
	• Water and energy conservation systems	327/ 328
	• New landscape	

Solar Office, Duxford International Business Park

This office is designed to incorporate 900 sq. m (9,660 sq. ft) of photovoltaic cells into the south-facing glass facade inclined at 60 degrees. This array is capable of generating a peak output of 73kW, equivalent to 55,000kWh per annum, meeting between one-third and one-quarter of the expected energy needs of the building. The solar-powered system is complemented by a natural stack ventilation system with sun-shading louvres,

both systems being controlled and monitored by computer. Potential overall energy savings of two-thirds are anticipated compared with a conventional office building.

✏️	Akeler Developments plc, UK	304
⚙️	Akeler Developments plc, UK	304
📜	Various, including photovoltaic array, monitoring systems	
♻️	• Energy conservation	327
	• Solar power	327

Suan – a space of comfort

Suan is a self-assembly transportable kit room or hut. It can be erected in any suitable location within an existing building to provide a multifunctional space for meditation, study, recreation or as a tearoom. Constructed of reclaimed timber from old housing and sustainably harvested materials, it is dry-fixed without nails. A protective coating of Koshibu, a varnish made from persimmons, repels insects, is antiseptic and resists water ingress. It is a space of comfort; every office block should have such a refuge, a time-out zone where weary desk-bound personnel can rest their minds and bodies.

✎	Osamu Okamura, Japan	308
✿	One-off	
▣	Reclaimed timber, Koshibu varnish	
♺	• Reused and renewable, sustainably harvested, materials	325
❢	IDRA award 2000–2001	330

Weobley Schools sustainable development

Weobley Schools energy management system is a test-bed to extend the sustainable energy initiatives of a local authority in response to Local Agenda 21. A holistic approach led to a wood-fuel boiler, using locally harvested coppice roundwood, which was chosen on the grounds that it was the most sustainable system. The coppice suppliers are paid according to the heat output of the wood (supplied as chips) rather than the quantity, encouraging quality supplies. Insulation is to very high standards coupled with computerized monitoring of the under-floor heating and internal environment of the building that work in tandem with passive design features, including solar shading and natural ventilation. The net effect is a very energy-efficient public building using local resources.

✎	Hereford & Worcester County Council, UK	307
✿	Various contractors	
▣	Biomass fuel from coppice	
♺	• Energy conservation	327
	• Renewable energy	327

Model Buildings

2.0 Objects for Working

Oxstalls Campus, University of Gloucester, UK

Low energy consumption and responsive regulation of the internal environment were the priorities for these new university buildings on the mixed brownfield and parkland campus site. The development comprised the Learning Centre, including 300 workstations, teaching rooms and a 200-seat lecture theatre, and the Sports Science building containing laboratories, teaching spaces and staff offices. The buildings are linked by a glazed 'bridge' over a water feature. To

(warm) stale air by using a thermal wheel. Exhaust air from the Termodeck heats the buffer zone and entrance areas. Lighting in the library is provided naturally by north-facing windows and by high-frequency T5 fluorescent fittings controlled by occupant and illuminance sensors. Similar strategies are used in the Sports Science building and photovoltaic panels are installed on a 'waveform' roof, providing 65% of energy requirements.

meet an overall energy consumption target of 110kWh per square metre per year for the Learning Centre (one-third of current best practice in the UK) a 'Termodeck' thermal mass floor system with special diffusers was used for heat storage and redistribution of warm/cold air according to the season. At the top of the atrium incoming cool air is warmed by heat extracted from rising

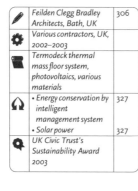

✏️	Feilden Clegg Bradley Architects, Bath, UK	306
⚙️	Various contractors, UK, 2002–2003	
📜	Termodeck thermal mass floor system, photovoltaics, various materials	
🎧	• Energy conservation by intelligent management system	327
	• Solar power	327
💬	UK Civic Trust's Sustainability Award 2003	

Norddeutsche Landesbank

A complex, seemly chaotic, arrangement of glass and steel buildings sits in a transition zone between the old residential area and historic centre of Hannover. This is a self-contained mini-metropolis, a landscape of transparent walkways, atria, and meeting places for 1,500 employees for Norddeutsche Landesbank. The key principles of the design focused on reducing energy use, increasing natural daylight and maximizing worker comfort. Insulation was the initial priority to reduce energy loss by exceeding the 1997 German insulation regulations by 10%. Appropriate air circulation designs by consultants Transsolar Energitechnik harnessed natural ventilation and the chimney-stack effect rather than using air-conditioning to deal with summer temperatures. Balustrade panels fitted with air shutters control incoming air; double-glazed windows fitted with highly reflective aluminium

reduce incoming solar gain while increasing natural light, and radiant slab cooling in the floors uses polyethylene tubes embedded in concrete in which to pump cool, geothermally heat-exchanged water at night. This entire system is controlled by office workers in their own part of the building so the comfort level is personalized. The architects demonstrate that an eco-tech strategy delivers reductions in energy expenditure and high aesthetic and architectural standards.

✏️	Behnisch, Behnisch & Partner, Stuttgart, Germany	304
⚙️	Various contractors, Germany, 2002	
📜	Steel, aluminium, glass, concrete	
🎧	• Natural ventilation	
	• Improved natural lighting	327
	• Energy conservation	327
	• Shared private/public spaces	
	• Personal control over comfort levels	

Four Horizons
'Autonomous' House

Four Horizons is an on-going design exercise in refining the concept of an autonomous house to suit the particular demands of its location. Lindsay Johnston, a Brit who settled

and the need for it to be an economical self-build. A significant aesthetic and structural feature of the house is the double 'fly' roof with curved inner trusses that is completely

the middle of each module to receive winter morning sun, and the breezeway permits through-flow of cooling north-east summer winds. South, east and west elevations are

reserve. Grey water is recycled for use in the walled vegetable garden. Energy is supplied by a combination of a BP Solar battery storing photovoltaic energy, 'Solahart' solar hot water, LPG gas and a back-up diesel generator. While not totally autonomous, the house uses half the energy of a conventional house in New South Wales. Measures to counter the threat from bushfires include cleared vegetation zones, concrete slab ground construction, 'Mini-orb' roofs and steel structure and non-combustible concrete block. This is 'Factor 2' sustainable design and the lessons learnt contribute to improving building specifications for new builds.

in Australia, designed the original house in 1994, self-built it in 1995–96, added several nearby studio houses and continues to evolve new autonomous technologies. Lacking all the usual utilities, the site is located at the north end of Watagans National Park on the eastern seaboard of Australia in a forested area. Temperatures range from 4°C to 38°C, rainfall averages 110 cm (43 in) per annum and prevailing winds are from the south to south east. Key features had to address the temperature range, bush fires, wood-eating termites

separate from the living modules underneath. This feature is revived from some of the early pioneer buildings in Australia. The roof frame comprises 50 x 50 mm (2 x 2 in) steel tube and 20 mm- (¾ in-) diameter bar covered in silver coloured 0.55 mm BHP Zincalume corrugated steel. This protects the two 9 m x 9 m (29 ft 6 in x 29 ft 6 in) dwelling modules, one including living, dining and kitchen activities and the other bedrooms, bathrooms, laundry and study. A curved thermal mass wall of solid concrete blockwork runs through

protected from excessive solar gain by recycled polyester wool insulation and the north side is clad with local blue gum planks, although use of timber is kept to a minimum throughout the building to preserve existing forests. Exposed concrete floors were treated with Livos hard-wearing finishes. The roof captures between 382,000 and 585,000 litres (84,000–128,000 gals) of water per year into Aquaplate steel tanks. Annual consumption is 220,000 litres (48,000 gals) per annum , so there is a six month water

✏️	Lindsay Johnston, New South Wales, Australia	307
⚙️	Lindsay Johnston, New South Wales, Australia	
🗒️	Concrete, steel, recyled polyester insulation, blue gum timber, photovoltaics	
♻️	• Low-energy	327
	• Passive and active solar heating	327
	• Autonomous water systems	328

✏	Bill Dunster Architects, Mark Lovell, and Oscar Faber, UK	304
⚙	Various contractors	
📜	Various	
🎧	• Integrated energy-efficient control system for home and domestic transport	327
	• Water conservation system	328
	• Solar power	327
✪	Design Sense award, 1999	

Fred

Although the concept is not new, Fred is a portable building with some special features. The basic room unit is 3 x 3 x 3 m (27 cu. m, 953 cu. ft) but the floor area can be doubled to 18 sq. m (194 sq. ft) by taking advantage of sliding wall and roof elements, which are electronically controlled. Each unit is equipped with a kitchen, toilet and shower and an area available for multipurpose use, but the basic utility services have to be connected. A fully glazed wall provides excellent natural light, and thick insulation in the walls and roof minimizes energy consumption.

✏	Johannes & Oskar Leo Kaufmann, Austria	307
⚙	Johannes & Oskar Leo Kaufmann, Austria	307
📜	Timber, metal, glass	
🎧	• Multi-use space	327
	• Low-embodied-energy of fabrication, transport and construction	325/ 326

Hope House

Hope House is a home, office, energy generator and leisure zone. Passive solar design combined with photovoltaic generation is sufficient to maintain an ambient internal climate and to run a Citroën electric car for up to 8,500 km (5,300 miles) per year, resulting in a net saving of about 4.13 tonnes of carbon dioxide per year. Mains water usage is minimized by using rainwater for the toilets and laundry room. All grey-water is reused to irrigate the garden after it has been passed through a sand filter. This project is a blueprint for a seventy-six-unit urban village with sun terraces planned for London by the Peabody Trust, a charitable organization that has been concerned with raising the social and environmental standards of British urban housing for over a century.

Exhibition hall

Imagine a building of 3,600 sq. m (38,750 sq. ft) floor made mainly of paper and cardboard. Impossible? Not in the hands of Shigeru Ban, the Japanese architect and designer with over two decades' experience of working with these materials to produce furniture and housing for disaster relief projects. The building premiered at EXPO 2000 in Hannover, where it became the first public building in the world to feature a 35 m- (115 ft-) span paper/plastic textile roof supported with a latticework of tubular cardboard. The building is designed to be demountable and reused.

✏	Shigeru Ban, Japan	304
⚙	Various contractors	
📜	Paper, cardboard, various	
🎧	• Renewable materials	325
	• Reusable building	325

Hooke Park Training Centre and Westminster Lodge

Untreated roundwood, of diameter 50 mm to 250 mm (about ⅕ to 1 in) – the thinnings from forestry management – forms the basic construction material for unique organic forms of architecture that take advantage of the natural properties of the timber. the Hooke Park Training Centre is a large, free-span space housing workshops for The Parnham Trust, whose college provides training in furniture design with emphasis on using indigenous timber.

✎	John Makepeace and others, Hooke Forest (Construction) Ltd, UK	308
⚙	Hooke Forest (Construction) Ltd, UK	308
▤	Roundwood timber	
🎧	• Locally sourced, renewable materials	324/ 325

SU-SI

Many people associate mobile or trailer homes with holiday parks and dubious lifestyles. Not so this customizable twenty-first-century modular home system, which can be erected on site within a few hours and is easily disassembled and reused in another location. The factory-produced modules measure 12.5 m x 3.5 m x 3 m (41 ft x 11 ft 6 in x 9 ft 10 in), each one interlocking with the next to create versatile domestic, office or exhibition spaces.

✎	Johannes & Oskar Leo Kaufmann, Austria	307
⚙	Johannes & Oskar Leo Kaufmann, Austria	307
▤	Timber, metal, composites, glass	
🎧	• Reusable buildings • Low-embodied-energy of fabrication, transport and construction	325 325/ 326
✹	iF Design Award, 2000	330

Airtecture

Weighing just 6 tonnes and easily packed on to a road vehicle for transport, Festo's portable building comprises a protected floor space of over 357 sq. m (3,810 sq. ft). This is achieved by supporting an inflatable cross-beamed roof on two rows of inflatable, Y-shaped columns. Stiffness is given to thin cavity wall panels by tensioning them with pneumatic muscles, which contract to oppose the effect of the wind. Air is the main insulator to assist with internal climate control.

✏️	Festo & Co, Germany	314
⚙️	Festo & Co, Germany	314
📃	Various	
♻️	• Reduction of resource consumption	325
	• Reusable and portable building	325/ 327
	• Mult-use single space	327

Ecover factory, Oostmalle, Belgium Project

Growth of Ecover's business in the early 1990s required an expansion of the existing factory near Antwerp, Belgium. Using an ecological grading system devised by the University of Eindhoven building materials were selected for their minimal environmental impact. Structural timber was obtained from sustainably managed forests and bricks from a clay-based residue from the coal industry provided high-insulation material. A huge multi-ridged turf roof covers the 5,300 sq. m (57,050 sq. ft) building, providing excellent insulation, controlling storm-water run-off and helping integrate the factory into the local landscape. In line with the company's philosophy of balancing commerce with social and environmental concerns, the factory has been developed to enhance conditions for the workforce. Many roof-lights create natural lighting and there are solar-powered showers for the workforce.

✏️	University of Eindhoven (Building Initiative Environmental Standards), the Netherlands, with Ecover, Belgium	310/ 314
⚙️	Various contractors	
📃	Various natural materials, turf roof, bricks from clay-residue	
♻️	• Energy conservation	327
	• Local materials from sustainable sources	324/ 325
	• Natural lighting	327

BedZED Housing

BedZED is a pioneering mixed-use and mixed-tenure development of housing, work space and public areas, which is being constructed on an old sewage works, a 'brownfield' site, in Beddington, Sutton, south of London. The whole scheme is designed to meet exacting environmental, social and financial requirements. Architect Bill Dunster and environmental consultants BioRegional have, in collaboration with the client, the Peabody Trust, adopted a holistic view of the local needs of the intended community, including a green transport system that was actually built into the planning permission and ratified by the local authority. BedZED hopes to cut total fossil fuel consumption to about half that of a conventional development by reducing the need to travel between living, work, health-care, shopping and recreational facilities. Reduced transport impacts are also encouraged by promoting good networking with existing train, bus and tram services and by providing decent bicycle storage facilities, attractive pedestrian links and on-site charging points for electric vehicles. There is a ten-year target to produce enough solar electricity on-site to power forty electric vehicles. Materials for the eighty-two flats and houses for sale and rent have been selected from natural, renewable or recycled sources, mainly near by. Each dwelling is an energy-efficient design using passive solar gain and a high insulation specification, including triple-glazed windows. A central combined heat and power-generation facility

✏️	Bill Dunster and BioRegional	305
⚙️	Various contractors	
📷	Various, including locally sourced	
♻️	• Energy neutral development	327
	• 'Carbon neutral'	
	• Integrated transport plan	327
	• Socially mixed housing	
🔍	RIBA award, 'best example of sustainable construction', 2000	330

will utilize on-site tree waste to provide all the development's heat and electricity requirements. Further on-site generation from photovoltaics will make this the first large-scale 'carbon neutral' development in Europe. Water conservation will be encouraged by providing up to 18% of on-site consumption from stored rainwater and recycled water and by installing water-efficient appliances.

Mobile Theatre

Real-time, real-life entertainment is becoming a rarer event in our lives as the Information Age delivers 'reality TV', the Internet offers 'live shows' and 'video phones' send facsimiles of friends and family. Although theatreland appears to be flourishing in large metropolitan centres those in smaller conurbations or rural locations do not often have the facilities to enjoy the theatre. This conceptual design takes the theatre to the people using three 7.5-tonne trucks, with attached trailers. Everything runs on bio-diesel, from the motors for the trucks to the generators for lighting. Once unfurled the theatre has a capacity for 500 people in a raked auditorium surrounding a 75 sq. m (800 sq. ft) stage.

✏	Springboard Design Partnership, UK	309
⚙	Conceptual prototype	
▤	Unspecified	
⏏	• Renewable energy	327
	• Transportable building	327
	• Low-cost community entertainment	

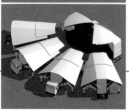

House in a suitcase

Inspired by Louis Vuitton travelling suitcases that fold open, then unpack again to create a travel wardrobe and chest-of-drawers, Prats and Flores conceived 'house in a

suitcase' as a temporal home perched on top of an existing apartment block. There are a series of trunks that unfold within the 9 x 3 x 3 m (29 ft 6 in x 9 ft 10 in x 9 ft 10 in) space to create seating, a kitchen and a bed area according to the time of day and specific needs. Each time the occupants visit they bring minimum belongings into the room. This confined multi-use space is intended to discourage accumulation of objects and paraphernalia. This is a game of chess between the fixed and mobile objects (hand luggage) that encourages reassessment of real needs. Spatial scales cascade from the whole room, lit by a linear skylight, to the two main trunks and, as these are unfolded, reveal further nested spaces and sliding trays where smaller objects can nestle. The bedroom closets unfold to define the sleeping zone, where one can change in privacy. The entire room is a dynamic space, a quest to explore the minimum spaces that one can live in. It questions the need for weekday city workers or city weekenders to have huge apartments, with their consequent resource use. It challenges us to review our spatial needs continually. In these days of inflated prices for city apartments and houses in many parts of Europe, it also offers hope for urban nomads and those seeking work.

✏	Eva Prats and Ricardo Flores with Frank Stahl, Spain	309
⚙	Prototype, Spain, 2001	
▤	Plywood, veneer, metal	
⏏	• Spatial economy	328
	• Reduced resource use	325
	• Multi-use space	325

The Clam House

This experimental structure is built as a prototype for sustainable techniques using only natural biosphere or abundant geosphere materials. This eccentric building represents what Bernard Rudolfski meant by the term 'architecture without architects'. Yet this building is architecture, it is a realized conceptual experiment in delineation of form and space and it bears the mark of its maker. A beautiful complex curved roof, the inspiration for the house's name, hints at biomimicry or 3D digital influence yet was probably realized by taking advantage of the natural warping of the thin rafters. Walls are lime-plastered, straw bale infilled between the frame, supported on a gravel/concrete trench foundation. The roof is wooden roofing shingles and the floor is poured earth. The structure is breathable and well suited to a passive solar heating system.

	Jacques Abelman, France	304
	One-off	
	Timber, straw, lime, gravel, concrete	
	• Low embodied energy dwelling	325
	• Low energy construction	325
	• Energy conservation	327
	• Renewable and abundant materials	324/ 325
	IDRA Award 2001	330

Moon House

Most garden sheds in the UK are factory built using inexpensive, insect-repellent pressure-treated softwood harvested from non-certified sources. The lifespan, serviceability and lifecycle impacts of these buildings leave a lot to be desired. This is not the case for Moon House, a one-room garden retreat, designed and built by Guy Martin. His sustainable philosophy is apparent in the careful selection of materials – local, weather-resistant Western Red Cedar for the frame, deck and roof shingles; walls infilled and insulated with local straw bales then finished with two coats of lime-based render plus coloured lime wash; underfloor insulation with pulped paper 'Warmcell'; 'Thermafleece' wool batts for roof insulation. A small woodburning stove provides energy efficient heating. A wavy barge board encourages rapid water movement away from the structure.

	Guy Martin Furniture, UK	306
	Guy Martin Furniture, UK	306
	Western Red Ceder, straw bales, lime render, insulation	
	• Renewable and compostable materials	324/ 325
	• Low embodied energy	325
	• Energy conservation	327

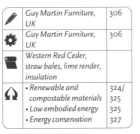

Ecology hotel

As one would expect from the ambitiously named Ecology hotel, Shimanto no yado, in Kochi, Japan, the fulfilment of the concept is in the detail and interconnected systems. All the rooms in this thirty-room hotel are furbished with natural materials, including wood, Japanese paper, bamboo, stone and mortar. Energy conservation, water and waste management are reflected in the use of natural ventilation and lighting, solar heating, photovoltaics, green roofing, double-glazed windows with heat

reclamation, high-efficiency lights and appliances, and kitchen waste fermentation. Non-ozone-depleting chemicals are used for construction, maintenance and cleaning materials.

	Takatoshi Ishiguro and PES Kenchiku Kankyo Sekkei, Japan	307
	Various contractors	
	Various	
	• Water-conservation	328
	• Solar power	327
	• Energy reclamation	327
	• Natural, recycled and non-ozone depleting materials	324
	IDRA award, 2002–2003	330

Project 19, Project 23

In the 1990s Germany saw the emergence of an eco-tech aesthetic expressing new forms of solar architecture. Thomas Spiegelhalter, an architect and town planner, cleverly mixed high-tech photovoltaics with passive systems and created a modern eco-tech design. Both Project 19, a new settlement at Rieselfeld near Freiburg, and Project 23, solar row houses at Ihringen, were intended to demonstrate low-cost, low-energy-technology houses with a plan area of 250–300 sq. m (2,700–3,200 sq. ft) requiring less than 30 kWh/ sq. m per annum to operate. Both projects deployed prefabrication, Internet-based collaboration, lifecycle assessment, and energy performance simulation. Most importantly, the local, social, cultural and landscape features were meshed with climatic data to create an exciting solar vernacular architecture. So

the curved north-facing roofs of Project 19 and materials selection reflect traditions yet experiment with new materials. Passive solar gain through deployment of south-east-facing glasshouses and use of cooling northerly air with natural stack ventilation ensures a maintenance-free system for comfort in winter or summer. Spiegelhalter's vision was holistic, extending to soft landscaping, rainwater harvesting systems, use of recycled materials and the need to create flexible spaces for living and working. To reduce costs the Rieselfeld project was based on a segmented assembly method from basement to fourth floor. Service systems for water, PV electricity, heating, telecommunications and sewage with integrated heat recovery, were centralized in each house. Prefabricated, insulated panels of reinforced concrete or wood–steel

modules were carefully jointed to prevent thermal bridging to maintain a U-value equal to 0.1W/m2K. Double- and triple-pane low-e glazing with Argon gas maintains U values of between 0.7 to 1.5 W/m2K while maximizing transmission of visible light for day lighting. Rooms are orientated to provide natural lighting and ventilation, and simple controls enable people to select their level of comfort. Twenty-kilowatt photovoltaic arrays generate

sufficient electricity to meet 70–78% of the entire low temperature heat, cooling and electricity demand, with excess power at daytime peak hours exported to the national grid. The PV panels also provide shading to the south-facing facades to prevent excessive solar gain. These projects demonstrate the eco-efficiency levels attainable for temperate European houses and set valuable new standards for lowering the total carbon dioxide emissions of future housing stock. The eclectic mix of technology, technosphere materials and quirky features create low-cost houses with individuality.

✏	Thomas Spiegelhalter, Germany	309
✿	Various	
▦	Concrete, steel, wood, mineral insulation, glass, photovoltaics	
⌂	• Low-energy housing	327
	• Affordable housing	326

Quarkhome

Young German designer Rouven Haas, has created an innovative transportable fold-out prototype mini-home. Diverging from the usual rectilinear semiotics of such buildings, Haas's three curvilinear spaces – being and cooking, living, and sleeping – emerge concertina-like from a cuboid structure, which, now empty of its curved walls, becomes a dual-function wet space for washing and an entrance lobby. This use of space has many ramifications for eco-efficiency. The ratio of materials consumed to space generated is high. Areas that are used temporarily (such as bathrooms) can be multifunctional. With such small air volumes to heat, cool or ventilate, the entire house becomes less demanding of energy and water. Micro-homes such as Quarkhome could provide factory jobs in constructing prefabricated units while restoring dignity to their occupants. Transportable homes are also ideal for temporary emergency shelters.

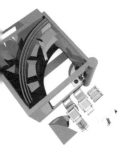

✏	Rouven Haas, Vienna, Austria	306
⚙	Conceptual design	
▤	Various technosphere and biosphere materials	
♻	• Pre-fabricated, disassemblable, recyclable units	324
	• Reusable, transportable building	325
	• Spatial economy	328

Villa Eila, Mali, Guinea

This is a private residence for Eila Kivekäs, founder of the Indigo Association responsible for promoting vocational training for women in Guinea, West Africa. Built in 1995, it is a powerful example of 'intermediate' or 'appropriate' technology using local resources and skills to develop a distinctive aesthetic. All the bricks and floor tiles were made of local soil mixed with 5% cement, manually pressed and then dried in the sun. Roof tiles were made in a similar fashion but sisal fibre was added to a higher cement–soil mix to form a stronger composite. Woven bamboo screens provide dappled light and shade to protect internal walls from heat absorption. The mono-pitch roof covers a variety of simple rectilinear or circular rooms, the simple geometry creating mini-arenas for the natural materials to speak quietly. Villa Eila is more than a vernacular building – although it does embrace all the virtues of thinking locally and being true to local materials, it is a statement of self-help and self-belief in local, rather than Western industrial solutions.

✏	Heikkinen Komonen Architects, Helsinki, Finland	306
⚙	Various	
▤	Soil, clay, cement, sisal, bamboo	
♻	• Local renewable or geosphere materials	324
	• Low-embodied-energy materals and construction	324/ 325
	• Naturally ventilated	

Loftcube

Look out over the rooftops of any major city and acres of unused space come into view. Aisslinger's Loftcube aims to colonize these sunny plots of prime urban spaces. Two prototypes, a 'home' and an 'office' version, each providing 36 sq. m (388 sq. ft) floor space, were launched at Designmai in Berlin in 2003 on the roof of the premises of Universal Music Deutschland. This prefabricated, modular structure utilizes honeycomb-type wooden modules protected with white laminate plystyrol with durable Bankirai wood for exterior features. The structural 6.6 x 6.6 x 3 m (21 ft 8 in x 21 ft 8 in x 9 ft 10 in) body rests on supporting legs and offers customizable window spaces (transparent, translucent wooden slats) according to clients' needs. Interior panels and partitioning, made of DuPont Corian® Glacier White, is also based on a modular structure and offers considerable flexibility. Flooring is the uniform, tough quartz-crystal composite of DuPont Zodiac® in wet areas or those needing frequent cleaning. In the office version DuPont Antron® Excel nylon-based fibre is used in tough-wearing 'Nandou' carpet from Vorwerk Teppichwerke, but a fluffy, warmer look is achieved in the home version by using a new loopware carpet, Antron® Supergloss.

Aisslinger believes his 'flying building' will appeal to young urbanites, business people and those wanting to be in the heart of the city but who need a refuge for time out. Whether the rooftops of cities will be colonized with Loftcubes depends upon investors' perception of economic returns to rent roof space and develop appropriate infrastructure.

✏	Werner Aisslinger, Studio Aisslinger, Berlin, Germany	304
⚙	Prototype	
📋	Wood, plystyrol, DuPont Corian®/Zodiac®/Antron® and other materials	
🎧	• Modular, pre-fabricated	327
	• Design for disassembly	325
	• Recyclable technosphere materials	324
	• Conserves land resources	328

Minibox

Designers Holz Box have tested Minibox on the citizens of Innsbruck in Austria. At the heart of the home – which is not for the claustrophobic – is the multifunctional kitchen/dining/work table, with a wood-burning stove neatly inserted underneath. A single and double sleeping platform, top-lit by a roof-light, occupy the top third of the transportable prefabricated 2.6 m- (8 ft 6 in-) cube unit. Every object has multiple functions, so the ply benches have steel storage shelves underneath and a storage unit doubles as steps to the beds. Another storage area contains a shower and camping-style toilet. Every detail of the larch and softwood laminate/plywood panels, stainless steel stove pipe and metal shelving contributes to a modern, functionalist aesthetic. Four units fit on a standard EU lorry so Minibox is adaptable to a variety of needs from temporary housing, autonomous extensions, holiday chalets or roof-top urban living.

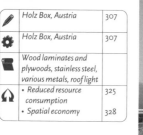

✏	Holz Box, Austria	307
⚙	Holz Box, Austria	307
📋	Wood laminates and plywoods, stainless steel, various metals, roof light	
🎧	• Reduced resource consumption	325
	• Spatial economy	328

Watervilla

The low-lying Netherlands, Belgium and Denmark may face an increasing battle to keep out the sea as climate change takes its course. Architects in the Netherlands are really enjoying the challenge. Sitting in the port of Middelburg, Watervilla revels in the creative possibilities opened up by living on water. The main house structure is a prefabricated steel skeleton covered in a metallic skin and braced with a concrete and steel plate floor. A variety of floor plans permits customization of glazed and decking areas or a roof terrace. An atrium brings light and air to a dynamic arrangement of sub-forms. Water is used

to cool the house in summer but the facility to rotate the house by ninety degrees on its platform means it can be oriented to facilitate internal climate control, obtain privacy or change the view. Its unique platform, made of steel pipes from the offshore oil and gas industry,

	Architetuurstudio Herman Hertzberger, Amsterdam, The Nethelands	304
	Various	
	Steel, metal sheet, concrete, glass, wood	
	• Re-usable and recyclable materials	324/ 326
	• Passive and active interior climate control	
	• Flexible housing	
	• Customizable	327

minimizes swaying movements induced by wind and waves. Re-locating is a simple matter of getting a tow from a motorized boat. Watervilla points the way to using the skills of the boatbuilder and the architect to create a new industry for modern

houseboat construction. The prefabricated structures minimize waste and can be assembled and disassembled for recycling at end-of-life. Here's an opportunity to make house building as streamlined as an automobile factory.

M House

This transportable house looks like a chaotic abstract puzzle. It is in reality a prefabricated bolt-together structure of support legs, a platform and a series of interconnected rectangular frames of square-section steel to which an array of

panels is attached. Its overall form mutates as panels are moved in or out, in vertical or horizontal mode, to encourage ventilation or provide furniture. All this happens according to the requirements of the

occupants. Within the rectilinear frame, climate-controlled spaces can be created by incorporating glass windows and doors. Frames can be made to bespoke sizes, but the inherent flexibility of the design makes the structure customizable. Designed as a vacation retreat, studio or guest-house, the basic M House offers about 37 sq. m (400 sq. ft) of insulated

controlled environment and 80 sq. m (860 sq. ft) of decking. Panels can be adapted for photovoltaics. There is space to store rainwater, grey-water and sewage under the platform. A prototype under construction is using painted steel and Viroc® cement and wood-fibre panels.

	Michael Janzen, Human Shelter Research Institute, California, USA	307
	Conceptual prototype	
	Steel, Viroc® panels, glass	
	• Modular, transportable design	327
	• Potential low-energy operation	327
	• Design for disassembly and recyclability	325

Criss Cross

Making a welcome change from the ubiquitous rectangular paving block, the Criss Cross paving system comprises four different forms that can be interlocked in regular or random patterns. Glindower Ziegelei still fires these blocks in a kiln dating from 1870. Natural variation in the clay minerals yields a range of colours and textures.

Finding Form in Forming Cardboard

Pulped cardboard using recycled and a small percentage of virgin paper fibre is a versatile material. Lightweight yet strong, it can be moulded into complex shapes, such as egg boxes and protective packaging. Its characteristics have been deployed to make a unique load-bearing roof structure, inspired by the architecture of Antoni Gaudí and Buckminster Fuller. Conjuring up the columns and ceilings of the Grand Mosque at Cordoba in Spain, this modular construction celebrates the tactility and aesthetic pleasure of the material while suggesting new possibilities for interior design.

🖊	Henk van Dijke, graduate student 2003, Design Academy Eindhoven, the Netherlands	305
⚙	Prototype with Fiberform Engineering, Arnhem, the Nethelands	
📃	Pulped cardboard	
☊	• Lightweight strong materials	324
	• Recycled and recyclable materials	324

🖊	Ecke: Design, Germany	305
⚙	Glindower Ziegelei GmbH, Germany	315
📃	Clay minerals	
☊	• Abundant geosphere materials	324
✑	iF Design Award, 2000	330

LockClad terracotta rainscreen

Combining the aesthetics and durability of fired clay tiles with ease of installation, this rainwater cladding on aluminium rails is a cost-effective method of protecting the exterior of a building from the elements. Each clay tile is locked in place on an extruded aluminium rail, LockRail, which meets all UK and Ireland wind loadings. This minimal-maintenance, lightweight cladding permits extra insulation materials to be applied to the outer skin of the building's structure, improving energy conservation. Natural ventilation behind the clay tiles and protection from the sun reduce temperature variations in the load-bearing structure.

🖊	Red Bank Manufacturing Company, UK	320
⚙	Red Bank Manufacturing Company, UK	320
📃	Clay, aluminium	
☊	• Durable, recyclable materials	324
	• Improved energy conservation for buildings	327

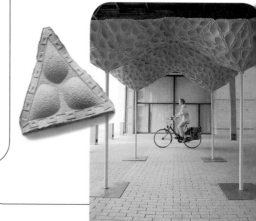

✏	K-X Industries, USA	317
⚙	K-X Industries, USA	317
📦	Waste wood and fly ash, Portland cement	
♻	• Partially recycled and renewable content	324/ 325

Faswall®

A post-and-beam structural grid is created by filling wall forms with reinforced concrete. Wall forms are manufactured using K-X® recycled wood waste chips. The entire wall structure, known as Faswall®, comprises up to 85% K-X Aggregate (from waste wood) bound with Portland cement (containing up to 15% fly ash content by volume). A finished Faswall shows good R-values (thermal insulation) of between 18 to 24 and it is an excellent sound barrier and substrate for dry-wall or direct finishes. Standard blockmaking equipment permits local manufacturing of Faswall® components.

Eco-shake®

Made of 100%-recycled materials, reinforced vinyl and cellulose fibre, eco-shake® shingles are available in four colour shades designed to mimic weathered wooden shakes. The shakes qualify under strict fire-rating, wind and rain resistance and impact tests.

✏	Re-New Wood, USA	320
⚙	Re-New Wood, USA	320
📦	Recycled wood, recycled plastics	
♻	• Recycled materials	324

Parallam®, Timberstrand®, Microllam®

TrusJoist produce a range of patented engineered timbers made by drying short or long veneer 'strands' or sheets, bonding them with adhesives or resins and subjecting them to high pressure and/or heat. They produce three 'timbers', Parallam® PSL, Timberstrand® LSL and Microllam® LVL, and a special composite structural timber floor joist, the Silent Floor® Joist. They claim improved strength and avoidance of defects such as cracking and warping for all their timbers. Furthermore, thanks to the raw veneer ingredients, they can use virtually the whole diameter of a sawn log and small-diameter second-growth trees. This results in a considerable saving on raw materials to produce the same amount of structural timber as sawn wood. For example, the Silent Floor® Joist system uses one tree to every two to three trees for a conventional sawn-wood joist flooring system. Microllam® LVL uses 30% more of the timber from each tree and, being stronger than solid timber, provides almost double the structural value per unit volume of raw material than sawn wood. However, quite a lot of energy is needed to make these composite timbers, so detailed examination of the embodied energy of TrusJoist versus sawn timber should be made on a case-by-case basis.

✏	Trus Joist, USA	322
⚙	Trus Joist, USA	322
🗐	American softwoods and hardwoods, waterproof adhesives, polyurethane resin	
♻	• Efficient use of raw materials	325

Wedge, Cube & Pyramid tiles

It is almost a certainty that the trauma of moving house is exacerbated by having to live with someone else's idea of tasteful kitchen or bathroom wall tiles. Inevitably the garish, naff, sad or downright nasty things have to be hacked off the wall, usually necessitating re-plastering before re-tiling. One way of preventing this regular wasteful cycle is to move away from contentious areas of taste with colours and patterns to create a more neutral, less fashion-oriented surface. Dene Happell has done just that with his geometric Wedge, Cube and Pyramid tiles. Suddenly a tiled wall becomes a frieze or mural, the angled or raised surfaces of the standard 150 mm x 150 mm (6 in x 6 in) tiles catching the light. The purity of the individual geometry of each tile 'typeform' goes beyond mere pattern. Will these tiles slow consumption? Time will tell.

	Happell, UK	316
	Happell, UK	316
	Ceramic, glazes	
	• Abundant geosphere materials	324
	• Anti-obsolescence	324

2Zones2®

The 1974 'Marche en Avant' (walk ahead) regulations changed catering practices in France by encouraging nuclearization of activities according to its temperature and hygiene requirements (cold/clean, cold/soiled, warm/clean, warm/soiled), create a

to avoid cross-contamination of raw, cooked or waste food. This led to the development of mono-functional rooms, resulting in a significant expansion in floor space required. The 1989 Hazard Analysis and Critical Control Point (HACCP) regulation enforced new cleaning and disinfecting procedures, which meant increased investment and time was required. Recently 3bornes Architectes re-examined the parameters needed to maximize health and safety while balancing business and eco-efficiency. Their vision was simple: divide any activity series of parallel climate-controlled zones and utilize standard GastroNorme (GN) containers and a new trolley, Transfert, to transfer food between zones, which are separated by the temperature-controlled cabinets that can be accessed from adjacent zones. Food enters the reception and is stored in GN containers in the first cabinet, the chef retrieves the container and places it in a steel canal – in effect a multifunctional stainless steel work surface plumbed with water and waste drainage below. Food is unpacked, vegetables and fruit are peeled and disinfected and rinsed above the canal. At the end of the canal prepared food is placed in another GN container and the trolley is wheeled into the cabinet adjacent to the transformation zone (for cutting, mixing, cooking, assembly) on a similar canal system. The prepared food is then placed in the cabinet accessible from the distribution zone. 2Zones2 is a modular stainless steel building with in-built ventilation, lighting, plumbing and waste disposal, producing a 50% reduction in floor space to a traditional kitchen layout. A final innovation is the use of directed temperature-controlled air streams, such as down onto each canal, to force bacteria toward the soiled areas that are easily disinfected. Local chilling makes better working temperatures for the staff and substantial energy savings. 2Zones2 kitchens have been installed for various institutions, including the Association National des Directeurs de Restaurants Municipaux.

	François Tesnière & Anne-Charlotte Goût, 3bornes Architectes, France	310
	S C Bourgeois with Halton (ventilation), Electricité de France (electrics)	320
	Stainless steel, electrics, ventilation & refrigeration units	
	• Energy efficiency	327
	• Improved health & safety	327
	• Improved ergonomics	327
	• Modular design	327
	• Durable	327

SunPipe

Natural daylight provides a more relaxing spectrum of light for human vision than artificial light sources but, more importantly, reduces energy consumption in work spaces. SunPipe is a system of conveying natural sunlight from rooftops into buildings. Eight different versions are available in the SunPipe range but the components are similar – a transparent dome of UV-protected polycarbonate is held on the roof by an ABS/acrylic universal flashing. Below the dome is a tube made of Reflectalite 600, silvered coated aluminium sheeting with 96% reflectance. Four standard-diameter tubes, 330 mm up to 600 mm (13–24 in), and a range of elbow joints permit light to be directed into the required space. A 200 mm-(8 in-) diameter version is being developed for domestic spaces. A vertically oriented SunPipe of 330 mm can deliver 890 Lux in full summer sun and 430 Lux in overcast conditions in the temperate British climate, which is sufficient to provide natural daylight to an are a of approximately 14 sq. m (150 sq. ft). Doubling the diameter of the pipe roughly doubles the Lux delivered.

✏️	Terry Payne, Monodraught Ltd, UK	308
⚙️	Monodraught Ltd, UK	308
📦	ABS/acrylic, polycarbonate, aluminium	
♻️	• Natural lighting	327

power glass® walls, balustrades

LED technology is driving diverse creative ideas for illumination. Two projects by Glas Platz demonstrate the unique features of their transparent panels of glass, conductive transparent materials and LED electronic systems. A complete power glass® facade was installed in Tour Europe. The spatial dynamics within the building are transformed by the wall of light. Another project at the Galeries Lafayette, Paris, is the installation of balustrades over three floors. It indicates the graphical and communications potential of power glass® systems, which can provide economical to run lighting and dramatic new spatial interpretations for interior designers.

✏️	Glas Platz, Germany	315
⚙️	Glas Platz, Germany	315
📦	Glass, conductive materials, LEDs and electronics circuits	
♻️	• Low energy lighting and wall systems	327

Papertex

A cotton-like yarn is produced from wood fibre, forming the warp in the woven Papertex carpet. This tightly bound weave is hard-wearing and easy to clean and sits well with the modern Scandinavian aesthetic.

✏️	Ritva Puotila, Finland	309
⚙️	Woodnotes Oy, Finland	323
📦	Wood (cellulose) fibre	
♻️	• Renewable and recyclable materials	324 325

Dalsouple

Dalsouple manufacture standard and bespoke rubber flooring tiles in a huge variety of colours and surface textures that are 100% recyclable.

✏	Dalsouple Direct, UK	313
⚙	Dalsouple Direct, UK	313
🗋	Synthetic and natural rubbers	
☊	• Recyclable	324
	• Clean, chlorine-free production process	325

All Dalsouple rubber is free from PVC, CFCs, formaldehyde and plasticizers. Production waste is virtually all recycled within the manufacturing plant and emissions meet local statutory requirements. Service partners to Dalsouple include Uzin Adhesives, who offer water-based and solvent-free adhesives including polyurethane and epoxy resins.

Ecoplan/ecoment

Freudenberg manufacture a diverse range of rubber flooring under the 'nora' range but 'ecoplan' and 'ecoment' are the only two made with up to 75% factory and post-installation waste. Granite and marbled-effect patterns are available. All products are free of PVC, plasticizers, formaldehyde, asbestos, cadmium and CFCs and production follows stringent waste-management procedures, minimal packaging and a zero-emissions environment for the workforce.

✏	Freudenberg Building Systems, UK	315
⚙	Freudenberg Building Systems, UK	315
🗋	Rubber, recycled rubber	
☊	• Recycled and recyclable materials	324
	• Clean production	325

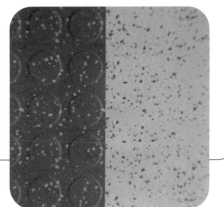

composite bathroom floor tiles

Combining the insulation properties of wood with the moisture resistance of plastic, these composite bathroom tiles are formed from waste wood and plastic. Experimentation with surface textures and shapes creates new tactile sensations for the feet to enjoy and explore. The highly competitive domestic and industrial flooring markets would benefit from such imaginative innovations. Collaborative projects with universities might help close recycling loops within other manufacturing industries by using factory floor waste.

✏	Antoinette Klawer, graduate student 2003, Design Academy Eindhoven, the Netherlands	305
⚙	Prototypes	
🗋	Waste wood, waste plastic	
☊	• Lightweight strong materials	324
	• Recycled materials	324

Re:Source Floor Care

Interface was an early US advocate of adopting sustainable thinking into a company's business strategy and practices. Chief Executive Ray Anderson was an early adopter of the Natural Step philosophy and today the company has its own 'Seven Steps to Sustainability'. Re:Source Floor Care is a service system (SS), although it is also available as a product-service-system (PSS) if Interface flooring

products (such as Entropy™ carpet tiles) are purchased. The core focus is to create bespoke cleaning solutions for clients' carpets on the basis that no two stains and traffic wear patterns are alike, and that maintenance requirements vary with clients. A range of non-abrasive, low-moisture, easy-to-remove products includes Intercept® anti-microbial and IMAGE care products. Any replacement carpet

tiles can be stuck down with Factor 4 Spray Adhesive System requiring less than 25% of the volume of a conventional adhesive. Re:Source service providers provide the first

pan-European network within the floorcovering industry so the benefits of lower environmental floor care impact can reach further afield.

✏	Interface, Inc., USA	316
⚙	Interface, Inc., USA	316
📜	Various cleaning and adhesive materials	
♻	• Reduction in use of consumables	327
	• Improved product lifespan	327

Silencio 6

This is a special 6 mm-thick (about ¼ in) fibreboard composed of 100%-waste softwood fibre, which provides good attenuation against impact sound and insulation as an underlay for wooden and laminate floating floors.

PLYBOO®

The basic component of all this company's products is strips of bamboo measuring 0.5 x 1.9 x 183 cm (³⁄₁₆ x ¾ x 72 in), which are extremely durable. These strips can be bent, woven and laminated as required for flooring, interior-decoration fittings and furniture. Four-, two- and one-ply laminates are available.

✏	Smith & Fong Company, USA	321
⚙	Smith & Fong Company, USA	321
📜	Bamboo, adhesives	
♻	• Renewable material	325

✏	Hunton Fiber AS, Norway	316
⚙	Hunton Fiber AS, Norway	316
📜	Recycled softwood fibre	
♻	• Waste materials	325
	• Recycled, recyclable materials	324

Marmoleum® Real/Fresco

True linoleum is a heavy-duty, durable, non-allergenic floor covering containing at least 30% linseed oil from flax or similar renewable oils from plants. Linoleums are predominantly constituted from natural raw materials such as linseed oil, rosin, wood flour and chalk, which are bonded under heat and pressure to a backing of jute (or occasionally polyester). Forbo-Nairn is the largest linoleum manufacturer in Europe, supplying up to 25 million sq. m (269 million sq. ft) per annum. Marmoleum® linoleums have much lower emissions and lower acidification output than PVC floorings or carpeting. The Marmoleum® Real is available in thirty-six colourways and Marmoleum® Fresco in twelve marbled colourways.

✏	Forbo-Nairn Ltd, UK	315
⚙	Forbo-Nairn Ltd, UK	315
📃	Linseed oil, pine resins, wood flour, cork, mineral fillers, jute	
↩	• Renewable materials and abundant non-renewables	324/325
	• Reduction in emissions and toxins	326
	• Safe, non-toxic	327

Stratica

Stratica is a laminated flooring product comprising a very tough, durable layer of chlorine-free, ionomer coating, DuPont Surlyn®, a printed layer, a backing layer to the print and a final bottom layer. Surlyn® is the finishing material on golf balls. Stratica is naturally flexible but doesn't use plasticizers and is free of volatile organic compounds (VOCs). Over forty-five different 'natural' surfaces can be mimicked in the printing process, from stone to marbles, granites, terrazzos and woods, plus over twenty solid colours. Abrasion resistance is very high and maintenance costs are low. With certification to ISO 14001, recycling pre-consumer waste, preventing pollution and saving energy in production are high priorities.

✏	The Amtico Company, UK	311
⚙	The Amtico Company, UK	311
📃	Dupont Surlyn®, mineral-filled ethylene copolymer	
↩	• Clean production	325

Acousticel Acoustic Underlay (A10)

This underlay is suitable for reducing noise transmission for wood, concrete and asphalt floors. It comprises a top layer of high-grade felt (jute, hair mixture) fixed to a layer of recycled rubber crumb, giving a nominal thickness of 10 mm and nominal weight of 2.455 kg per square metre. Supplied in 1.37 x 11 m (4 ft 6 in x 36 ft) rolls it is easy to lay and is recommended on uneven floors.

✏	Sound Service (Oxford), UK	321
⚙	Sound Service (Oxford), UK	321
📃	Felt, rubber	
↩	• Recycled and renewable materials	324/325 326
	• Reduced noise	

Colorette, Linorette, Marmorette, Uni Walton

This diverse range of linoleum products is fabricated from natural ingredients, such as linseed oil, with minerals including chalk bonded with heat and pressure to a jute or hemp backing.

✏	Armstrong World Industries, Inc., USA	311
⚙	Armstrong World Industries, Inc., USA	311
📃	Linseed oil, flax, pine resins, wood flour, cork, mineral fillers	
↩	• Renewable materials	325
	• Abundant geosphere materials	324

Flooring

Coconutrug®

While most people associate coconut fibre with uninspiring doormats, Deanna Cornellini has worked with the natural characteristics of the fibre to produce an inspiring range of textured rugs dyed with non-toxic colours. As well as standard ranges, including warm spectrum colours, blues, purples and greens, there are a number of limited edition runs, such as 'Signs', 'Views', and 'Horizons', with juxtaposed colour panels and hand-stitched lines. Ultra-pure fibre is hand-woven on traditional looms to produce distinctive finishes to each product ·

line. These tough rugs are suitable for hard-wearing areas, from children's playrooms to offices. Each carpet arrives rolled up in a jute bag. Cleaning is easily achieved by brushin with a broom or using a carpet beater – good old-fashioned but time-tested methods that require only human energy.

✏	Deanna Cornellini, Italy	30
⚙	G T Design, Italy	31
🗋	Coconut fibre, non-toxic colouring dyes	
♻	• Renewable materials	32
	• Safe, non-toxic	32

Super Duralay

Over 60,000 used car tyres are processed each week at a new Duralay plant to provide the raw material for a range of rubber crumb underlays suitable for carpets and wooden flooring. Super Duralay is rated for heavy domestic use but other grades are suitable for contract usage. Bacloc, a woven backing of paper and synthetic thread,

gives extra strength to the underlay. Other grades use a mixed backing of jute and plastic.

✏	Duralay, UK	314
⚙	Duralay, UK	314
🗋	Recycled tyres, latex, Bacloc or polyjute	
♻	• Recycled materials	324
	• Conservation of landfill space	328

Recycled glass tiles

Hand-made glass tiles possess an individualistic quality, the marks of the maker imperceptibly solidified in this vitreous medium. A diverse range of colours is created entirely from 100% recycled glass sources. Knowing a batch of glass tiles is unique, since it reflects the nature of the glass source, is another source of pleasure from these durable products.

✏	Terri Raudenbush, USA	309
⚙	Sandhill Industries, USA	320
🗋	100% recycled glass	
♻	• Recycled materials	324
🔍	IDRA award, 2000–2001	330

Entropy™

As the name suggests, Entropy™ carpet tiles embody principles of energy movement in natural systems. Tiles are made of DuPont Antron Lumena® yarn, a recyclable nylon using the tufted tip-sheared production process. All tiles are designed for installation in random patterns in any direction. This means that Entropy™ provides near-zero installation waste, ages gracefully and has an extended lifecycle because worn tiles can be swapped or replaced. Using one dye lot ensures even colouring. Worn-out tiles are taken back and recycled. This product embraces Interface's 'Seven Steps to Sustainability' – elimination of waste, benign emission, renewable energy, closing the loop, resource-efficient transport, sensitivity hook-up (creating synergistic communities to work with the company) and redesign of commerce. This is further reflected in a new scheme Interface have introduced, with the Climate Neutral Network, called Cool Carpet™ – buy 15 sq. m (18 sq. yds) of carpet, plant a tree. Interface have calculated that new tree planting absorbs the same amount of carbon dioxide as is emitted by their carpets. Paying for new forestry plantings with one's purchase may become a regular feature of retailing.

✏	Interface, Inc., USA	316
⚙	Interface, Inc., USA	316
📜	DuPont Antron Lumena® yarn	
🎧	• Recyclable product	324
	• Reduction in emissions	326
	• Ease of repair	327

Loomtex

Lloyd Loom is renowned for its patented process for creating furniture from metal wire and twisted paper. The technology has been further developed to create a highly durable woven paper carpet with 100% cotton welt. A wide range of patterns is available in widths up to 2.5 m (8 ft 2 in) and any length can be made to special order.

✏	Lloyd Loom of Spalding, UK	317
⚙	Lloyd Loom of Spalding, UK	317
📜	Paper, cotton	
🎧	• Renewable materials	325
	• Durable material	324

Composite lumber

This decking is a composite material using oak fibre and recycled polyethylene with foaming compounds and additives. Containing over 90%-recycled materials by weight, it is very moisture-resistant and durable. It also weathers like conventional wood but without any associated rotting.

✏	Smartdeck Systems/ Plastic Lumber, USA	321
⚙	Smartdeck Systems/ Plastic Lumber, USA	321
📜	Composite wood	
🎧	• Recycled materials	324

Dal-lastic

Dalsouple's huge range of coloured, textured floorcoverings for domestic and heavy industrial use is well-known. Dal-lastic is a new variant, a thick chunky tile suitable for outdoor use and made from 100%-recycled rubber fibre. It has a springy impact and sound absorbent surface that is self-draining and feels warm underfoot. Tiles are loose-laid but jointing is invisible. Plain, flecked or terrazzo patterning plus a large range of colours gives plenty of choice for patios, terraces, roof gardens and other landscape features.

✏	Dalsouple® Direct, UK	313
⚙	Dalsouple Direct, UK	313
📜	100% recycled rubber	
🎧	• Recycled materials	324

Aquair 100, Aquair U.W.

The Aquair 100 is a water turbine that can be towed behind a sailing vessel. The Aquair U.W. is designed to be stationary in a moving body of water. At 6 knots (3 m or 10 ft/second) the Aquair U.W. generates 6 amps at 12V whereas the Aquair 100 generates 5 amps continuous charge. Durable marine-grade materials are used with double 'O' seals and hydraulic fluid in the alternator body to provide maintenance-free turbines. An Aquair U.W. in a fast-running stream can generate up to 2.4kW per day.

🖊	Ampair Ltd, UK	311
⚙	Ampair Ltd, UK	311
📃	Marine-grade metals and plastics	
⟲	• Renewable, water-driven, power	327

Furlmatic 1803

This three-phase alternator generates 340W at wind speeds of 10m/s (35 km/h or 22 mph) but is capable of generating electricity from wind speeds as low as 3 m/s. It provides sufficient output for off-grid domestic lighting and any remote site requiring power for lighting, pumping water or low-voltage equipment. An automatic system produces a furling point at 15 m/second to protect the generator from damage by excessive winds. Other automatic features include an overcharge battery protection device, which stalls the turbine, and a 12V or 24V controller unit. This twin-blade turbine, diameter 1.87 m (6 ft), is mounted on a minimum 6.5 m (21 ft) tower. All components are manufactured at an ISO 9001-compliant factory.

🖊	Marlec Engineering Company, UK	318
⚙	Marlec Engineering Company, UK	318
📃	Stainless steel/aluminium, GRP	
⟲	• Renewable wind power	327

Farm 2000 'HT' boilers

A range of high-temperature (HT) boilers has been designed to accommodate typical biomass fuels available on the farm, such as circular or 1-tonne rectangular straw bales, as well as woodchips, cardboard and other combustible wastes. Heat outputs vary from 20kW to 300kW depending on the boiler and the equivalent electricity costs per kWh are between 25% and 33% of those of kerosene oil or natural gas. An upper refractory arch encourages complete burning of gases, improves the overall efficiency and minimizes atmospheric emissions. Annual or short-rotation crops for biomass fuels absorb carbon dioxide that is released on burning, so this cycle is neutral and makes no net contribution to the greenhouse effect.

🖊	Teisen Products, UK	321
⚙	Teisen Products, UK	321
📃	Steel	
⟲	• Reduced emissions, carbon-dioxide neutral, heating system	326
	• Renewable, local biomass fuels	327

Filsol solar collector

Water heating in buildings in temperate climates can be readily supplemented by installation of a solar collector. Sunlight enters the acrylic collector, which transmits 89% of incident light, and heats a 'Stamax' absorber plate made of specially coated stainless steel. Colourless oxides of chromium, nickel and iron provide an absorption of 0.93 of the incident energy, transferring it to an aqueous antifreeze mixture running in channels in the absorber plate. This hot aqueous mixture is pumped whenever its temperature is higher than the water in the hot-water tank, where further heat exchange ensues.

✏	Filsol, UK	314
⚙	Filsol, UK	314
🎞	Stainless steel, aluminium, alloys, acrylic, poly-isocyanurate foam	
⚘	• Solar power (passive)	327

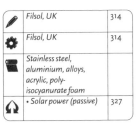

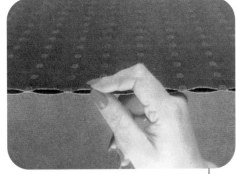

Rutland 913 Windcharger

Weighing just 13 kg (28 lb), this wind generator has a twenty-year pedigree and has been well proven in a wide variety of climates by yachtsmen and scientific researchers and for military and telecommunications operations. Continuous electrical generation starts at wind speeds of 5 knots (5.75 mph). Durable marine-grade materials are combined with quality engineering, units being manufactured to ISO 9001.

✏	Marlec Engineering Company, UK	318
⚙	Marlec Engineering Company, UK	318
🎞	Stainless steel / aluminium, glass-reinforced polymer	295
⚘	• Renewable wind power	327

Thermomax®

at-plate solar collectors ow a reduction in ficiency from 60% at an perating temperature of ℃ down to about 40% s the temperature doubles. ot so with vacuum-tube lar collectors, which can aintain efficiencies of over 0% at temperatures of ℃. A semi-conductor bsorber plate sits within evacuated glass tube. he special liquid-filled heat pe is in intimate contact ith the absorber plate here heat from the sun uses the liquid to aporate to the top of a ondenser unit. Between e pipe and the condenser s a spring made of ape-memory metal, hich limits heat transfer through the pipe when pre-set temperatures are reached, so preventing overheating. Water surrounding the condenser absorbs heat as it flows. Energy conversion even on cloudy days is very efficient, with an overall annual efficiency of over 70%. The reduction of gas or electricity heating costs for an average household is about 40%. All Thermomax manufacturing plants comply with ISO 9001.

✏	Thermo Technologies, USA	322
⚙	Thermo Technologies, USA	322
🎞	Low-iron soda glass, copper, shape-memory metal	
⚘	• High-efficiency solar-powered hot-water system	327

GFX

The heat within waste water from domestic showers, baths and sinks can be reclaimed to heat the incoming cold-water supply to the hot-water tank. GFX is an insulated spiral coil of copper tubing carrying the cold supply, which is in intimate contact with a falling-film heat exchanger through which the waste hot water travels by gravity. The system is capable of saving up to 2kW of power from each shower.

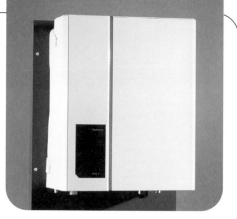

Logamax plus GB112-19 (Linea)

This wall-mounted boiler unit offers an output capacity of between 9.6kW and 19.1kW. It uses an efficient ceramic burner to provide more complete combustion of the gas fuel, resulting in emission levels well under those specified in the German Blue Angel eco-label scheme.

✏	Buderus Heiztechnik GmbH, Germany	312
⚙	Buderus Heiztechnik GmbH, Germany	312
▤	Steel, various metals including aluminium-silicon alloy, ceramics (burner unit)	
♻	• Energy efficient • Blue Angel eco-label	327 328

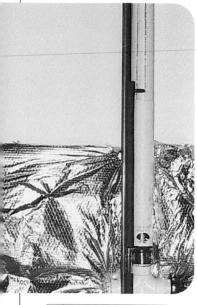

✏	WaterFilm Energy Inc, USA	322
⚙	WaterFilm Energy Inc, USA	322
▤	Copper, insulation	
♻	• Energy conservation	327

Smart Roof© Solar Shingles

Why install a photovoltaic panel on top of an existing roofing surface when the job of keeping out the weather and generating energy can be combined in one product – the Solar Shingle? These 2 m- (7 ft-) long , 30 cm- (12in-) wide, photovoltaic panels can be simply nailed on to the roofing substructure instead of roofing shingles, slates or tiles. Each panel is subdivided into 12 x 30 cm (5 x 12 in) sections that visually mimic traditional roofing materials. Generating 17W each at 6V, panels can be wired together to produce the required capacity.

✏	Uni-Solar, USA	322
⚙	Uni-Solar, USA	322
▤	Photovoltaics, Tefzel glazing, stainless-steel backing	
♻	• Solar-power generation • Reduced construction materials used	327 325

NSS (Non-Stop Shoes)

Translating the expenditure of human energy into power is well understood in the context of sport, but how much energy expended in everyday activities can be harnessed to power appliances in our daily lives? Emili Padrós suggests that even the process of walking could generate electricity, which could be utilized to power lamps and radios once we return home. Developing the mechanisms to generate and store energy could redefine the shoe of the future.

	Emili Padrós, Barcelona, Spain	308
⚙	Conceptual design	
🔋	Various, battery energy storage	
🎧	• Renewable (human) energy generator	327

Logano G124

This free-standing boiler unit offers an output capacity of between 9kW and 34kW, making it suitable for heating single or multiple dwellings. A key feature is the efficient ceramic burner, which provides more complete combustion of the natural or liquid gas fuel, reducing emission levels of nitrous oxides and carbon monoxide below the levels set by the German Blue Angel eco-label scheme.

	Buderus Heiztechnik GmbH, Germany	312
⚙	Buderus Heiztechnik GmbH, Germany	312
🔋	Cast iron, various metals including aluminium-silicon alloy, ceramics (burner unit)	
🎧	• Energy efficient • Blue Angel eco-label	327 328

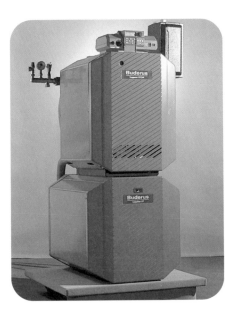

NSD (Non-Stop Doors)

Any repetitive human movement in our everyday lives expends energy, some of which can be captured and stored for later use. In public buildings the act of opening and passing through doors is repeated hundreds of times each day. This human energy is transferred to stored energy, which lights up the frame of the door. Potentially this improves the eco-efficiency of each individual as some of the energy acquired from primary food reserves is recycled. But the real advantage of this concept is that renewable energy is generated without requiring any behavioural changes.

	Emili Padrós, Barcelona, Spain	308
⚙	Conceptual design	
🔋	Various, battery energy storage	
🎧	• Renewable (human) energy generator	327

Multibrid wind energy converter

This wind turbine is designed to work offshore exposed to high-speed, salt-laden winds, so all components are sealed to prevent ingress of water. Unique aspects of the design include slow rotational movement to ensure that the unit can be operated without maintenance for the first three years. This massive generator, with individual blades spanning 50 m (164 ft), is an innovative rotor with excellent aerodynamics.

✏	Bartsch Design Industrial Design Gbr, Germany	304
⚙	aerodyn Energiesysteme GmbH, Germany	311
📜	Various	
🎧	• Renewable energy	327

Paradigma CPC Star

This solar collector is based upon a modular design allowing easy separation of components and materials and facilitating almost 100% recycling. Material usage has been kept to a minimum, giving a lightweight structure with a high efficiency in low sunlight and at ambient temperatures below freezing point.

✏	Büro für Produktgestaltung, Germany	305
⚙	Ritter Energie- und Umwelttechnik GmbH KG, Germany	320
📜	Aluminium, glass	
🎧	• Design for disassembly and recycling	325
🏆	iF Ecology Design Award, 2000	330

POWER Cell

Over 16% of incident sunlight is converted into electrical power with these Sunway solar cells, a very efficient ratio compared with conventional solar cells. Various versions of the POWER cells are manufactured, including those offering up to 30% transparency. The transmitted light is white, yet there is a range of external colours for the cells, employing a process of texturing that avoids the use of chemicals. Now solar cells can be integrated into any aperture intended to introduce light into a building, such as windows and roof lights, thereby reducing overall construction costs.

✏	Roland Burkhardt, Sunways, Germany	305
⚙	Sunways, Germany	321
▣	Silicon polycrystalline wafers	
🎧	• Dual-function	327
	• Solar power	327

Solar-powered service station canopy

Forecourts of most service stations protect customers from the weather by means of a canopy. BP Amoco, in line with their long-term aim of becoming a clean and responsible energy company, have designed special photovoltaic arrays for installation on canopies to generate all the electricity needed to pump fuel and for lighting and so on. This energy-neutral installation will eventually be incorporated into all new service stations.

✎	BP Amoco, UK	312
✿	BP Amoco, UK	312
▤	Photovoltaic panels	
⌂	• Energy-neutral building	327
	• Solar power	327

Topolino

Traditional wood-burning stoves are stoked with timber in a haphazard fashion, causing rapid, uncontrolled combustion with significant heat loss up the chimney stack. GAAN's range of wood-burning heaters encourage optimal combustion because wood is stacked vertically and burns from the top, like the wick of a candle, producing a fuel efficiency of 85%. As the warm combustion gases rise they are forced through a double swan-necked constriction where heat is absorbed into the surrounding materials. Immediate space heating is provided by radiated heat from the toughened glass door, while the remaining heat passes into the surrounding cast stone, granite or steatite body panels, where 60% of the total combustion energy is stored and emitted over the next six to eight hours. Emissions are significantly lower than required under existing EU and Swiss regulations.

✎	GAAN GmbH, Switzerland	306
✿	Tonwerk Lausen AG, Switzerland	322
▤	Steel, glass, granite	
⌂	• Improved energy generation	327
	• Durable construction	327

Watercone®

Paradoxically for a planet whose surface is 70% covered in oceanic water and has two major ice caps, the availability of fresh drinking water is extremely limited in many parts of the world. Ground water and surface water from rivers and lakes comprise less than 1% of the earth's total water reserves. These potable reserves are unequally and unequitably distributed, so almost 2.5 billion people, or 40% of the world's population, don't have access to adequate supplies of safe drinking water. This self-help solar still provides a functional, low-cost solution to in-situ provision of safe drinking water. It can provide 1–1.5 litres of water per day, a basic daily water supply. Contaminated, dirty water or seawater is placed in the central reservoir of the base section. Sunlight causes evaporation from the water source, which condenses on the underside of the vacuum-

formed polycarbonate (PC) cone and trickles down inside the cone to the collection trough. A screw-thread apex on the cone allows it to be used to empty fresh condensate and water from the trough into any suitable receptacle. Weighing less than 2 kg the Watercones are easily stacked for transportation and are ideal for generating water for poor communities and disaster relief schemes. With a duration of three

to five years the US $50 investment pays back well within the lifetime of the product, so it is attractive to not-for-profit and non-governmental organizations (NGOs) in terms of its economic effectiveness and eco-efficiency. Not only is the UV resistant polycarbonate robust but it is non-toxic, non-flammable and 100% recyclable. The black base section is made from 100% recycled PC. In the future controlled

manufacturing and the provision of servicing and take-back facilities could be localized; both of these activities could create local employment.

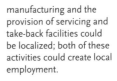

✏️	Stephan Augustin, Augustin Produktentwicklung, Germany	304
⚙️	Disc-O-Bed GmbH, Germany	313
🧻	Polycarbonate	
🔁	• Water generation	328
	• Improved health	327
🏆	iF Design Award, 2003	330

Vulcan Ram

A ram is a water-driven pump using a natural head of water to force water in a small-diameter pipe uphill. Water flows through a pipe taken from a stream or lake feed and is accelerated as it passes through a pulse valve. This valve snaps shut when sufficient pressure builds up in the 'input' chamber in the ram, with the result that a proportion of the water is forced through a delivery valve

into the 'output' chamber. Typically there are between forty and ninety open/shut cycles in the pulse valve each minute. Air under pressure in the output chamber converts the pulsing water through the delivery valve into a steady flow to a header tank. Rams are capable of raising water up to 100 m (330 ft) above the ram and pumping 250,000 litres (55,000 gals) in twenty-four

hours. Using traditional cast-iron and gunmetal production techniques, Green & Carter have been

✏️	Originally patented by Pierre Montgolfier in 1816	
⚙️	Green & Carter, UK	315
🧻	Cast iron, metals, rubber	
🔁	• Energy and water conservation	327/ 328
	• Durable	327

manufacturing rams since 1928 and export worldwide. They still repair rams designed and made by Josiah and James Easton, who installed water-pumping schemes for many eighteenth- and nineteenth-century landowners. The ram is an example of Industrial Revolution technology still proving durable, reliable and economical.

Axor Starck two-handled basin mixer

This functional, easy-to-clean mixer tap limits water output to 7.2 litres (1.6 gal) a minute, eliminates limescaling and has a special stop valve.

✏	Philippe Starck, France	309
⚙	Hans Grohe GmbH, Germany	316
▬	Chromium-plated steel	
☊	• Water conservation	328
✪	iF Design Award, 1999	330

Axor Starck showerhead

This free-standing shower unit is sparing in its use of materials as parts of the frame also act as hot- and cold-water pipes to deliver to the overhead and hand-held roses. Pleating the polyester curtain prevents it from clinging to the user. The unit is easily plumbed in, provides excellent access for maintenance and can be repositioned when moving house. Stainless steel would be a preferred substitute for the chromium-plated steel to minimize the impacts of this unit even further.

✏	Philippe Starck, France	309
⚙	Hans Grohe GmbH, Germany	316
▬	Enamelled and chromium-plated steel, polyester, polymer base	
☊	• Reduction in materials used	325
✪	iF Ecology Design Award, 2000	330

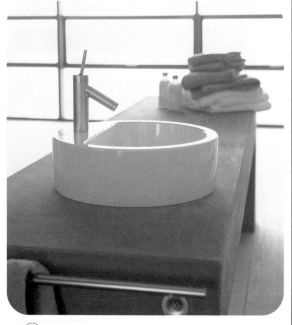

Clivus Multrum composting toilets

This company has been manufacturing composting toilets since 1939. This particular model, made of 100%-recycled polyethylene, provides adequate sanitation for a three-bedroomed house. An integral moistening system ensures biomass volume reductions of 95%. Water vapour and carbon dioxide are the only emissions.

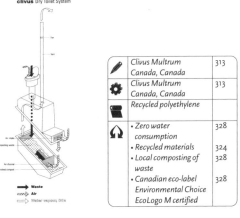

clivus Dry Toilet System

Waste
Air
Water vapours, Gills

Plush tap

Conservation of resources in public buildings ought to be a high priority but it often needs an innovation to encourage capital investment before tangible results can be achieved. Plush Tap is such an innovation. It allows existing cross-head taps to be converted into push taps by using an adaptor to fit on the old tap body. Water is conserved, as taps cannot be accidentally left running. The advantages of push taps are especially felt where water is metered.

✏️	Flow Control Water Conservation, UK	315
⚙️	Flow Control Water Conservation, UK	315
📇	Brass, seals, stainless steel	
🎧	• Water conservation	328

✏️	Clivus Multrum Canada, Canada	313
⚙️	Clivus Multrum Canada, Canada	313
📇	Recycled polyethylene	
🎧	• Zero water consumption	328
	• Recycled materials	324
	• Local composting of waste	328
	• Canadian eco-label Environmental Choice EcoLogo M certified	328

Ifö Cera range

The humble toilet bears the hallmark of a couple of hundred years of traditional industrial design but few people know what happens inside the water cistern. In traditional toilet designs extravagant volumes of water are used to flush even small quantities of human effluent. Today, to meet the need for water conservation, sanitary-ware manufacturers such as Ifö Sanitär have introduced dual-flush cisterns offering two- or four-litre (0.4 or 0.9 gal) water delivery and, more recently, adjustable flushing volumes, from three to eight litres (0.7–1.8 gal) in the Ifö Cera range. Polypropylene or duroplastic seating is

available, ergonomically designed to fit a variety of posteriors, and with hygienic surfaces in a typically clean, sculptural, Scandinavian form.

✏️	Ifö Sanitär AB, Sweden	316
⚙️	Ifö Sanitär AB, Sweden	316
🗞️	Ceramics, polypropylene or duroplastic	
🎧	• Water conservation • Improved ergonomics	328 327

Excel NE

With over twenty-five years' experience of designing composting, waterless toilets, Sun-Mar Corporation have developed a range of self-contained and central composting toilet systems. Most models are equipped with an electrically driven fan to provide an odour-free atmosphere but the Excel NE is totally non-electric, using a vent chimney instead. The operating principle in all Sun-Mar toilets is identical. A mixture of peat, some topsoil and/or 'Microbe Mix' is added to the Bio-drum™ and a cupful of

peat bulking mix is added per person per day. After use the Bio-drum is mechanically turned between four and six revolutions every third day or so to aerate the mixture. Fully degraded compost is removed from a bottom finishing drawer as required. An evaporating chamber at the rear of the drum ensures excess moisture is removed. So confident are the manufacturers in the robust design of their toilets that they offer free parts for three years and a twenty-five year warranty on the fibreglass body.

✏️	Sun-Mar Corporation, Canada	321
⚙️	Sun-Mar Corporation, Canada	321
🗞️	Fibreglass	
🎧	• Water conservation • Local composting of waste	328 328

Naturum

Scandinavia has a well-developed market for composting toilets, so expectations of standards are high. The Naturum is a very practical and efficient rotary-drum composter suitable for installation indoors in any bathroom serving up to five people daily. Whether used in a seated or standing position, the urine is separated by the shape of the bowl. As the compost space is closed with a shutter-seal, a shower hose turns the bowl into a bidet. Solid wastes fall directly through the trap door into the drum where an absorbent, such as unfertilized milled peat, is placed. Depressing the pedal rotates the drum

✏️	Biolan Oy, Finland	312
⚙️	Biolan Oy, Finland	312
🧻	Glass fibre, polyethylene, stainless or acid-resistant steel	
🎧	• Water conservation	328
	• Local composting of waste	328
	• No chemical use	327

and instantly turns old compost over the new waste. Each time the pedal is depressed more mixing occurs. As the digestion and decomposition process proceeds the compost mass of about 30 litres (6–7 gals) is kept constant and rises to its operating temperature. Any excess mass is shaken out in sterile, odourless particles by the rotation motion into a 10-litre (22-gal) separate container, which is emptied when required – about once a month if used by four people. The front and bowl are fibreglass but inner parts are recyclable polyethylene and any metal parts are high-grade stainless or acid-resistant steel.

Oxfam bucket

Jerrycans holding about 22 or 45 litres (5 or 10 gals) are the normal means of carrying water for aid or disaster relief work by agencies such as Oxfam. But rigid jerrycans take up a lot of valuable space on aid work planes, so Core Plastics developed a stackable bucket with a removable lid, which incorporates a filler hole/spout with snap-on top. An indentation in the base helps reduce the risk of spinal injuries when the bucket is carried on the head. The bucket design improves the efficacy of relief efforts.

✏️	Oxfam/Core Plastics, UK	313
⚙️	Core Plastics, UK	313
🧻	Lightweight, UV-resistant plastic	
🎧	• Reduction in transport energy	326

'W' High-Efficiency Motor

Brook Crompton is a major supplier of heavy-duty electric motors to UK industry. The 'W' High-Efficiency Motor uses a new type of steel with improved magnetic characteristics, which increases electrical efficiency by 3%. Electric motors account for over 65% of the energy usage by UK industry, so this new motor can potentially reduce carbon dioxide emissions in the UK by up to 2.5 million tonnes per annum as existing motors are replaced. Use of this special steel also allows a reduction of 30% in weight compared with conventional steels.

✎	Brook Crompton with Sheffield and Cambridge Universities, UK	312
⚙	Brook Crompton, UK	312
▤	Steel, copper and other materials	
🎧	• Energy efficiency • Reduction in materials used	327 325

ureprint

onventional web-offset rinting processes use ater with about 10% dustrial alcohol, such as PA, to ensure ie plates stay wet so the ks can flow. IPA is ghly mobile as it readily vaporates and 'dissolves' water. It is also a arcinogen and therefore creates a potentially toxic environment for workers. Beacon Press avoid using water or alcohol and instead use silicon rubber to ensure appropriate 'wetting' of the plates and sharper resolution. Nor are any chemicals used in preparation of filmwork, and a strong corporate environmental policy ensures that Beacon Press operate a clean technology printing plant in all aspects, from supply-chain management to car-sharing for employees.

✏	Originally developed in Japan and USA	
⚙	Beacon Press, UK	312
▤	Silicon rubber	
♻	• Water conservation • Avoidance of use of toxic substances	328 326

Erosamat Type 1, 1A, 2

Woven mats of natural fibres placed on the surface of the soil absorb raindrop impact and significantly reduce water run-off and consequent soil erosion. Erosamat Type 1 and 1A are made from jute fibres whereas Erosamat Type 2 is a heavier duty geotextile of coir fibres extracted from the husks of coconuts. The latter takes longer to decompose but affords greater protection to soils more at risk from erosion. All types of mat can be seeded to create a dense sward of vegetation, which further bonds the surface of the soil.

✏	Unknown	
⚙	Various for ABG Ltd, UK	311
▤	Jute, coir	
♻	• Renewable materials • Protection against soil erosion	325 328

Volvo Articulated Hauler A 35/40D

Volvo has a considerable reputation in the construction and mining industries for their quality of build and serviceability of their heavy-duty vehicles. Any downtime for these machines, or undue stress on the operator, imposes a burden on the contractor, so this new model has focused on reliability, high capacity per load and an optimal safety environment for the driver. All mechanical components and systems are easily accessed to enable repairs or refits. Vision from the driving seat is improved by a steeply raked hood, while instruments and other controls are optimized for ease of use when making repetitive actions. Easy disassembly at end of life ensures that 95% of the vehicle is recyclable.

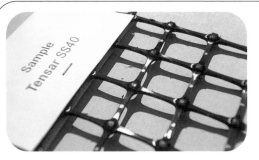

Tensar® range

Steep construction slopes can be reinforced with Tensar® 80RE, a uni-axial grid of polypropylene with elongated apertures, which improves the shear strength of the exposed surface layers. Extruded sheets of special polyethylene are punched with regularly shaped apertures, then stretched under heat to create a high-strength grid. When this geomat is laid on the surface it reduces soil erosion by absorbing the energy of raindrop impact and providing an anchor for plant roots.

✏️	Tensar International, UK	322
⚙️	Tensar International, UK	322
📜	Polypropylene or polyethylene	
🎧	• Protection against soil erosion and slope failure	328

✏️	Hans Philip Zachau, Nya Perspektiv Design AB, Sweden	310
⚙️	Volvo Articulated Haulers, Sweden	322
📜	Various materials	
🎧	• Ease of maintenance	327
	• Improved health and safety	327
	• Recyclable	324
🏆	iF Design Award, 2002	330

Agil-Hubgerät, lightweight stacker lifting truck

In effect this is equivalent to a hand-operated mini forklift truck capable of lifting diverse objects of up to 80 kg (176 lb). It is highly manoeuvrable in tight corners owing to its small turning circle, making it ideal for maximizing use of valuable warehouse space. It can lift a wide variety of objects by attaching accessories to the vertical axis and, if greater stability is required, is easily adjusted to widen the tracking between the wheels. A rechargeable lead gel battery powers the electric motor and a low-maintenance cog drive train.

Mimid

Land mines planted during military and civil conflicts during the twentieth century kill or maim innocent civilians every day. An estimated 100 million undiscovered mines form a lethal legacy for future generations, so this portable, compact mine detector is a useful addition to the tools available to mine-clearance personnel.

✏	Prof. Gerhard Heufler, Germany	307
⚙	Schiebel Elektronische Geräte AG, Austria	320
▤	Various	
↻	• Improved health and safety	327

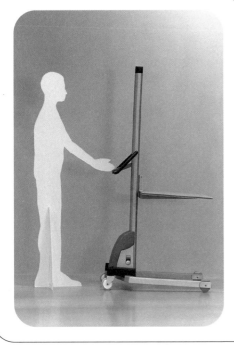

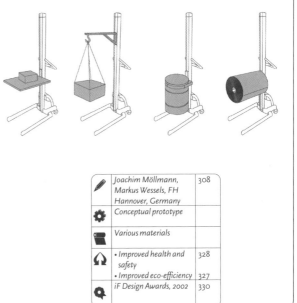

✏	Joachim Möllmann, Markus Wessels, FH Hannover, Germany	308
⚙	Conceptual prototype	
▤	Various materials	
↻	• Improved health and safety	328
	• Improved eco-efficiency	327
✦	iF Design Awards, 2002	330

3.0 Materials

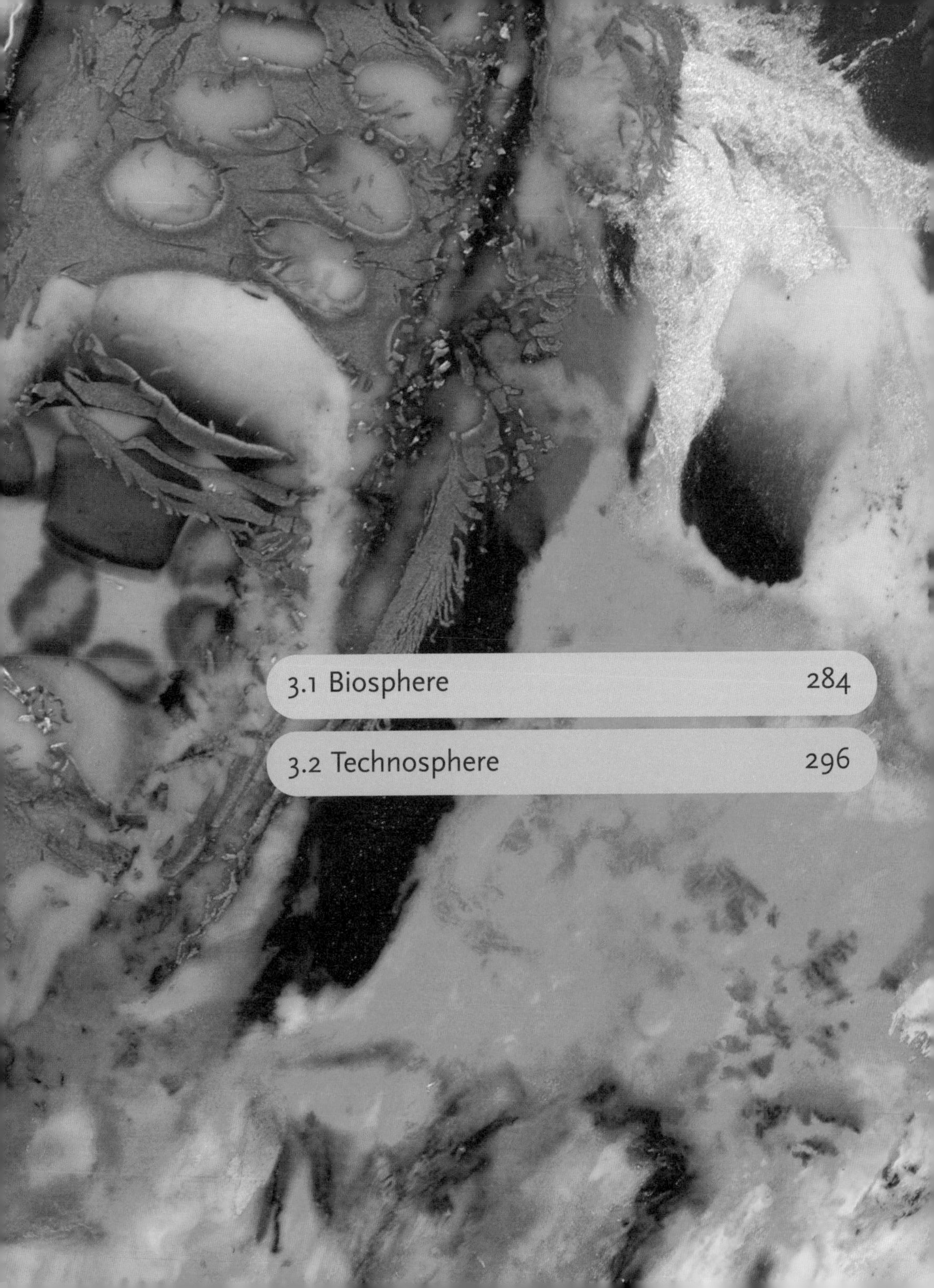

It's A Material World

People first developed methods to synthesize materials from nature at the time of the earliest civilizations in Mesopotamia. The Industrial Revolution and two world wars accelerated the synthesis of new man-made materials. Today the designer is faced with a mind-boggling array of hundreds of thousands of materials, some of which have no or little impact on the environment while others generate a rucksack of environmental impacts including depletion of non-renewable resources, toxic or hazardous emissions to air, water or land, and the generation of large quantities of solid waste.

While designers have traditionally selected materials on the basis of their physical, chemical and aesthetic properties, as well as by cost and availability, other parameters, such as resource depletion, are now proving important. Designers are now obliged to observe legal restrictions on the use of materials from endangered species, as listed in the 1973 Convention on International Trade in Endangered Species (CITES). Various voluntary certification schemes, such as the Forest Stewardship Council and SmartWood schemes, ensure that materials originate from sustainably managed forests. Unfortunately, designers have few published guidelines about criteria for selecting materials in relation to environmental, social and ethical issues. The checklist in Table 1 offers a method of considering the potential impacts of a material.

Ecomaterials

An ecomaterial is one that has a minimal impact on the environment but offers maximum performance for the required design task. Ecomaterials are easily reintroduced into cycles. Ecomaterials from the biosphere are recycled by nature and ecomaterials from the technosphere are recycled by man-made processes.

Embodied energy

One measure of eco-efficiency is the degree of efficiency of use of energy within an ecosystem, that is, the energy captured, energy flows within the ecosystem and energy losses. All materials represent stored energy, captured from the sun or already held in the lithosphere of the earth. Materials also represent or embody the energy used to produce them. One tonne of aluminium takes over a hundred times more energy to produce than one tonne of sawn timber, so the embodied energy of aluminium is comparatively high. Materials with a low embodied energy are generally those with a smaller rucksack of environmental impacts. Materials extracted directly from nature and requiring little processing tend to be low-embodied-energy materials, while man-made materials tend to possess medium to high embodied energy (Table 2).

In complex products, such as a car, involving application of many materials, the calculations of embodied energy are more involved. For instance, using lightweight aluminium as opposed to steel in the chassis of a car will ensure greater fuel efficiency and so reduce the total energy use over the lifetime of the product. Selection of high-embodied-energy materials, which are durable and extend product life, may be preferred to lower-embodied-energy materials, which have a short product life. So a very important consideration is the embodied energy of the material over the lifespan of the product.

Materials from the biosphere and lithosphere

Materials derived from the living components of the planet, the biosphere, are renewable and originate from plants, animals and micro-organisms. Biosphere materials include special groups of man-made materials such as compostable biopolymers and biocomposites derived from plant matter. Such materials are readily returned to the cycles of nature. Materials derived from the lithosphere (geological strata of the earth's crust) fall into two main categories. The first category is widely distributed or abundant materials such as sand, gravel, stone and clay, while the second category includes materials whose distribution is limited, such as fossil fuels, metal ores and precious metals/stones. Materials

Table 1: A checklist for selecting materials

Material attribute	Low environ-mental impact	High environ-mental impact
Resource availability	Renewable and/or abundant	Non-renewable and/or rare
Distance to source (the closer the source the less the transport energy consumed) km	Near	Far
Embodied energy (the total energy embodied within the material from extraction to finished product) MJ per kg	Low	High
Recycled fraction (the proportion of recycled content) per cent	High	Low
Production of emissions (to air, water and/or land)	Zero/Low	High
Production of waste	Zero/Low	High
Production of toxins or hazardous substances	Zero/Low	High
Recyclability, reusability	High	Low
End-of-life waste	Zero/Low	High
Cyclicity (the ease with which the material can be recycled)	High	Low

from the biosphere or lithosphere are often processed by synthesis or concentration to create technosphere materials.

Materials from the technosphere

Technosphere materials are generally non-renewable. Synthetic polymers (plastics, elastomers and resins) derived from oil, a fossil fuel, are technosphere materials. Embodied-energy values tend to be much higher than in biosphere materials. Most technosphere materials are not readily returned to the cycles of nature and some, such as plastics, ceramics (glass, glass/ graphite/carbon fibres) and composites (ceramic, metal), are inert to microbial decomposition and will never re-enter the biosphere. In a world of finite resources we need to be aware of the need to recycle technosphere materials.

Recycling

Three exhibitions in the 1990s encouraged designers to focus on the potential of using recycled materials from the technosphere. Rematerialize (1994), collated by Jakki Dehn of Kingston University, UK, displayed a diverse selection of contemporary materials made using recycled content. 'Mutant Materials', curated by Paola Antonelli at the Museum of Modern Art (MoMA) in New York in 1997, examined the application of recycled thermoplastics alongside new material developments such as specialist polymers, foamed

alloys, foamed ceramics and unusual composites.

An exhibition called Recycling, organized by Craftspace Touring in the UK in 1996, revealed the beauty of hand-crafted products made from recycled materials.

Materials from the biosphere are readily taken back into nature's cycles by the process of biodegradation, or composting, by the action

of microbes and by water and weather. Nature recycles all its materials but humans recycle only certain man-made materials. Materials of low monetary value tend to have low volumes of recycling. Thus relatively expensive ferrous metal and light alloys often include a recycle fraction of between 70 and 80%, non-ferrous metals between 10 and 80% and precious

Table 2: Embodied energy values for common materials

Material type	Typical embodied energy (MJ per kg)
Biosphere and lithosphere materials	
Ceramic minerals, e.g., stone, gravel	2–4
Wood, bamboo, cork	2–8
Natural rubber (unfilled)	5–6
Cotton, hemp, silk, wool	4–10
Wood composites, e.g., particleboards	6–12
Technosphere materials	
Ceramics – bricks	2–10
Ceramics – glass	20–25
Ceramics – glass fibre	20–150
Ceramics – carbon fibre	800–1,000
Composites – titanium-carbide matrix	600–1,000
Composites – alumina fibre reinforced	450–700
Composites – polymer – thermoplastic – Nylon 6 (PA)	400–600
Composites – polymer – thermoset – epoxy matrix – Kevlar fibre	400–600
Foam – metal – high-density aluminium	300–350
Foam – polymer – polyurethane	140–160
Metal – ferrous alloys – carbon steel	60–72
Metal – ferrous alloys – cast iron – grey (flake graphite)	34–66
Metal – light alloys – aluminium – cast	235–335
Metal – non-ferrous alloys – copper various alloys	115–180
Metal – non-ferrous alloys – lead various alloys	29–54
Metal – precious metal alloys – gold	5,600–6,000
Polymer – elastomer – butyl rubber	125–145
Polymer – elastomer – polyurethane	90–100
Polymer – thermoplastic – ABS	85–120
Polymer – thermoplastic – nylon	170–180
Polymer – thermoplastic – polyethylene	85–130
Polymer – thermoplastic – polypropylene	90–115
Polymer – thermoset – melamine	120–150
Polymer – thermoset – epoxy	100–150

Adapted from Cambridge Engineering Selector, version 3.0, Granta Design Ltd, UK

metal alloys (gold, platinum, silver) between 90 and 98%. Relatively inexpensive polymers (plastics), on the other hand, exhibit recycle fractions of between zero and 60%, the most commonly recycled plastics being PET (20–30% recycle fraction), polypropylene (25-35 per cent), polyethylene as LDPE or HDPE (typically 50–60%) and polystyrene (35–40%). Specialist technosphere materials, especially composites, for example, thermosets and reinforced thermoplastics, often have less than 1% recycle fraction.

Closed-loop recycling of materials from the technosphere significantly reduces environmental impacts. Metals made entirely of recycled content and recycled plastics have an embodied energy that is often only half or even as little as 10 % of that of virgin metals. Increasing the recycle fraction in more materials, by re-evaluating the idea of 'waste', will bring savings in energy.

Green procurement

Designers can also reduce the impacts of materials they use if they specify sources of materials and minimum recycle fractions and if they insist on compliance with certain standards, such as eco-labels or voluntary industry schemes (see Green Organizations, p. 329). Specifiying suppliers or manufacturers that comply with internationally recognized environmental management systems, such as ISO 14001 or EMAS, is also desirable.

Boards and Composites
BIOFIBER™

This decorative, rich, golden-yellow, smooth-surfaced, fine-fibre composite panel is made from wheat straw, an abundant agricultural residue. The patterning runs throughout the thickness of the material, which has a consistent density. Out-gassing solvents are not added during manufacturing so there are no emissions to the atmosphere during the lifetime of the product. BIOFIBER™ is intended for commercial and domestic interiors only.

⚙	Phenix Biocomposites, USA	319
♻	• Renewable materials	325
	• Utilizes agricultural waste	325

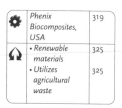

Environ™

Environ™ is possibly the first example of a mass-produced biocomposite using a plant-based resin to bond recycled materials. It is manufactured from recycled paper and soy flour into sheets and floor strips and is claimed to be harder than oak wood and suitable for interior decoration and furniture.

⚙	Phenix Biocomposites, USA	319
♻	• Renewable and recycled materials	324/325

General-purpose particleboard

Some 14,000 tonnes of waste wheat straw annually go to make this half-inch- (1 cm-) thick, wheat-based particleboard suitable for furniture, construction and interior design.

⚙	Prairie Forest Products, USA	319
♻	• Renewable material	325

Kronospan®

This Swiss company manufactures a diverse range of particle boards, tongue-and-groove panels, MDF, Kronoply (an orientated strand board, OSB), laminate flooring and post-formed panels and surfaces for interior use. Laminated flooring sheets are FSC certified, the formaldehyde-free, panel-type 'Hollywood' qualifies for a Blue Angel eco-label and the company is certified to ISO 14001. Timber is generally sourced locally.

⚙	Kronospan AG, Switzerland	317
♻	• Renewable materials	325
	• Clean production	325
	• Certification of various products to FSC or Blue Angel eco-label	324

Masonite

This tough, dense board is made from long-fibre wood compressed to attain a very high density of 940 kg/sq. m (192 lb/sq. ft), which makes it especially suitable where structural strength is required. It is manufactured to ISO 9001 and guaranteed and approved by the Swedish National Board of Housing.

⚙	Masonite Corporation, USA	318
♻	• Renewable materials content	325

MeadowBoard™ panels and sheeting

Compressed panels of ryegrass straw are suitable for all interior design, exhibition and furniture production.

⚙	Meadowood Industries, Inc., USA	318
♻	• Renewable material	325

Medite ZF

Medite ZF is the trade name for an interior-grade, medium-density fibreboard (MDF) manufactured using softwood fibres bonded with formaldehyde-free synthetic resin. Free formaldehyde content of Medite ZF is less than 1 mg/100g (one part in 100,000), which is equivalent to or less than natural wood, and formaldehyde emissions are well below general ambient outdoor levels. All other Medite MDF boards are manufactured to Class A EN622 Part 1, complying with free formaldehyde content of less than 9 mg/100g (nine parts per 100,000). The company has applied for FSC certification for Medite.

⚙	Weyerhaeuser Europe Ltd, Ireland	322
♻	• Renewable materials	325
	• Reduction in toxic ingredients and emissions	326

Moulded strandwood (MSW)

Thin strands of aspen pulpwood are rebonded in moulds for furniture and automobile parts, such as legs, shells and seat backs.

⚙	Strandwood Molding, Inc., USA	321
↻	• Renewable materials	325

Multiboard

Composite sheeting and boards are made from a diverse range of recycled materials, of which 55% originates from used PE-coated milk cartons, newsprint and corrugated paper, the remaining 45% from the industrial or production waste streams. This company is registered to ISO 9001, ISO 14001 and EMAS.

⚙	Fiskeby Board AB, Sweden	314
↻	• Recycled materials	324
	• EMAS policy	328

N.C.F.R. Homasote®

Homasote claim to be the oldest manufacturer of building board from 100%-recycled post-consumer paper in the USA, with a pedigree stretching back to 1909. All the products in their range are free of asbestos and formaldehyde additives. For each tonne of recycled paper there is a net reduction of 73% emissions to air, 40 to 70% less water consumption and 70% less energy than virgin wood pulp fibre. N.C.F.R. Homasote®, a multipurpose interior or exterior board, is a good insulator and a barrier to moisture, noise and fire (when impregnated with fire retardants).

⚙	Homasote Company, USA	316
↻	• Recycled materials	324
	• Low-embodied energy	324

Pacific board™

Wheat straw and Kentucky bluegrass are the main fibre constituents of the particleboard manufactured by this company.

⚙	Pacific Northwest Fiber, USA	319
↻	• Renewable material	325

Pacific Gold Board

This is a straw-based building board suitable for interior uses.

⚙	BioFab LLC, USA	312
↻	• Renewable material	325

Panda Flooring

Bamboo is botanically classified as a grass yet it is harder, tougher and more elastic than many temperate hardwoods. The genus bamboo shows phenomenal growth with reportedly one metre per day for some species. Panda bamboo flooring is made from three-year growth canes of the Chinese bamboo, *Phyllostachys pubescens*, a straight, regular-diameter cane with a very dense fibre and lower carbohydrate content that improves resistance to insect and fungal attack. Canes are harvested from managed forests. They are then treated by steaming, heating, dehydrating and oven-drying followed by the application of anti-insect and mildew treatments. Lengths of bamboo are cut and prepared before laminating into tongue-and-groove boards 920 x 92 x 15 mm (36 x 3½ x ½ in) and receiving UV-resistant lacquer protection. Colours vary from very light natural beige to dark carbonized browns.

⚙	Panda Flooring Co., USA	319
↻	• Renewable resources	325
	• Selective harvesting from managed forests	325
	• Durable	324
	• Ease of maintenance	327
		327

PrimeBoard®

Annually in the USA 60 million tonnes of straw are burnt off or ploughed back into the soil. This raw material could provide a valuable alternative to wood-based fibre boards, and relieve pressure on existing temperate softwood and tropical hardwood forests. Instead of the traditional urea formaldehyde for bonding, PrimeBoard® uses a new adhesive called MDI, which produces no off-gassing and therefore has Emission Free Board (EFB®) status. It is a light, moisture-resistant board with good laminating properties.

⚙	Primeboard, Inc., USA	319
↻	• Renewable materials	325
	• Non-toxic	324

Quikaboard

Quikaboard is a versatile range of lightweight honeycomb boards and panels. Reconstituted paper is used for the honeycomb core varying from under 10 mm to over 60 mm (about ½–2½ in). Thickness of the boards and panels varies according to honeycomb depth and the thickness and nature of the external facings. Typical facing materials include hardboard, composites, laminates, plywood, chipboard and MDF. Boards and panels can be bent to a range of radii for curvilinear designs. Quikaboard has high strength-to-weight ratio, is durable and is suitable for a wide range of applications, such as signage, point of sale, shopfittings, exhibitions, office interiors and caravan fixtures. One of the more ambitious applications of Quikaboard was the waterproofed laminates for the external and roofing panels used at the Cardboard School at Westcliff-on-Sea, Essex, UK designed by Cottrell and Vermeulen Architecture (p. 238).

⚙	Quinton & Kaines Ltd, UK	320
♻	• Recyclable	324
	• Durable	324

Resincore™

Resincore is formaldehyde-free particleboard composed of sawdust, phenolic resin and wax.

⚙	Rodman Industries, USA	320
♻	• Renewable and recycled materials	324/ 325

Shortstraw

Although compressed straw panels are a familiar product in the USA they are just beginning to receive attention in the UK. Shortstraw is introducing small-scale production of standard and bespoke panels. Straw is shredded, mixed with MDI binder and then pressed into 850 x 850 mm (33½ x 33½ in) or smaller panels. Striking visual diversity is achieved using wheat, barley or oat straw, but the most dramatic feature of these panels is the introduction of other plant materials (such as leaves, flowers and branches) to create varied abstract patterns, colours and textures. Nigel Pearce, the designer at Shortstraw, has also experimented with including fabrics, metal and plastics into the straw boards. This allows designers to work with the manufacturer to achieve specific structural and aesthetic qualities.

⚙	Shortstraw, UK	321
♻	• Renewable material	325
	• Waste agricultural material	325

Thermo-ply

These fibreboards are made of 100%-recycled materials including cardboard, office waste, mill waste and production scrap.

⚙	Ludlow Coated Products, USA	317
♻	• Renewable and recycled materials	324/ 325

WISA® plywoods

Birch, spruce and pine from managed forests in Finland are used to manufacture a range of plywoods suitable for interior, exterior and concrete formwork and as laminboard. Special tongue-and-groove panel plywood laminated floorings include Schauman Birchfloor, Sprucefloor and Spruce Dek. The company is certified to ISO 14001.

⚙	WISA Wood Products, Finland	323
♻	• Renewable materials	325
	• Stewardship sourcing	325

Timbers

Bamboo

Bamboo sourced from Vietnam is the principal material for strip and laminated flooring manufactured by the company, but poles and bamboo for structural purposes can also be supplied.

⚙	Bamboo Hardwoods, Inc., USA	312
♻	• Renewable material	325

Bendywood®

A wide range of sections and profiles of Bendywood® are available in oak, beech, ash and maple, offering designer–makers and manufacturers a ready-made material for forming tight- or wide-radius curves for furniture and other product applications. Bendywood® is created by subjecting timber to heat and directional pressure, which restructures the cellulose fibres, allowing them to concertina when subject to bending. This enables Bendywood® to be bent in a cold dry state to a radius of about ten times the thickness of the section or profile. Tighter bends can be achieved by raising the moisture content prior to bending then securing the work and letting it dry in situ. Available stock sizes are up to 100 x 120 x 2,200 mm (about 4 x 4½ x 86½ in) depending on the type of wood.

⚙	Mallinson, UK	318
♻	• Renewable material	325

Certified timber

This company supplies high-density prepared boards from sustainably harvested palm trees. The timber is guaranteed 100% chemical-free and is suitable for structural, flooring and furniture applications. A wide range of North American and tropical hardwoods is supplied from certified sources.

	Eco Timber International, USA	314
	• Renewable materials	325
	• Non-toxic	324
	• Certified sources	324

Iron Woods®

Diniza, Purpleheart, Greenheart and Macaranduba are tough, exceptionally dense and durable, tropical hardwoods. Promoted as THL Iron Woods®, these sawn and planed woods are certified by the FSC and the Rainforest Alliance's SmartWood® schemes as originating from sustainably managed forests. They do not require any chemical treatments to prolong life.

	Cecco Trading, Inc., USA	313
	• Renewable materials	325
	• Certified sources	324

Microllam® and Intrallam®

With minimal wastage from the forest roundwood, TJM bond layers of aspen wood with resin to form high-strength timber composites – Microllam™ comprising thin even layers and Intrallam™ formed from more irregular layers and chips.

	Trus Joist, USA	322
	• Reduced production waste	325
	• Efficient use of materials	325

Rubber wood

There are over 7.2 million hectares (17.8 million acres) of cultivated rubber trees worldwide, of which over 5.2 million hectares (12.8 million acres) are in Malaysia, Indonesia and Thailand. Latex yields decline twenty-five to thirty years after planting. These older trees are now being harvested for *Hevea*, or rubber wood, the dominant species being *Hevea braziliensis*. In 1990 the ASEAN (the Association of south-east Asian Nations) produced about 17 million cu. m (600 million cu. ft) of rubber wood. The timber is suitable for a wide range of applications, such as flooring, particle boards, kitchen utensils and furniture.

	Numerous manufacturers in tropical countries	
	• Renewable material	325

Timber from sustainable sources

Harwood Products supplies timber and products from sustainably managed forests certified by the FSC. A member of CERES.

	Harwood Products, USA	316
	• Renewable materials	325
	• Certified sources	324

Timberstrand® LSL

TJM produce a range of engineered timbers composed of strands or sheets of veneer bonded with adhesives or resins at high pressure and heat. Timberstrand® TSL is a general-purpose structural timber. TJM products encourage better resource usage than sawn timber since almost all the sawn log is used in the composite timber.

	Trus Joist, USA	322
	• Renewable material	325
	• Efficient use of materials	325

Biopolymers

Biocorp: biopolymer

Corn starch is the main ingredient of the compostable biopolymers made by Biocorp. Products include plastic bags and cutlery.

	Biocorp, USA	312
	• Renewable materials	325
	• Compostable	324

Bioplast®

Biotec specializes in biodegradable plastics using vegetable starch as the raw ingredient. Trade products include Bioplast®. It has similar properties to polystyrene, so it is suitable for making disposable cups for vending machines and catering companies.

	Biotec, Germany	312
	• Renewable materials	325
	• Compostable	324

Biopolymers

Biopolymers and industrial starches are extracted and processed from the corn (maize) plant.

⚙	Cerestar, USA	313
⟲	• Renewable material	325
	• Compostable	324

Biopolymers and resins

This company specializes in the manufacture of starch-based biopolymers and resins suitable for injection moulding. Clean Green is loose-fill packaging that is water-soluble.

⚙	Starch Tech, Inc., USA	321
⟲	• Renewable materials	325
	• Compostable	324

depart® and adept™

Derived from polyvinyl alcohol, depart® is a water-soluble and biodegradable plastic.

⚙	Adept Polymers Ltd, UK	311
⟲	• Compostable	324

Eco-Foam®

Polystyrene chips, often made by injecting chorofluorocarbon (CFC) gases, can now be substituted with biodegradable chips of foamed starch polymer, Eco-Foam®, where steam is used as the blowing agent. This biopolymer is made of 85% corn starch, so it is biodegradable, water-soluble and reusable. It is also free of static, which makes the packaging process easier.

⚙	National Starch & Chemical Co., USA	318
⟲	• Renewable materials	325
	• Reusable packaging	326
	• Free of CFCs	326

Eco-Flow

Eco-Flow is an extruded packaging material primarily composed of wheat starch.

⚙	American Excelsior Company, USA	311
⟲	• Renewable material	325
	• Compostable	324

Flo-Pak Bio 8

A loosefill packaging material in a figure-of-eight shape made by extruding corn, wheat and/or potato starch biopolymer, it is suitable for using in all standard commercial packaging systems. Like all starch-based biopolymers it readily dissolves in water and is biodegradable.

⚙	FP International, USA	315
⟲	• Renewable materials	325

Natural rubber (NR)

Today over 70% of rubber production, 5.2 million hectares of a world total of 7.2 million hectares (12.8 million acres and 17.8 million acres respectively), centres around Malaysia, Indonesia and Thailand. Trees have a productive lifetime of up to thirty years, after which latex production declines. Plantations also act as a sink for absorbing carbon dioxide. Natural rubber is used pure or mixed with synthetic rubbers and fillers to manufacture a huge range of products from tyres and tubes, industrial components and medical goods to footwear and clothing. Special grades of NR produced by Malaysia include SUMAR (non-smelly rubber), ENR (epoxidized NR), DPNR (deproteinized NR), and PA/SP (superior processing rubber).

⚙	Many in tropical countries	
⟲	• Renewable material	325
	• Versatile natural polymer	

PHA

PHAs are biodegradable plastics derived from plants or bacteria that are water-soluble and easily recycled. PHAs are suitable for medical and food-packaging uses.

⚙	Metabolix, Inc., USA	318
⟲	• Renewable materials	325
	• Compostable and recyclable	324

PLA

Plastics manufacturers all over the world are examining the commercial viability of making plastics using renewable resources. In 2000 Cargill Dow Polymers announced that their 'NatureWorks Technology' had created a new bioplastic called polyactide (PLA), derived from the maize plant. A new factory in Blair, Nebraska, is producing up to up to 150,000 tonnes per annum. PLA has attracted interest from the packaging and computer industries. Questions remain over the lifecycle benefits of PLA and the use of GM maize as source material.

⚙	Cargill Dow, USA	313
♻	• Renewable materials	325
	• Compostable	324

Tenite Cellulose Acetate

Eastman manufactures a range of biodegradable polymers from cellulose acetate.

⚙	Eastman Chemical Company, USA	314
♻	• Renewable materials	325
	• Compostable	324

Fillers/Insulation

Chanvrisol, Chanvribat

Loose-fill and blanket insulation is made by combining cellulose (wood) fibres with hemp fibre. Chanvrisol is loose-fill insulation, Chanvribat is supplied in a roll and has a thermal conductivity of 0.049W/K.

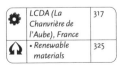

⚙	LCDA (La Chanvrière de l'Aube), France	317
♻	• Renewable materials	325

Heraflax

Long and short flax plant fibres are separated; the former are used for weaving linen and the latter are manufactured into insulation battens and quilts. In the Heraflax WP battens and Heraflax WF quilt the fibres are integrated with polyester fibres to form standard 60 mm- or 80 mm- (2¼ or 3⅓ in-) thick products. Both materials are good insulators with a thermal conductivity of 0.42 W/sq. m.

⚙	Deutsche Heraklith GmbH, Germany, and Österreichische Heraklith Gmbh, Austria	313
♻	• Renewable materials	325
	• Energy-saving product	327

Hypodown

Fibres from the milkweed plant provide the raw materials for this company's range of hypoallergenic down products, Hypodown, suitable for use in bedding and upholstery.

⚙	Ogallela Down Company, USA	319
♻	• Renewable materials	325
	• Improved health	327

Thermafleece™

Thermafleece™ is made from sheep's-wool fibres that can absorb and desorb water vapour without changing their excellent insulation properties. Manufactured in batts of 50, 75 or 100 mm (2, 3, or 4 in) thickness, 1,200 mm (47 in) length and 400 mm or 600 mm (15½ or 23½ in) width, they can be layered according to insulation needs in walls, roofs or floors. Water absorption is 100%, relative humidity is 40% and the thermal conductivity (K-value) is 0.039 W/m.K, similar to other natural-fibre insulation materials. Thermafleece™ requires much less energy to manufacture than glass-fibre insulation. The embodied energy is 5.45 MJ/kg – 14% of that needed to make glass fibre batts. Fire resistance of wool is also higher than cellulose and cellular plastic insulants. Naturally derived additives are used for insect proofing and fire resistance, so Thermafleece™ does not contain permethrin, pyrethroids, pesticides or formaldehydes. It can thus be readily re-used and recycled.

⚙	Second Nature UK Ltd, UK	320
♻	• Renewable material	325
	• Non-toxic, safe	324

Thermo-Hanf®

Hemp is an ideal crop for all aspiring organic farmers. It does not require the application of any herbicides or insecticides, it is a good weed suppressant, helping to clean the land, and it is a prolific producer of biomass and fibre, growing up to 4 m (13 ft) high in 100 days. Hemp cultivars with minimal active 'drug' chemicals have been grown in Germany since 1996 specifically for the production of this new insulation material. Fibres are extracted from the harvested plants and reworked into panels using 15% polyester for support and 3–5% soda for fireproofing. It is suitable for insulating between stud walls and roofing timbers. Thermo-Hanf® (thermo hemp) conforms to all DIN-Norm standards and has a thermal conductivity of 0.039 W/mk for DIN 52612. It also has in-built resistance to insect attack from the plant's own natural defences.

⚙	Hock Vertrieb GmbH/ Swabian ROWA, Germany	316
♻	• Renewable material	325
	• Clean production	325

Paints/Varnishes

Auro paints, oils, waxes, finishes

Auro manufactures an extensive range of 'organic' paints, oils, waxes, stains and other finishes without the use of fungicides, biocides or petrochemicals including white spirit (an isoaliphate). Oils originate from renewable natural sources such as ethereal oils, balm oil of turpentine or oil from citrus peel, so waste from the manufacturing process is easily recycled and the potential health hazard of the finished products is less than in petrol or isoaliphatic-based manufacturing systems. Emulsion paints for interior use include white chalk and chalk casein paints, which can be tinted using pigments from a range of 330 colours. Exterior-grade gloss paints and stain finishes are suitable for applying to wood, metal, plaster and masonry.

	Auro Pflanzenchemie AG/Auro GmbH, Germany	311
	• Non-toxic ingredients	324
	• Clean production	325

BioShield paints, stains, thinners, waxes

BioShield Paint Company manufactures a diverse range of paints, stains, thinners and waxes from natural ingredients such as oils from linseed, orange peel and soy bean.

	BioShield Paint Company, USA	312
	• Renewable materials	325

Earthnes

This company bulk-manufactures dyes from natural sources to supply other industries with alternatives to synthetic colourants.

	Color Trends, Inc., USA	313
	• Renewable materials	325

Livos

In 1975 Livos developed techniques for dispersing ingredients in natural resins. Its range includes natural-based primers (with linseed oil), hardening floor agents (pine tree resins), transparent glazes (phytochemical oils such as citrus), wood polishes (beeswax), wall glazes (beeswax and madder root) and varnishes. Livos URA Pigment Paint comprises organic beeswax, linseed/

stand oil, orange-peel oil and dammar mixed with water, methylcellulose, isoaliphate, ethanol, iron oxide, mineral pigments, borax and boric acid. Colour varies according to the amount of pigment.

	Livos Pflanzenchemie, Germany	317
	• High content of natural, renewable materials	325
	• Low or no VOC content	326
	• EU eco-label for some products	328

Milk-based paints

Traditional milk-based paints, suitable for interior design, restoration and furniture production, are made by this company. These paints follow authentic recipes and are free of synthetics.

	Old Fashioned Milk Paint Co., USA	319
	• Renewable materials	325

The Natural Choice

All paints and finishes in the Natural Choice collection utilize natural oils and solvents from citrus peel or seeds, resins from trees, waxes from trees and bees, inert mineral fillers and earth pigments. Oils are extracted

by cold pressing or with low heat and all products are packaged in biodegradable or recyclable containers.

	BioShild Paint Co., USA	312
	• Renewable materials	325
	• Reduced pollution manufacturing	326
	• Recyclable packaging	327

Nutshell®

Nutshell produces a full range of adhesives, paints, herb and resin oils, varnishes and stains with natural pigments.

	Nutshell Natural Paints, UK	319
	• Renewable materials	325
	• Non-toxic	324
	• Clean production	325

OS Color

Waxes from carnauba and candelilla plants and oils from sunflower, soy bean, linseed and thistle are the raw ingredients of a wide range of natural stains and protective finishes for exterior and interior wood surfaces. OS Color wood stain and preservative is a natural oil-based, microporous, water-repellent treatment for timber exposed to the

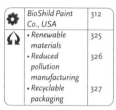

weather. The natural oils, water-repellent additives and lead-free siccatives (drying agents) form the binder, which comprises almost 85% of the solids content. This binder is mixed with the active (bacteria and fungi) protective ingredients, low-odour solvents (benzole-free, medical-grade white spirit) and pigments (iron oxide, titanium dioxide). Floor treatment, such as the OS Color hardwax-oil, is an oil-based application. It doesn't contain biocides or preservatives. Manufacturing plants are covered by ISO 9001 and ISO 14000.

	Ostermann & Scheiwe, Germany	319
♻	• Reduced toxins (solvents, VOCs, biocides, preservatives and citrus oils)	326
	• Natural, renewable raw materials	325

Papers/Inks

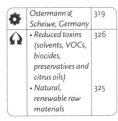

Continuum

Old denim jeans, worn-out money notes and industrial cotton waste are recycled in a diverse range of papers. Zenus Crane's mill has been recycling waste textiles and paper since 1801. The tradition

continues with the Continuum brand of tree-free papers using 50% cotton fibre and 50% hemp fibre.

	Crane & Company, USA	313
♻	• Renewable resources	325
	• Conservation of forest resources	

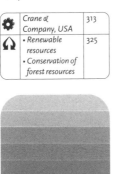

'Context' and other Paperback papers

There are tens of paper manufacturers and distributors in the UK that make recycled papers but Paperback offers the most extensive range of gloss- and matt-coated papers, uncoated offsets, letterheads and speciality grades manufactured from recycled waste paper. This process consumes less than half the energy required to make paper from virgin wood pulp. The company was set up in 1983 and is committed to encouraging use of recycled paper to decrease the disposal of six million tonnes of waste paper annually in the UK. Boards range in weight from 225 gsm to 300 gsm with a variety of finishes from smooth white watermarked to natural-coloured micro-fluting. All 'Context' papers and boards contain 75%-de-inked used waste to a

NAPM approved grade and 'Context FSC' is made from 75%-de-inked fibre and 25%-FSC-endorsed pulp.

	Paperback, UK	319
♻	• Recycled materials	324
	• Reduction in embodied energy	325
	• NAPM approved	336
	• Stewardship sourcing, FSC	325

ECO range

Curtis Fine Papers produce a wide range of uncoated paper made with 80% Forest Stewardship Certified (FSC) virgin fibre and 20% fibre derived from post-consumer waste. The virgin fibre bleaching process at the pulp mill is totally chlorine free (TCF – no chlorine gas or chlorine dioxide) or elemental chlorine free (ECF). The company has ISO 14001 certified mills, EMAS accreditation and yearly environmental reports. The ECO range includes Smooth (Retreeve, Conservation), Detail (Scotia), Clear (Scotia, Retreeve) and Hot (seasonal collections), so there's a great choice of finishes and 28 colours available in 100 g/sq. m to 280 g/sq. m. All ECO papers are NAPM (National Association of Paper Merchants) approved recycled papers.

	Curtis Fine Papers, UK	313
♻	• Recycled content	324
	• NAPM certified	336
	• Certified sources	324

Office and sanitary paper

Over four hundred tree-free papers, made from plant fibres and recycled waste paper, are available from this manufacturer.

	New Leaf Paper, USA	318
♻	• Renewable and recycled materials	324/ 325
	• Conservation of forest resources	

Tree-free paper

A range of papers is made from natural plant fibres, such as cotton and hemp, and post-consumer paper waste.

	Green Field Paper Company, USA	316
♻	• Renewable and recycled materials	324/ 325
	• Encourages forest resource conservation	

Treesaver range

This company produces a vast range of craft, packaging grade and printing papers. The Treesaver range uses 100%-waste paper to create recycled papers such as MG Greentreesaver

Kraft, MG Green Envelope and MG Treesaver Plus Kraft used in the manufacture of envelopes.

⚙	Smith Anderson & Co. Ltd, UK	321
♻	• Recycled content	324

Vanguard Recycled Plus™

This tree-free, bond-quality paper is manufactured from 10% hemp/flax and 75% post-consumer waste paper.

⚙	Living Tree Paper Company, USA	317
♻	• Renewable and recycled materials	324/325
	• Conservation of forest resources	

Vision® and Re-vision® printing paper

Kenaf fibre is the principal raw material for the manufacture of a range of 100% tree-free and chlorine-free printing papers.

⚙	Vision Paper, KP Products, USA	322
♻	• Renewable materials	325
	• Conservation of forest resources	

Printing inks

Alden & Ott manufacture a range of heat-set soy-based inks with about 20–25% soy content and colour pigments avoiding the use of heavy metals.

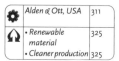

⚙	Alden & Ott, USA	311
♻	• Renewable material	325
	• Cleaner production	325

EcoPure

EcoPure is a range of inks derived from soy beans. The company also produces a diverse range of water-based flexographic inks and specialist inks for printing on metal.

⚙	Inx International Ink Co., USA	317
♻	• Renewable materials	325

Printing inks

An extensive range of vegetable-based inks is available for offset and lithographic printing.

⚙	Flint Ink, USA	315
♻	• Renewable materials	325

Soy bean inks

This company manufactures a diverse range of inks derived from soy beans.

⚙	Ron Ink Company, USA	320
♻	• Renewable materials	325

Textiles

Argyll Range

Designer Jasper Morrison has built on a long

Scottish tradition of weaving woollen textiles by creating a new range of furnishing fabrics for Bute Fabrics in vibrant, contemporary colours, yet the durability and warm surface textures associated with traditional crafted products are retained. Bute Fabrics source much of their raw materials locally and adopt clean production, minimizing the use of harmful substances during processing, as an integral part of their environmental policy. These fabrics are suitable for restoration projects and for new furniture.

⚙	Bute Fabrics Ltd, UK	312
♻	• Renewable materials	325
	• Clean production	325

Bincho-Charcoal Border

Charcoal has long been known as an agent to filter and purify air and water. It also has good insulating characteristics. This predominantly wool–silk fabric has a border of charcoal fibres incorporated to take advantage of its positive properties. (Wool 39%, silk 29%, nylon 11%, polyurethane 6%, wood charcoal 15%.)

⚙	Reiko Sudo & Nuno Corporation, Japan	319
♻	• Natural purifying agents	
	• Some renewable fibre	325

Cantiva™

Hemp is a very strong natural fibre, naturally resistant to salt water, mould, mildew and UV light, and its use in China is documented through ten thousand years. Tens of different pure hemp or hemp/ natural-fibre fabrics are designed by Hemp Textiles International using the Cantiva™ brand hemp fibre Hemptex®. Fabrics range from heavy-duty pure hemp canvas weighing 620 g per sq. m (18.3 oz per sq. yd) to lightweight hemp/silk or hemp/cotton mixtures weighing between 92 and 193 g per sq. m (2.7 and 5.7 oz per sq. yd). Bulk or wholesale orders are produced in contractual arrangements with a Chinese mill.

⚙	Hemp Textiles International Corporation, USA	316
♻	• Renewable materials	325

Climatex® LifeguardFR™

Following the success of Climatex® Lifecycle™ series of upholstery fabrics, Rohner Textil has continued setting high standards in ecological textile design by developing a range of fire-retardant upholstery fabrics called Climatex® LifeguardFR™. Eco-design tends to involve much larger groups of stakeholders than conventional industrial design. Rohner Textil recognized this by collaborating with the independent German environmental institute, EPEA, Clariant, a leading chemical producer, and fibre manufacturer Lenzing AG, undertaking extensive laboratory trials to understand the full environmental impacts of the proposed flame retardants. Climatex® LifeguardFR™ emerged as probably one of the most advanced ecological textiles with fire retardant certification meeting standards worldwide. Climatex® LifeguardFR™ is made of wool and the cellulose fibre Redesigned LenzingFR™ is extracted from beech trees. The Colors Kollektion, using Climatex® LifeguardFR™ uses environmentally sound chemicals and 16 dyes from Ciba developed

with EPEA for the Climatex® Lifecycle™ series. Now any designer or manufacturer of public seating has no excuse to specify furnishing textiles that create high negative environmental impacts – there are viable alternatives.

⚙	Rohner Textil AG, Switzerland	320
♻	• Avoidance or toxic and hazardous manufacturing	325
	• Safe chemicals and dyes	327
	• Durable	324

Green Cotton®

Well before 'organic' became the adjective of the late 1990s, companies such as Novotex were re-examining the sustainable features of their business. Sources of raw materials were analyzed and it was discovered that hand-picked cotton from pesticide-free South-American sources required less cleaning than intensively grown 'commercial' cotton. Long-fibre cottons were selected to provide a yarn that could be woven to facilitate dyeing with water-based dyes and reduce chemical additives throughout the production process. As a result Green Cotton is free of chlorine, benzidine and formaldehyde. Waste water generated in processing

is chemically and biologically cleaned in situ. Supply-chain management, cleaner and quieter production have also created a healthier environment for employees.

⚙	Novotex A/S, Denmark	319
♻	• Clean production of 'organic' textiles	325
	• Reduced toxins	326

Green fabric (Eco-green fabric)

This is a new, fully biodegradable, maize-starch fibre developed by Mitsui Chemical and Kanebo Synthetics in Japan. It is fully compostable by micro-organisms to release water and carbon dioxide. Using a Dobby loom, threads of the fibre are 'overspun' to create a delicate crepe in 800 mm (31½ in) widths.

⚙	Reiko Sudo and Nuno Corporation, Japan	319
♻	• Biodegradable textile	324

hemp textiles

The hemp plant is said to have over 50,000 documented uses, although the media would have you thinking its primary use was ingestion or smoking for its narcotic effects. Industrial varieties of hemp, however, contain no THCs, the active narcotic ingredient, and so are safe to grow. Industrial hemp is, in fact, a fantastic cash crop that requires little or no fertilizer, suppresses weeds and produces 5–10 tonnes of fibre per hectare. It is a durable fibre suitable for rope making, webbing and canvas. In the fashion arena it has long been associated with 'hair shirts' and hippies, which has led to a significant failure to recognize the full potential of hemp textiles. Hemp fibre is often mixed with cotton, silk and wool. The range of textures, colours and drapeability of hemp textiles is diverse, as the catalogue of textiles at Hemp Traders attests.

⚙	Hemp Traders, USA	316
♻	• Renewable resource	325

Ingeo fibre

Cargill Dow is one of the leading manufacturers of polylactide acid (PLA), a biodegradable polymer. Ingeo is a new textile fibre made from corn (maize) plants. After harvesting the corn is fermented to release the plant sugars, which are then converted into polylactide acid through the patented

NatureWorks™ PLA process. This PLA is extruded as Ingeo™ fibre that can be applied to the manufacture of furnishing fabrics, fibrefill, carpets and non-wovens. Currently Ingeo™ is co-operating with Diesel and the Its# International fashion show to create a range of modern apparel. At end-of-life Ingeo is completely biodegradable.

⚙	Ingeo/Cargill Dow, USA	316
↻	• Renewable annual crop fibre	325
	• Compostable	324

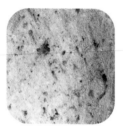

TENCEL®

TENCEL® is a modern textile that uses natural raw materials in the form of 'lyocell' cellulose fibre derived from wood pulp harvested from managed forests. This lyocell fibre is processed through the unique TENCEL® 'closed loop' solvent spinning process, which is economical in its use of water and energy and uses a non-toxic solvent that is continuously recycled. The resultant TENCEL® fibre is soft, breathable, absorbent and fully biodegradable. Luxurious surface finishes are achieved by abrading the wet fibres, a technique called fibrillation. A wide

variety of fibrillated or non-fibrillated (TENCEL A100) finishes is achievable. TENCEL® filament is suitable for knitted and woven fabrics, is softer in feel yet stronger than cotton and provides a good surface for printing and dyeing. Many of the world's leading fashion designers have taken advantage of the versatility of fabrics woven with TENCEL® yarn.

⚙	Acordis Fibres (Holdings) Ltd, UK	311
↻	• Renewable, compostable materials	324/325
	• Clean production	325
	• Low-energy production	325

Terrazzo Felt 'Colour Chips'

This non-woven, needle-punched, blanket-type fabric fuses dye-chips into a 100%-natural-coloured alpaca-wool felt over a core of polyester organdy to produce unique pieces of material.

⚙	Nuno Corporation, Japan	319
↻	• Renewable and recyclable materials	324/325

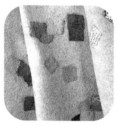

Terrazzo Felt 'Nuno'

Industrial-waste snippets of various Nuno fabrics and 'outtakes' in raw wool

are combined in a needle-punched technique to create an interesting textured terrazzo effect. The constituents are 85% alpaca wool with 15% Nuno production waste.

⚙	Nuno Corporation, Japan	319
↻	• Renewable and recycled materials	324/325

Miscellaneous

Bean-e-doo™

Franmar Chemical, Inc. manufacture a multipurpose, industrial-strength cleaner, Bean-e-doo™, derived from soy beans.

⚙	Franmar Chemical, Inc., USA	315
↻	• Renewable materials	325
	• Reduction in toxic chemicals and VOCs	326

Bio T®

Bio T® is a general-purpose cleaner derived from terpene, which is suitable for use in the manufacturing industries and public-sector maintenance.

⚙	BioChem Systems, USA	312
↻	• Derived from renewable materials	325

ECOSTIX™

A whole new family of starch adhesive biopolymers called ECOSPHERE™ has been created by Ecosynthetix by redesigning starch molecules to form high solids dispersion in water without resorting to the use of heat and caustic chemicals. Tack and drying times are much faster than traditional starch adhesives. ECOSTIX™ are water-borne pressure sensitive 'smart' adhesives (PSAs) whose adhesive properties turn off when subject to specific external conditions. Using patented ECOMER™ technology these smart adhesives solve the problem of recycling millions of discarded 'stickies'. The United States Postal Service has approved ECOSTIX™ as part of its Benign Stamp Program. The environmental footprint of using ECOSTIX™ with paper is reduced by improving recycling pulpability, lower VOCs and a biodegradable content.

⚙	Ecosynthetix, USA	314
↻	• Renewable, compostable materials	324/325
	• Reduced VOCs	326
	• Improved recyclability	328

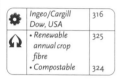

Fern vine (Yan lipao)

This abundant jungle fern is found in southern Thailand where it has supported the vibrant Yan lipao basketry industry for over 150 years. After removal of the outer layer, the pith is dried in the shade, then polished and smoothed, giving a characteristic black or brown colour. It is tough, durable and versatile as a weaving material. In recent years, Italian companies, such as Lino Codato, have created new furniture collections using the woven vine over hardwood frames thereby encouraging continuity of the skill and knowledge while offering European consumers contemporary furniture in natural materials. Any significant increase in world demand may require careful management of jungles to prevent depletion of stocks, so Yan lipao production should be monitored and sustainable harvesting adopted where possible.

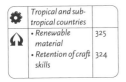

⚙	Tropical and sub-tropical countries	
🎧	• Renewable material	325
	• Retention of craft skills	324

Soy Clean

Soy Clean is a range of biodegradable, non-toxic cleaners and paint removers derived from soy beans.

⚙	Courtney Company, USA	313
🎧	• Renewable materials	325
	• Non-toxic	324

SOY Gel™

Removal of paints, urethanes and enamels from timber surfaces is facilitated by application of SOY Gel™, made from 100% American-grown soy beans. This stripper is odourless and easy to use. Simply apply it to the surface to be treated, allow it time to work into the paint, then strip it off with hand-operated scrapers. Like most strippers it is a caustic and potentially harmful substance but it does not have the high VOC content of synthetic-based strippers and originates from annually grown soy beans.

⚙	Franmar Chemical Inc., USA	315
🎧	• Renewable material	325
	• Reduced off-gassing	

Sundeala and Celotex Sealcoat

Sundeala is a soft board manufactured from unbleached recycled newsprint available in a range of natural colours. Celotex Sealcoat Medium Board is also made from recycled newsprint and is coloured with natural mineral pigments. Both boards are suitable for interior applications, pinboards, noticeboards, exhibition displays and furniture.

⚙	Sundeala Ltd, UK	321
🎧	• Recycled materials	324
	• Clean production	325

Water hyacinth (Eichornia crassipes)

Originating from tropical and sub-tropical regions of South America, the water hyacinth has invaded the southern United States and many other areas of the globe. This invasion results in the extinction of local flora and fauna and causes flooding by blockage of natural and man-made water systems. It is remarkably productive, one plant having the ability to produce 1,200 daughter plants in just four months. This biomass is now being harvested, prepared and dried to produce a tough, light but biodegradable weaving material. It is suitable for use as natural undyed material but can be dyed for different aesthetic effects. Italian companies, such as Lino Codato, have created new furniture collections using water hyacinth and, in doing so, encourage local conservation work and provide employment.

⚙	Tropical and sub-tropical countries	
🎧	• Renewable material	325

Boards and Sheeting

Bottle range, type d

Commingled recycled plastic sheeting, of varying thicknesses, is becoming a familiar material to designers. One is high density polyethylene (HDPE) from post-consumer bottles. Typically the re-manufacturing process produces flow patterning and streaks towards the edges of the sheet. Colour palettes vary according to the supply of recycled bottles and between batches. Type d is known as a Regular UK mix, as collected and unsorted, but more pastel or natural colours are also possible.

⚙	Smile Plastics, UK	321
⍥	• 100% recycled HDPE	324

ChoiceDek®

This product is manufactured from a mixture of recycled polyethylene (HDPE and LDPE) plastics and waste wood fibre. ChoiceDek are plank sections suitable for decking.

⚙	Advanced Environmental Recycling Technologies (AERT), USA	311
⍥	• Recycled materials	324

FlexForm®

This hybrid composite combines natural fibres, kenaf and hemp, with synthetic polymers, polypropylene or polyester, in custom-blended compositions. FlexForm® is a non-woven composite available as a mat, a low-density board (LD) for 2D and 3D applications ideal for variable geometrical forms and high-density hardboard (HD) at densities of 450 g/sq. m to 2,400 g/sq. m. The manufacturing plant permits continuous sheet production or rolls to 3.5 m (11½ ft) width. 2D lamination multi-layer compositions of mat, LD and HD products are available between 2,000–5,000 g/sq. m and up to 1.5 m (5 ft). Polymer content is increased where specific plasticity is required. This is a versatile group of composites with flexible moulding, variable specifications and lower VOC emissions. FlexForm® is recyclable and factory trim is recovered and reused.

⚙	FlexForm Technologies, USA	314
⍥	• Renewable materials fraction	325
	• Partly biodegradable	324
	• Recyclable	324

Stokbord™ and Centriboard

Stokbord™ is a smooth or embossed low-density polyethylene (LDPE) sheet available in standard thicknesses of 6, 9, 12 and 14 mm (about ⅕ to ½ in). It is constituted from 40–50% post-consumer waste and 50–60% industrial/commercial waste. Centriboard is available in three grades: L – a smooth LDPE sheet, 1.5 mm to 18 mm thick (¹⁄₂₀ to ⁷⁄₁₀ in); H – smooth HDPE sheet, 2 mm to 6 mm thick (about ¹⁄₁₂ to ¼ in); and P – smooth polypropylene sheet, 2 mm to 6 mm thick.

⚙	Centriforce Products Ltd, UK	313
⍥	• Recycled materials	324
	• Reduction in embodied energy (compared with virgin plastics)	325
	• Encouraging conservation of timber resources	

Origins Solids

Commingled, recycled post-consumer polythene plastics are extruded into sheets suitable for a wide range of uses from packaging to laminates. A huge range of colours are available.

⚙	Yemm & Hart, USA	323
⍥	• Recycled content	324

Origins Patterns

Recycled HDPE is used to manufacture new plastic boards suitable for a wide variety of applications.

⚙	Yemm & Hart, USA	323
⍥	• Recycled content	324

Tectan

Used drinks cartons are mixed with industrial scrap from the carton-manufacturing plants under the Duales System Deutschland scheme to provide the ingredients for this tough, water-resistant board. The raw material is shredded, then compressed under heat and pressure, causing the polyethylene fraction to melt and bond the particles. It is suitable for building and furniture manufacturing.

⚙	Tetrapak Ltd, UK	322
⍥	• Recycled materials	324

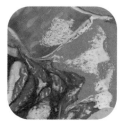

Wellies

There is a substantial market for children's wellingtons. They are produced in bright primary colours and many are branded products related to toys or other children's products. As such, many wellingtons lead very short lives before being discarded. Smile Plastics brings thousands of these wellies back into the recycling loop. Old wellies are shredded and subject to heat and up to 1,000 tonnes of pressure to produce flexible, soft, rubber 2 mm-thick water-resistant sheets suitable for a wide variety of applications.

⚙	Smile Plastics, UK	321
♺	• Recycled rubber	324

Dapple

Industrial foodstuff containers are shredded and mixed with Smile Plastics' own factory waste.

The 'black veins' in this commingled recyclate are added during the process. There is variation between batches but special mixes can be made to order.

⚙	Smile Plastics, UK	321
♺	• 100%-recycled plastics	324

Plastic profiles

DPR Plaswood

Reclaimed polythene and polypropylene – 30% waste from supermarkets and 70% production factory waste – are reblended into extruded profiles suitable for uses requiring tough, rot-free materials.

⚙	British Polythene Industries, UK	312
♺	• Recycled materials	324
	• Reduction in embodied energy (compared with virgin plastics)	325
	• Encouraging conservation of timber resources	

DURAT®

DURAT® is a smooth, silky surfaced, waterproof polyester-based plastic containing 50% recycled material. It is available in a standard range of forty-six colours including pure bright hues and terrazzo- and granite-like abstract patterns. Tonester use DURAT® for a wide range of bathroom and kitchen fittings and fixtures. It is fully recyclable.

⚙	Tonester, Finland	322
♺	• Recycled content	324
	• Recyclable	324

Durawood

Durawood is a high-density material available in a range of rectangular profiles, which is made entirely from recycled plastics. It is especially suited to the manufacture of street and outdoor furniture.

⚙	Durawood, UK	314
♺	• Recycled content	324

DUROSAM®

Jute is an important natural fibre for agriculture and industry in India, Bangladesh and China, often used as an eco-friendly fibre for sacking, packaging materials and twine. Like hemp, sisal and coir, it is durable and strong but, for a number of years, these fibres have been replaced by synthetic fibres. AB Composites set themselves the task of finding new applications for jute. The outcome is Natural Fibre Thermoset

Composite DUROSAM®, a composite of jute and polymer and a unique system of PREPEG (a method of preparing and treating the polymer during manufacturing). A fire-proof material from jute composite is being used for the manufacture of fire-proof coaches for Indian Railways.

⚙	AB Composites Pvt Ltd, India	311
♺	• Encourages agricultural cash crop	
	• Based on renewable material	325

Epoch

Commingled, recycled HDPE plastics are extruded to form rectangular, square or plank sections suitable for a multiplicity of uses in street and garden furniture.

⚙	Ultra Plastics Ltd, UK	322
♺	• Recycled content	324

Govaplast®

A range of square, round, rectangular and tongue-and-groove profiles is produced using recycled polyethylene and polypropylene plastics. Colours include charcoal grey, grey–green and mid-brown. The tongue-and-groove is used in everything from fabrication of equestrian buildings to outdoor planters.

⚙	Govaerts Recycling NV, Belgium	315
♺	• Recycled and recyclable content	324

Holloplas

To date Centriforce has supplied more than 150,000 tonnes of recycled finished products to construction, industrial, agricultural and recreational markets in over thirty countries. It offers an extensive range of hollow extruded profiles using a blend of recycled plastic from waste from retail distribution (40–50%) and industrial and commercial waste including film, pipe and packaging (50–60%). Standard sections are suitable for decking, tongue-and-groove flooring, fencing, railings and street furniture.

⚙	Centriforce Products Ltd, UK	313
♺	• Recycled materials	324
	• Reduction in embodied energy (compared with virgin plastics)	325
	• Encouraging conservation of timber resources	

Mobiles

There are now about 45 million mobile phones in circulation in the UK and an estimated 15 million are discarded annually, many ending up in shredders and then landfill sites. A phone fashion industry has emerged, encouraging users to customize their phones with colourful new plastic covers. New regulations, in the form of the WEEE (Waste Electrical and Electronic Equipment) Directive from the European Union, will shortly require all manufacturers of mobile phones and a wide range of electronic equipment to take responsibility for recycling and safe disposal of these products. Recycling the plastic covers to create new materials is one solution. This process requires no additives or resins, just the old phone covers, and it

⚙	Smile Plastics, UK	321
♺	• Recycled materials	324

produces an eclectic range of colours according to the nature of the waste streams.

Plastic lumber

Commingled, recycled plastics are extruded into a variety of rectangular sections, making an alternative material for traditional uses such as decking and outdoor furniture.

⚙	Yemm & Hart, USA	323
♺	• Recycled content	324

Plastic profiles

Profiles and stakes in a variety of round and square shapes are manufactured from recycled plastics.

⚙	WKR Altkunst-stoffproduktions-u. Vertriebsgesell-schaft mbH, Germany	323
♺	• Recycled content	324

Plastic profiles

A variety of round and square profiles and stakes are made from recycled plastics.

⚙	Hahn Kunststoffe GmbH, Germany	316
♺	• Recycled content	324

Polymers

Acousticel

Recycled rubber fibre and granules are bonded with a special latex to bitumenized felt-cardboard to make rolls or panels suitable for sound insulation on floors and walls. R10 is a 10 mm (½ in) thick roll weighing 3 kg/sq. m and M20AD is supplied in 1 sq. m panels weighing 15 kg/sq. m. Sound insulation of 43dB to 45dB is achievable when applying M20AD to single-thickness concrete block walls and fixing two layers of 12.5 mm plasterboard over the panels. Cork granules are added where better thermal insulation is needed.

⚙	Sound Service (Oxford), UK	321
♻	• Recycled and renewable materials	324/ 325

Correx Akylnx

This lightweight twin-walled PP sheet is made from 100% production and customers' returned waste. It is utilized for packaging, self-assembly storage systems and tree shelters.

⚙	Kaysersberg Plastics, France	317
♻	• Recycled materials	324

EcoClear®

EcoClear® is a resin and film made from recycled PET, which is suitable for beverage and food packaging.

⚙	Wellman, Inc., USA	322
♻	• Recycled content	324

RecyTop

This CFC-free protection and drainage sheeting,

22–35 mm (1–1½ in) thick is used in civil engineering. It is made of closed-cell polyethylene (PE) foam from recycling waste. Individual foam flakes are heat-bonded without further additives or adhesives.

⚙	Schmitz Recycling, the Netherlands and SSP, Specialised Sports Products, UK	320
♻	• Recycled materials	324

Rubber granulate

This company manufactures rubber granulate, 0.5 mm to 30 mm (up to 1 in) diameter particles, from 100%-reclaimed scrap tyres. The granulate can be bonded with virgin natural or synthetic rubber and elastomers and is ideal for play surfaces or other uses to reduce impact damage.

⚙	Charles Lawrence Recycling Ltd, UK	313
♻	• Recycled materials	324

Flo-Pak® and Super 8®

These figure-of-eight loosefill packaging chips are made from 100%-recycled polystyrene. The form permits interlocking, improves packing ability and offers improved protection. Super 8® is the heavier grade.

⚙	FP International, USA	315
♻	• 100%-recycled	324
	• Recyclable	324

SoyOyl® Biosynthetic polymers

USSC manufactures a range of specialist polyurethane foams using soy-bean oil. Like synthetic PU, the USSC foams are suitable for everything from loose-fill packaging to panels and shoe parts.

⚙	Urethane Soy Systems Company (USSC), USA	322
♻	• Renewable content	325

Paints

Biora, Aqua range

Biora is a range of water-based acrylic resins suitable for application to walls, ceilings and other interior surfaces. Qualifying for the EU eco-label, these eight paints and varnishes have less environmental impact than conventional paints, especially VOCs and toxic ingredients. Teknos are also certified to ISO 9001 and ISO 14001 and are working with the Swedish Paint Makers Organization to develop tools, such as lifecycle analysis, to make further improvements.

⚙	Teknos Group Oy, Sweden	321
♻	• Cleaner production	325
	• EU eco-label ensuring low toxicity of constituents	328

Ecos

This company claims to manufacture the only solvent-free odourless paints and varnishes in the world, with zero VOC content, independently tested by the US EPA and the Swedish National Testing & Research Institute. The Ecos range is, however, based upon synthetic resins, albeit non-allergenic, harmless resins, processed from crude oil, so it is not from a renewable source.

⚙	Ecos Organic Paints, UK	314
♻	• Free of toxins (VOCs and vinyl chloride)	326

Innetak and Bindoplast

At the paint manufacturing plant at Malmö, Sweden,

Akzo Nobel produce over 30 million litres (6.6 million gallons) of decorative coatings and 16 million litres (3.5 million gallons) of industrial coatings. Since 1995 the company has set itself a series of environmental targets, such as reducing the emissions of solvents to the air by 50% between 1995 and 1999 and reducing the total energy consumption per litre of paint manufactured by 5% between 1995 and 2000. Innetak and Bindoplast are decorative, water-based emulsion paints, which were the first brand in Europe to receive the EU eco-label.

suitable for all mineral substrates, Concretal protects concrete against corrosion and Biosil is a water-borne, silicate-based paint suitable for interior applications. Ecosil is a recently introduced interior-quality paint, which is water-based, contains no chemical solvents and is VOC-free. Keim are certified to ISO 14001 and ISO 9001.

⚙	Keim Mineral Paints Ltd, UK	317
♻	• Abundant geosphere materials	324
	• Non-toxic	324

⚙	Nordsjö (Akzo Nobel Dekorativ), Sweden	319
♻	• EU eco-label ensuring low VOCs and general reduction in toxicity	328
	• Reduction in emissions to air	326

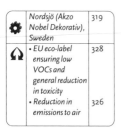

Keim paints

All the paints manufactured by Keim use inorganic materials that are abundant in the geosphere, including potassium silicate binders, mineral fillers and earth oxide colour pigments. Granital is an exterior paint with a range of 350 colours

Pinturas Proa

A range of water-borne, vinyl polymer interior paints containing less than 45% volatiles is certified with an EU eco-label. The company is also registered with the Spanish eco-label certification authority, AENOR, and participates in the Punto Verde (Green Dot) packaging disposal scheme.

⚙	Pinturas Proa, Spain	319
♻	• EU eco-label	328
	• Reduction in toxic substances	326
	• Recycled and recyclable packaging	327

Textiles

Elex Tex™

Conductive fibres are woven with traditional, natural yarns to create a flexible textile suitable for a variety of applications such as electronic clothing, roll-up keyboards and so on.

⚙	Eleksen Ltd, UK	314
♻	• Dual-function material	327

EcoSpun®

Wellman is one of the world's leading manufacturers of yarn and textiles using PET from recycled drinks bottles. Wellman supply the furnishing and clothing industries, including Patagonia, the outdoor clothing company. Ecospun® is a specialist fibre made using recycled plastics.

⚙	Wellman, Inc., USA	322
♻	• Recycled content	324

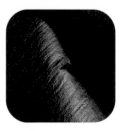

Otterskin

This 100%-polyester, non-woven, needle-punched fabric is made from recycled PET bottles. A surface coating of polyurethane provides wind- and waterproofing, yet the material is breathable and retains body heat.

⚙	Nuno Corporation, Japan	319
♻	• Recycled materials	324

Stomatex

Stomatex is a breathable fabric made of neoprene and polyethylene, which mimics transpiration, the process of evaporation of moisture from leaves. Perspiration vapour generated by the wearer is collected in small depressions on the inside of the fabric where tiny pores allow the vapour to escape to the outside. Stomatex is activated only by sufficient body perspiration, so this is a responsive, 'smart' textile.

	Stomatex, UK	321
	• Improved personal health with breathable fabric	327

Take/Bamboo Hexagonal Pattern

Personal hygiene is given a boost if you wear a garment made with Take, as it uses copper sodium-chlorophyll as a catalyst during the manufacture of this silk fabric to generate strong anti-bacterial and odour-suppressing properties. It is made from a mixture of natural and synthetic fibres (rayon 45%, silk 38%, nylon 10% and polyurethane 7%). The copper sodium-chlorophyll is extracted from brush bamboo (*Kuma sasa*) and is sold as a commercial digestive medicine, while the residues provide natural fibre. In a full lifecycle analysis Take will offer considerable reductions during the usage phase of any garment, since washing frequency can be reduced, saving water and detergent consumption.

	Kazuhiro Ueno, and Nuno Corporation, Japan	319
	• Natural anti-bacterial agents	
	• Some renewable fibre	325

Terratex

Made entirely of recycled PET recycled plastic bottles, Terratex is a tough, versatile, recyclable fabric for furnishing and similar applications.

	Interface, Inc., UK/USA	316
	• Recycled and recyclable materials	324

Trevira NSK/Trevira CS

This is a recyclable fabric made from two types of polyester yarn, Trevira NSK, which gives strength, and Trevira CS, which acts as a flame retardant. Being 100% polyester, it can be reworked by pleating, dyeing and printing and needs no flame-proof coating.

	Trevira GmbH & Co. KG, Germany	322
	• Recyclable	324
	• Cleaner technology	325
	iF Design Award, 2000	330

Tyvek

With its durability and high chemical resistance, Tyvek was originally developed by DuPont for protective clothing but has since been used for haute-couture fashion and as a paper substitute for envelopes, stationery and various printed media. Tyvek is fully recyclable.

	DuPont, USA	314
	• Recyclable synthetic material	324

Velcro®

Velcro® is a combination of two nylon fabrics, one woven with a surface of hooks and the other with a smooth surface with loops. When juxtaposed the two fabrics adhere, as the hooks take up in the loops, creating a strong 'adhesive' bond.

	Velcro, USA	322
	• Temporary bonding system allowing reuse of textiles	326

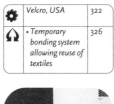

Waterfront, Messenger

Kvadrat is a manufacturer of high-quality modern woven and printed furnishing and curtain textiles for the contract and retail markets. Waterfront is a 100%-recycled polyester fabric with a wavy raised texture while Messenger is a close textured colourful range of fabrics using 78%-recycled polyester with 15% virgin polyester and 7% nylon. Kvadrat received ISO 14001 certification for environmental management in 1997.

	Kvadrat, Denmark	317
	• Recycled content	324

Miscellaneous

Syndecrete®

Syndecrete® is a chemically inert, zero out-gassing, concrete-like material composed of cement and up to 41% recycled or recovered materials from industrial or consumer waste. Typical wastes include HDPE, crushed recycled glass, wood chips and brass screw shavings. Pulverized fly ash (PFA), a waste residue from coal-fired power stations, is added to reduce the cement requirement by up to 15% and recovered polypropylene fibre scrap provides a 3D matrix to increase the tensile strength of this composite recyclate concrete. It is easily worked and polished to create a contemporary terrazzo look.

	Syndesis, Inc., USA	321
	• Recycled materials	324
	• Reduction in embodied energy of manufacture	325
	• Certified as a recycled product by the Californians Against Waste Foundation	328

4.0 Resources

Abalos & Herreros with Angel Jaramillo
C/ Gran Via, 16
3° Centro
28013 Madrid, Spain

Abelman, Jacques
Replican Design
134 bis rue de Charenton
75012 Paris, France
T +33 (0)672 435 407
E jabelman@ifrance.com

Accoceberry, Samuel and Antonio Cos
Viale Corsica 57/A
20133 Milan, Italy
E antonio-cos@tiscali.it

Aisslinger, Werner
Studio Aisslinger
Oranienplatz 4
D 10999 Berlin, Germany
T +49 (0)30 315 05 400
F +49 (0)30 315 05 401
E studio@aisslinger .de
W www.aisslinger.de

Akeler Developments plc
20 Berkeley Square
London W1X 5HD, UK
T +44 (0)20 7864 1800
F +44 (0)20 7864 1801/2
E info@akeler.com
W www.akeler.co.uk

Allard, Helena and Cecilia Falk
c/o Iform (p. 316)

Anthologie Quartett
D 49152 Bad Essen, Germany
T +49 (0)5472 94090
F +49 (0)5472 940940
E info@anthologiequartett.de
W www.anthologiequartett.de

Apotheloz, Christophe
Industrial Designer
Lörgernstrasse 27
CH 8037 Zurich,
Switzerland
T +41 (0)1 361 51 47
F +41 (0)1 361 51 97

Aquaball
c/o 21st Century Health
(p. 311)

Architetuurstudio Herman Hertzberger
Gerard Doustraat 220

1073 XB Amsterdam
the Netherlands
T +31 (0)20 676 58 88
F +31 (0)20 673 55 10
E office@hertzberger.nl
W www.hertzberger.nl

Arosio, Pietro
Studio Pietro Arosio
Via Gaetano Giardino 2/A
20053 Muggio (MI), Italy
T +39 (0)39 793 237
F +39 (0)39 278 1088
E studio@pietroarosio.it

Atfield, Jane
244 Grays Inn Road
London WC1X 8JR, UK
T +44 (0)20 7278 6971
F +44 (0)20 7833 0018

Atkins, Kelly
Carpet Burns
Britannia Mill
Mackworth Road
Derby D22 3BI, UK
T +44 (0)1332 594 044
E kelly@carpet-burns.com
W www.carpet-burns.com

Atwell, Tyson
c/o IDRA (p. 330)

Augustin, Stephan
Augustin
Produktentwicklung
Tengstrasse 45
D 80796 Munich, Germany
T +49 (0)89 2730690
F +49 (0)89 2730690
E stephan@augustin.com
W www.augustin.biz

Baas, Maarten
c/o Design Academy
Eindhoven (p. 305)

Ban, Shigeru
5-2-4 Matsubara Ban
Building 1Fl
Setagaya
Tokyo 156, Japan
T +81 (0)3 3324 6760
F +81 (0)3 3324 6789

Bär + Knell
7 Untere Turmgasse
D 74206 Bad Wimpen,
Germany
T +49 (0)7063 6891
F +49 (0)7063 6980
E Baerknell@aol.com
W www.baer-knell.de

Barlow-Lawson, Stephen
c/o Ground Support
Equipment (p. 316)

Baroli, Luigi
c/o Baleri Italia (p. 312)

Barron, Neil
E neil@gusto.co.uk
c/o Gusto Design (p. 316)

Bartsch Design
Industrial Design Gbr
Philipp-Müller-Strasse 12
D 23966 Wismar, Germany
T +49 (0)3841 758 160
F +49 (0)3841 758 161

Batchelor, Lucy
Hill House
Vicarage Road
Wigginton, Tring
Hertfordshire
HP23 6DY, UK
T +44 (0)794 100 6513
E lucy@lucyjanebatchelor.me.uk
W www.lucyjanebatchelor.me.uk

Behnisch, Behnisch & Partner Architekten
Christophstrasse 6
D 70178 Stuttgart,
Germany
T +49 (0)711 60 77 20
F +49 (0)711 60 77 299
E buero@behnisch.com

Bazzan, Vreni
Hochschule für Gestaltung
und Kunst
Zurich, Switzerland
W www.hglz.ch

Berger, Susi and Ueli
c/o Röthlisberger (p. 320)

Bergman, David
Fire & Water Lighting and
Furniture, USA
T +1 212 475 3106
F +1 212 677 7291
E bergman@cyberg.com
W www.cyberg.com

Bergne, Sebastian
Bergne Design for
Manufacture
2 Ingate Place
London SW8 3NS, UK
T +44 (0)20 7622 3333
F +44 (0)20 7622 3336
E bergne.dfm@mailbox.co.uk

Bernett, Jeffrey
c/o Cappellini SpA (p. 312)

Bernstrand, Thomas
Bernstrand & Co.
Skånegatan 51
s-116 37 Stockholm,
Sweden
T +46 (0)8 641 91 10
F +46 (0)8 641 91 20
E mail@bernstrand.com
W www.bernstrand.com

Berthier, Marc
Design Plan Studio
141 Boulevard St Michel
75005 Paris, France
T +33 (0)143 26 49 97
F +33 (0)143 26 54 62
E dpstudio@wanadoo.fr

Bey, Jurgen
Nikkelstraat 40
3067 GR Rotterdam,
the Netherlands
T +31 (0)10 425 8792
F +31 (0)10 425 9437
E studio@jurgenbey.nl
W www.jurgenbey.nl

Bill Dunster Architects, Mark Lovell and Oscar Faber
24 Helios Road
Wallington
Surrey SM6 7BZ, UK
T +44 (0)20 8404 1380
F +44 (0)20 8404 2255
E info@zedfactorcy.com
W www.zedfactory.com

Blanca, Oscar Tusquets
c/o Escofet 1886 SA (p. 314)

Blashki, Guy
c/o IDRA (p. 330)

Bocchietto, Luisa
c/o Serralunga (p. 321)

Boeing
W www.boeing.com

Boeri, Cini and Tomu Katayanagi
c/o Fiam Italia SpA (p. 314)

Bogdan, Lea
c/o IDRA (p. 330)

Boner, Jörg
Edenstrasse 16
CH-8045 Zürich, Switzerland
T +41 (0)1 201 79 34

F +41 (0)1 201 19 35
E mail@joergboner.ch
W www.joergboner.ch

Borgersen, Tore and Espen Voll
c/o Iform (p. 316)

Boym, Constantin and Laurene with Rebecca Wijsbeek
c/o Moooi (p. 318)

Bredahl, Pil
Njalsgade 19.6 sal
DK 2300 Copenhagen, Denmark
E pilbredahl@get2net.dk

Brown, Julian
Studio Brown
6 Princes Buildings
George Street
Bath BA1 2ED, UK
T +44 (0)1225 481 735
F +44 (0)1225 481 737
E julian@studiobrown.com

Burke, David
c/o IDRA (p. 330)

Burkhardt, Roland
c/o Sunways (p. 321)

Büro für Form
Hans-Sachs-Strasse 12
D 80469 Munich, Germany
T +49 (0)89 2694 9000
F +49 (0)89 2694 9002
E meek@bürofürform.de
W www.bürofürform.de

Büro für Produktgestaltung
Brendstrasse 83
D 75179 Pforzheim, Germany
T +49 (0)7231 442 115
E f-neubert@s-direktnet.de

Cahen, Antoine
Les Ateliers du Nord
Place du Nord 2
CH 1005 Lausanne, Switzerland
T +41 (0)21 320 58 07
F +41 (0)21 320 58 43
E antoine.cahen@
atelierdunord.ch

Campana, Fernando and Humberto
Rua Barão de Tatui 219
São Paulo 01226030, Brazil

T +55 (0)11 | 36 66 41 52
F +55 (0)11 825 3 408
E campana@
campanadesign.com.br

Cantono, Chiara
WELL-TECH
Via Malpighi 3
20129 Milan, Italy
T +39 (0)2 295 18792
F +39 (0)2 295 18189
E chiara.cantano@tiscali.net
W www.well-tech.it

Carduff, Ian and Hamid von Koten
VK & C Partnership
2/2 248 Woodlands Road
Glasgow G3 6ND, UK
T/F +44 (0)141 332 2049

Carlson, Julia
c/o IDRA (p. 330)

Coates, Nigel
Branson Coates Architecture
23 Old Street
London EC1V 9HL, UK
T +44 (0)20 7490 0343
F +44 (0)20 7490 0320

Colwell, David and Roy Tam
David Colwell Design
Station Building,
Llanidloes,
Powys SY18 6EB, UK
T +44 (0)1686 414 848
F +44 (0)1686 414 849
E info@davidcolwell.com
W www.davidcolwell.com

Cornellini, Deanna
c/o G T Design (p. 315)

Cottrell & Vermeulen
1B Iliffe Street
London SE17 3LJ, UK
T +44 (0)20 7708 2567
F +44 (0)20 7252 4742
E cva@cvaarchit.
freeserve.co.uk
W www.cottrellandvermeulen.
co.uk

Crasset, Matali
c/o Domeau & Perés
(p. 313)

Davids, Rik
c/o Goods (p. 315)

Day, Robin
c/o Magis SpA (p. 318)

De Carlo, Jacopo and Andrea Gualla with Raffaella Godi
DeCarlo Gualla
Studio Di Architettura
Via Palermo 12
20121 Milan, Italy
E decarlo.gualla1@tiscalinet.it

de Leede, Annelies
c/o Goods (p. 315)

DEKA Research & Development
340 Commercial Street
Manchester NH 03101, USA
W www.dekaresearch.com

Dell'Orto, Filippo
c/o Pallucco Italia (p. 319)

Design Academy Eindhoven
Emmasingel 14
PO Box 2125
5600 CC Eindhoven,
the Netherlands
T +31 (0)40 239 39 39
F +31 (0)40 239 39 40
E info@designacademy.nl
W www.designacademy.nl

Design Tech
Zeppelinstrasse 53
D 72119 Ammerbuch,
Germany
T +49 7073 91890
F +49 7073 918916
E info@designtechschmid.de
W www.designtechschmid.de

Deuber, Christian
N2 Büro
Breisacherstrasse 64
CH 4057 Basel, Switzerland
T +41 (0)61 693 4011

Dillon, Jane and Tom Grieves
c/o Lloyd Loom of Spalding
(p. 317)

Ditzel, Nanna
c/o Fredericia Furniture A/S
(p. 315)

Dixon, Robert
c/o Advanced Vehicle
Design (p. 311)

Dixon, Tom
Tom Dixon Design
The Shop
28 All Saints Road
London W11 1HG, UK
T +44 (0)20 7792 5335

F +44 (0)20 7792 2156
E michaela@tomdixon.net
W www.tomdixon.net

Dolphin-Wilding, Julienne
34 Cecil Rhodes House
Goldington Street
London NW1 1UG, UK
T +44 (0)20 7380 0950
F +44 (0)20 7252 1778
E dolphin@
julienne.demon.co.uk
W www.dolphinwilding.com

Douglas-Miller, Edward
c/o Remarkable Pencils Ltd
(p. 320)

Dranger, Jan
Dranger Design AB
Stora Skuggans Väg 11
115 42 Stockholm, Sweden
T +46 (0)8 153 929
F +46 (0)8 153 926

Dunster, Bill and BioRegional
BioRegional Development
Group
The Ecology Centre
Honeywood Walk
Carshalton
Surrey SM5 3NX, UK
T +44 (0)20 8773 2322
F +44 (0)20 8773 2878
E info@bioregional.com
W www.bioregional.com

Dyson, James
c/o Dyson Appliances
(p. 314)

Ecke: Design
Reuchlinstrasse 10–11
D 10553 Berlin, Germany
T +49 (0)30 347709 0
F +49 (0)30 347709 22
E berlin@eckedesign.de
W www.eckedesign.de

Eiermann, Professor Egon
c/o Wilde + Spieth (p. 322)

El Ultimo Grito
Studio 8
23-28 Penn Street
London N1 5DL, UK
T +44(0)20 7739 1009
F +44(0)20 7739 2009
E grito@btinternet.com

Ela, Adital
c/o Design Academy
Eindhoven (p. 305)

Eldøy, Olav
Stokke Gruppen (p. 321)

Emilio Ambasz & Assoc., Inc.
8 East 62nd Street
New York
NY 10021, USA
T + 1 212 751 3517
F + 1 212 751 0294
E info@ambasz.com
W www.emilioambasz.com

Enlund, Teo
c/o Simplicitas
Grevgatan 19
114 52 Stockholm, Sweden
T +46 (0)8 661 00 91
F +46 (0)8 661 00 97
W www.simplicitas.se

Enthoven Associates Design Consultants
Minderbroedersstraat 14
2000 Antwerp, Belgium
T +32 (0)3 2035300
E m.vogelzang@ea-de.com

Erik Krogh Design
Denmark
E erkr@dk-designskole.dk

Eurlings, Judith
c/o Design Academy
Eindhoven (p. 305)

FanWing
E peebles@flashnet.it
W www.fanwing.com

Farkache, Nina
c/o Droog Design (p. 313)

Fee, Brenda
Studio 2
The Watermark
Ribbleton Lane
Preston
Lancashire PR1 5EZ, UK
E brenda@brendafee.com

Feilden Clegg Bradley Architects
Bath Brewery
Toll Bridge Road
Bath BA1 7DE, UK
T +44 (0)1225 852 545
F +44 (0)1225 852 528
W www.feildenclegg.com

Feldmann & Schultchen
Timmermannstrasse 7
D 22299 Hamburg,
Germany

T +49 (0)40 510000
F +49 (0)40 517000
W www.fsdesign.de

Fleetwood, Roy
Roy Fleetwood Ltd
Office for Design Strategy
1 St John's Innovation Park
Cowley Road
Cambridge CB4 4NS, UK
T +44 (0)1223 240 074
E roy.fleetwood@
 fleetwoodinc.com
W www.fleetwoodinc.com

Force4 and KHRAS
Teknikerbyen 7
DK 2830 Virum, Denmark
T +45 (0)82 40 72 00
E cda@force4.dk
 khr@hkras.dk
W www.khras.dk

Formgestaltung Schnell-Waltenberger
Sponheimstrasse 22
D 75177 Pforzheim,
Germany
T/F +49 (0)7231 33364
E mail@formgestaltung.de
W www.formgestaltung.de

Förster, Monica
c/o OFFECCT (p. 319)

Fortunecookies
(Jacob Jurgensen Ravn)
Denmark
E jacob@fortunecookies.dk

Frazer, John-Paul
Bioinspiration
11 Harmood Street
London NW1 8DN, UK
T +44 (0)207 267 6227
W www.bioinspiration.com

GAAN GmbH
Sonneggstrasse 76
CH 8006 Zurich,
Switzerland
T +41 (0)1 363 52 00
F +41 (0)1 363 52 05
E info@gaan.ch
W www.gaan.ch

Geertman, Nine
c/o Design Academy
Eindhoven (p. 305)

Gehry, Frank O
Frank O Gehry &
Associates Inc

1520-B Cloverfield
Boulevard
Santa Monica
CA 90404, USA
T +1 310 828 6088
F +1 310 828 2098

Gismondi, Ernesto, Artemide Design
c/o Artemide SpA (p. 311)

Gluska, Natanel [sp?]
Renggerstrasse 85
CH 8038 Zurich,
Switzerland
E natanel@bluemail.ch
W www.natanelgluska.com

gmp – Architekten
von Gerkan Marg und
Partner
Elbchaussee 139
D 22763 Hamburg,
Germany
T +49 (0)40 88 151 0
F +49 (0)40 88 151 177
E hamburg-e@gmp-
 architekten.de
W www.gmp-architekten.de

Gomes, Patricia, João Cunna, Luis Temudo and Palmira Leinia
c/o IDRA (p. 330)

Grcic, Konstantin
Konstantin Grcic Industrial
Design
Schillerstrasse 40
D 80336 Munich, Germany
T +49 (0)89 55079995
F +49 (0)89 55079996
E mail@konstantin-grcic.com

Green Map System
PO Box 249
New York
NY 10002, USA
T +1 212 674 1631
F +1 212 674 6206
E info@greenmap.org
W www.greenmap.org

Grunert, Pawel
c/o Alicja Trusiewicz
Via Bramante 22/L
06100 Perugia, Italy
T/F +39 (0)75 572 6470
E alicjet@tin.it
W www.grunert.art.pl

Guimarães, Dr Luiz
GDDS, Federal University

of Campina Grande
Rua Agrimensor José de
Brito 419
Alto Branco 58102-560
Campina Grande PB, Brazil
T/F +55 (0)83 342 3703
E adocid@superig.com.br

Guixé, Marti
Calabria 252
E 8029 Barcelona, Spain
T/F +34 (0)93 322 5986
E info@guixe.com
W www.guixe.com

Gupta, Arvind
c/o IDRA (p. 330)

Guy Martin Furniture
Crown Studios
Old Crown Cottage
Greenham
Crewkerne
Somerset TA18 8QE, UK
T +44 (0)1308 868122
E guy_martin@mac.com
W www.guy-martin.com

Guynn, Mark D.
Systems Analysis Branch,
NASA Langley Research
Center
T +1 757 864 8053
W rasc.larc.nasa.gov

Haas, Rouven
RO Designment
Alserbachstrasse 33/16
A 1090 Vienna, Austria
E office@designment.cc
W www.designment.cc

Haberli, Alfredo & Christophe Marchand
c/o Röthlisberger (p. 320)

Haemmerle Linda, Unitec
c/o IDRA (p. 330)

Hanspeter Steiger Designstudio
c/o Röthlisberger (p. 320)

Hartgring, Aran
c/o Design Academy
Eindhoven (p. 305)

Heikkinen Komonen Architects
Antti Könönen
Kristianinkatu 11-13
FIN-00170 Helsinki,
Finland

E mk@heikkinen-komonen.fi
W www.heikkinen-komonen.fi

Hereford & Worcester County Council
Technical Services
Department
County Hall
Spetchley Road
Hereford and Worcester
WR5 2NP, UK
T +44 (0)1905 766 422

Hertz, David
c/o Syndesis, Inc. (p. 321)

Heufler, Professor Gerhard
Koröisistrasse 5
A 80101 Graz, Austria
T +43 (0)316 672 258
F +43 (0)316 672 258 4

Holz Box ZT GmbH
Colingasse 3
A 6020 Innsbruck, Austria
T +43 (0)512 561478
F +43 (0)512 561478 55
E mailbox@holzbox.at
W www.holzbox.at

Höser, Christoper
c/o Glas Platz (p. 315)

Houshmand, John
USA
T +1 212 965 1238
E johnhoushmand@yahoo.com

Human Factors with Oxo International
c/o Oxo International
(p. 319)

IDEO Japan
Axis Building, 4th floor
5-17-1 Roppongi
Minato-ku
Tokyo, Japan
T +81 (0)3 5570 2664
F +81 (0)3 5570 2669
W www.ideo.com

IDEO Product Development
Pier 28 Annex
The Embarcadero
San Francisco
CA 94105, USA
T +1 415 615 5000
W www.ideo.com

IDRA International Design Resource Awards
Design Resource Institute

347 NW 105th Street
Seattle
WA 98177, USA
T +1 206 289 0949
F +1 206 7893 144
E Designwithmemory@cs.com
W www.designresource.org

Interform Design
Am Wenderwehr 3
D 38114 Braunschweig,
Germany
T +49 (0)531 233 7810
W www.interform-design.de

Irvine, James and Steffan Kaz
via Sirtori 4
20129 Milan, Italy
T +39 (0)2 295 34532
F +39 (0)2 295 34534
E info@james-irvine.com

Ishiguro, Takatoshi and PES Kenchiku Kankyo Sekkei
c/o IDRA (p. 330)

Ito, Setsu
I.T.O. Design
Via Brioschi 54
20141 Milan, Italy
T/F +39 (0)2 8954 6007
E setsuito@micronet.it

Iverson, Jason and Shayan Rafie
University of Washington,
Industrial Design
Department
c/o IDRA (p. 330)

Jacobs, Camille
100 Taman Nakhoda
Villa delle Rose
257793 Singapore
T/F +65 475 3581

Jaensch, Peter
Behind-the-wheel
Product Design
Hofgärten 8c
D 70597 Stuttgart,
Germany
T +49 (0)711 72 80 622
F +49 (0)711 72 80 621
E pj@peterjaensch.de
W www.behind-the-wheel.de

Jakobsen, Hans Sandgren
Færgevej 3
DK 8500 Grenaa,
Denmark
T +45 (0)86 32 00 48
F +45 (0)86 32 48 03

E mail@hans-sandgren-
jakobsen.com
W www.hans-sandgren-
jakobsen.com

Jansen, Willem
c/o Design Academy
Eindhoven (p. 305)

Janzen, Michael
Human Shelter Research
Institute
E mjantzen@yahoo.com

Johnston, Eimeir
21 Sutton Downs
Bayside
Dublin, Ireland
T +353 (0)1 8393181
E eimeir@mash.ie
W www.mash.ie/eimeir/
index. html

Johnston, Lindsay
Four Horizons
Watagans National Park
Georges Road
Quorrobolong
PO Box 485
Cessnock
NSW 2325, Australia
T +61 (0)2 4998 6257
F +61 (0)2 4998 6237
E lindsay@fourhorizons.com.au
W www.fourhorizons.com.au/
lindsay

Jongerius, Hella
JongeriusLab
Schietbaanlaan 75b
3021 LE Rotterdam,
the Netherlands
T +31 (0)10 477 0253
E jongeriuslab@hotmail.com

Jørgesen, Carsten
c/o Bodum (p. 312)

Juanico, Juan Benavente
Juanico Design
Manuel Marti 5, 3
E 46201 Valencia, Spain
E juanico@juanico.net
W www.juanico.net

Karpf, Peter
Glentevej 8
DK 3210 Vejby, Denmark
T +45 (0)48 70 63 73
F +45 (0)48 70 63 79

Karrer, Beat
Zimmerlistrasse 6

CH 8004 Zurich,
Switzerland
T +41 (0)1 400 55 00
F +41 (0)1 400 55 04
E info@beatkarrer.net
W www.beatkarrer.net

Kaufmann, Johannes and Oskar Leo
Johannes Kaufmann
Architektur
Sagerstrasse 4
A 6850 Dornbirn, Austria
T +43 (0)5572 23690
F +43 (0)5572 236690 4
E office @jkarch.at
W www.jkarch.at

Kaz, Lorenz✭
Viale Tunisia 10
20124 Milan, Italy
T/F +39 (0)229 510 400
E info@lorenz-kaz.com
W www.lorenz-kaz.com

Kerr, Bernard and Pejack Campbell
c/o Solar Cookers
International (p. 321)

Klawer, Antoinette
c/o Design Academy
Eindhoven (p. 305)

Klug, Ubald
33 rue Croulebarbe
75013 Paris, France
T +33 (0)1 44 33 13 882
F +33 (0)1 45 35 31 54

Konings, Jan and Jurgen Bey
c/o Droog Design (p. 313)

Kotkas, Aki
Hamentie 130E
00560 Helsinki, Finland
T +358 50 5879077
T + 358 50 3673969
E akotkas@uiah.fi
akotkas@welho.com

Krier Sophie
c/o Moooi (p. 318)

Kuypers, Sonja
c/o Design Academy
Eindhoven (p. 305)

Lemmon, Michael
c/o IDRA (p. 330)

Lewis, David
c/o Vestfrost A/S (p. 322)

307

Liefting, Mander
c/o Design Academy
Eindhoven (p. 305)

Lino Codato Collection
Via del Credito 18
31033 Castelfranco
Veneto-Treviso, Italy
T +39 (0)423 722591
F +39 (0)423 724089
E lcc@lcc-collection.com
W www.lcc-collection.com

Lo, K. C.
31 Finsbury Park Road
London N4 2JY, UK

Looyen, Rick
c/o Design Academy
Eindhoven (p. 305)

Lovegrove, Ross
Lovegrove Studio X
21 Powis Mews
London W11 1JN, UK
T +44 (0)20 7229 7104
F +44 (0)20 7229 7032
E lovegroves_rmr@
  compuserve.com

M3 Architects
49 Kingsway Place
Sans Walk
London EC1R 0LU, UK
T +44 (0)20 7253 7255
F +44 (0)20 7253 7266
E post@m3architects.com
W www.m3architects.com

Maier-Aichen, Hansjerg
c/o Authentics artipresent
GmbH (p. 311)

Makepeace, John
Hooke Forest
(Construction) Ltd
Parnham House
Beaminster
Dorset DT8 3NA, UK
T +44 (0)1308 862 204
F +44 (0)1308 863 494
E info@hookepark.com
W www.hookepark.comv

Malcolm Baker Furniture Design
Rainscome Farm Buildings
Oare
Marlborough
Wiltshire SN8 4HZ, UK
T +44 (0)1672 564795
E studio@malcolmbaker
  .co.uk

Marczynski, Mike
c/o Business Lines (p. 312)

Mari, Enzo
c/o Alessi SpA
(p. 311)

Marriott, Michael
Unit F2,
2–4 Southgate Road
London N1 3JJ, UK
T/F +44 (0)20 7923 0323
E marriott.michael@virgin.net

Massachusetts Institute of Technology
77 Massachusetts Avenue
Cambridge MA 02139, USA
T +1 617 253 1000
W www.mit.edu

McCrady, Paul
AeroVironment
825 S. Myrtle Drive
Monrovia
CA 91016, USA
T +1 626 357 9983
F +1 626 359 9628
W www.aerovironment.com

Meda, Albeto
c/o Luceplan (p. 317)

Meller Marcovicz, Gioia
Gioia Limited
PO Box 381
30100 Venice, Italy
T +39 (0)41 795213
F +44 (0)20 7504 3771
E info@gioia.org.uk
W www.gioia.org.uk

Miles, J R
c/o Flexipal Ltd (p. 315)

Minakawa, Gerard
c/o IDRA (p. 330)

Mir, Ana and Emili Padrós
Emiliana Design Studio, s. l
Aribau, 240 8° d
Barcelona, Spain
T/F + 34 (0)93 414 34 80
E emiliana@
  emilianadesign.com
W www.emilianadesign.com

Mithun Architects
Pier 56, 1201 Alaskan Way,
Suite 200
Seattle, WA 98101, USA
T +1 206 623 3344
F +1 206 623 7005

E mithun@mithun.com
W www.mithun.com

Mitra, Vikram
National Institute of
Design, India
c/o IDRA (p. 330)

Moerel, Marre
182 Hester Street No. 13
New York
NY 10013, USA
T +1 212 219 8965
F +1 212 925 2371
E marremoerel@rcn.com

Möllmann Joachim, Markus Wessels
FH Hannover
Postfach 92 02 51
30441 Hannover,
Germany
W www.www.fh-hannover.de

Moore, Isabelle
22257 NE Inglewood
Hill Road
Redmond
WA 98053, USA
E isabellemoore@earthlink.net

Moriarty, Lauren
46 High Street
Rode
Bath BA11 6PB, UK

Morrison, Jasper
c/o Cappellini SpA (p. 312)

Mrasek, Sabine and Clemens Stüber
c/o Nils Holger Moormann
GmbH (p. 318)

N2
Switzerland
T/F +41 (0)61 693 4015
E n2@n2design.ch
W www.n2design.ch

NASA Dryden Flight Research Center
PO Box 273
Edwards, CA 93523, USA
T +1 661 276 331
W www.dfrc.nasa.gov

Navone, Paola
c/o Gervasoni SpA (p. 315)

Ng, Lian
c/o Publique Living
(p. 320)

Neggers, Jan
c/o Goods (p. 315)

Newson, Marc, Ross Lovegrove, Beatrice Santiccioli
c/o Biomega (p. 312)

Öjerstam, Carl
c/o IKEA (p. 316)

Okamura, Osamu
c/o IDRA (p. 330)

Omolo, Anouk
c/o Design Academy
Eindhoven (p. 305)

Opsvik, Peter
c/o Stokke Gruppen
(p. 321)

Padrós, Emili
Emiliana Design Studio
Aribau, 240 8° N
08006 Barcelona, Spain
T/F +34 (0)93 414 34 80
E emiliana@emilianadesign.com
W www.emilianadesign.com

Pakhalé, Satyendra
Atelier Satyendra Pakhalé
R. J. H. Fortuynplein 70
1019 WL Amsterdam
the Netherlands
T +31 (0)20-4197230
F +31 (0)20-4197231
E info@satyendra-pakhale.com
W www.satyendra-pakhale.com

Payne, Terry
c/o Monodraught Ltd
(p. 318)

Pearce, Nigel
c/o Shortstraw (p. 321)

Pellone, Giovanni and Bridget Means,
c/o Benza Inc.
(p. 312)

Piercy Conner Architects
Cairo Studios
4-6 Nile Street
London N1 7RF, UK
T +44 (0)20 7490 9494
F +44 (0)20 7490 9480
W www.piercyconner.co.uk

Pillet, Christophe
c/o Ceccotti Collezioni Srl
(p. 313)

**Porsche Design
Management GmbH
& Co. KG**
Porschestraße 1
74321 Bietigheim-
Bissingen, Germany
W www.porsche-design.com

**Prats, Eva and Ricardo Flores
with Frank Stahl**
Flores-Prats Arquitectors
Trafalgar 12, 3-1
E 08010 Barcelona, Spain
T +34 (0)93 268 4635
F +34 (0)93 268 2951
E flores-prats@coac.net

Prause, Philipp
Hietzinger Hauptstrasse
152/3/2
A 1130 Vienna, Austria
T +43 (0)1 876 4039
F +43 (0)1 876 4088
E office@diskleve.com
W www.disklev.com
W www.prausedesign.com

Produktentwicklung Roericht
c/o Wilkhahn (p. 323)

**PROFORM Design
Lener + Rossler +
Hohannismeier Gbr**
Seehalde 16
D 71364 Winnenden,
Germany
T +49 (0)7195 919100
F +49 (0)7195 919108
E proform_design@t-online.de
W www.industriedesign.com

Puotila, Ritva
c/o Woodnotes Oy
(p. 323)

Ratty, Sarah
Ciel Ltd
T +44 (0)1273 720042
E ciel.ltd@btopenworld.com

Raudenbush, Terri
c/o IDRA (p. 330)

Refsum, Bjørn
c/o Stokke Gruppen
(p. 321)

Reln
c/o Wiggly Wigglers(p. 322)

Remy, Tejo
c/o Droog Design (p. 313)

Rincover, Aaron
Mathmos Design Team
c/o Mathmos (p. 318)

Rizzatto, Paolo
c/o Serralunga (p. 321)

Rohland, Vibeke
Holbergsgade 19, 3
DK 1057 Copenhagen,
Denmark
T/F +45 (0)331 44406

Roije, Frederik
c/o Design Academy
Eindhoven (p. 305)

**Rossi, Diego and Raffaele
Tedesco**
c /o Luceplan (p. 317)

Roth, Antonia
FH Hannover
Postfach 92 02 51
30441 Hannover, Germany
E antonia@gmx.de
W www.www.fh-hannover.de

Salm, Jaimie
c/o MIO (p. 318)

Sandell, Thomas
c/o IKEA (p. 316)

Sanders, Mark
c/o Roland Plastics
(p. 320)

Scarff, Leo
Rear 91 Ballybough Road
Dublin 3, Ireland
T/F +353 (0)1 836 3135
E leoscarff@dna.ie

Schaap, Natalie
c/o Design Academy
Eindhoven (p. 305)

**Schneider, Professor Wulf
and Partners**
Schellbergstrasse 62
D70188 Stuttgart, Germany
T +49 (0)711 286 49 00
E bfg@profwulfschneider.de
W www.profwulfschneider.de

Schreuder, Hans
MOY Concept & Design
Klarendalseweg 530
6822 GZ Arnhem,
the Netherlands
T +31 (0)26 443 9977
E hans@moy.nl

Schwendtner, Gitta
Unit 25
Oxo Tower
Barge House Street
London SE1 9PH, UK

Sempé, Inga
c/o edra (p. 314)

Shiotani, Yasushi
c/o Canon, Inc. (p. 312)

Sodeau, Michael
c/o Gervasoni SpA
(p. 315)

Spiegelhalter, Thomas
WAH 204, 850 W 37 Street
Los Angeles,
CA 90089, USA
T +1 323 377 3685
F +1 213 363 7821
E spiegelh@usc.edu
W rcf.usc.edu/~spiegelh

**Springboard Design
Partnership**
15 Richmond Terrace
Clifton
Bristol BS8 1AA, UK
T +44 (0)117 973 9900
F +44 (0)117 973 2099
E info@springboard-
design.com
W www.springboard-
design.com

Starck, Philippe
Agence Philippe Starck
27 rue Pierre Poli
92130 Issy-les-Moulineaux,
France
T +33 (0)1 41 08 82 82
F +33 (0)1 41 08 96 65
E starck@starckdesign.com
W www.philippe-starck.com

Stark, Dr Herbert
Kopf Solardesign GmbH
+ Co (p. 317)

Startup Design
4 Albion Road
London N16 9PA, UK
T +44 (0)207 923 1223
F +44 (0)207 503 2123
E jasper@startupdesign.co.uk
W www.startupdesign.co.uk

**Steinmann, Peter and
Herbert Schmid**
c/o Atelier Alinea AG
(p. 311)

Stiletto DESIGN Vertrieb
Auguststrasse 2
D 10117 Berlin, Germany
T +49 (0)30 280 94 614
F +49 (0)30 280 94 615
E hallo@siltetto .de
W www.stiletto.de

Studio 7.5
c/o Herman Miller, Inc.
(p. 316)

**Studio Jacob de Baan and
Frank de Ruwe**
Nieuwevaart 128
1018 ZM Amsterdam,
the Netherlands
T +31 (0)20 77 600 18
F +31 (0)20 77 600 19
E info@jacobdebaan.com
W www.jacobdebaan.com

**Stumpf, Bill and Don
Chadwick**
c/o Herman Miller, Inc.
(p. 316)

Sugasawa, Mitsumasa
c/o Atelier Alinea AG
Zähringerstrasse 14
CH 4007 Basel ,
Switzerland
T +41 (0)61 690 97 97
F +41 (0)61 690 97 90

**Suppanen, Ilkka and Pasi
Kohhonen**
Studio Ilkka Suppanen
Punavuorenkatu 1 a 7 b
00120 Helsinki, Finland
T +358 9622 78737
F +358 9622 3093
E suppanen@kolumbus.fi

Sutton Vane Associates
Dimes Place
106-108 King Street
London W6 0QP, UK
T +44 (0)20 8563 9370
F +44 (0)20 8563 9371
W www.sva.co.uk

Sylvania Design Team
Sylvania Lighting
International
20 Route de Pré-Bois
1215 Geneva 15,
Switzerland
T +41 (0)22 717 0895
W www.sylvania.com

Takahashi, Koki
c/o IDRA (p. 330)

Teppich-art-Team
Zumbühl + Birsfelder
CH 7012 Felsberg,
Switzerland
T +41 (0)81 252 86 89
E info@teppich-art-team.ch
W www.teppich-art-team.ch

**Tesnière, François and
Anne-Charlotte Goût**
3bornes Architectes
70 rue Jean Pierre Timbaud
75011 Paris, France
T + 33 (0)1 47 00 78 27
F + 33 (0)1 47 00 78 57
E 3bornes@magic.fr

Thomas, Deborah
323(B) Grove Green Road
Leytonstone
London E11 4EB, UK

Threadgold, Tiffany
Tiffany Tomato Designs
539 S 5th Ave, Studio 1
Ann Arbor
MI 48104, USA
E tiffanytomato@yahoo.com

Timpe, Jacob
c/o Nils Holger Moorman
GmbH (p. 318)

Tolstrup, Nina
47 Warwick Mount
Montague Street, Brighton
East Sussex BN2 1JY, UK
T/F +44 (0)1273 570 179
E tolstrup@studiomama.com
W www.studiomama.com

Tomet, Nya Vatten
Sony Ericsson (p. 321)

Topen, Paul
c/o Designed to a 't' Ltd
(p. 313)

Tortel Design
6, Villa du Clos de Malevart
75011 Paris, France
T +33 (0)1 43550370
F +33 (0)1 43550349
E tortel.design@infonie.fr
W www.tortel.design.fr

Trubridge, David
44 Margaret Avenue
Havelock North
New Zealand
T/F +64 (0)6 877 46 84
E trubridge@clear.net.nz
W www.davidtrubridge.com

**Tsujimura Hisanobu Design
Office**
c/o IDRA (p. 330)

University of Eindhoven
c/o Design Academy
Eindhoven (p. 305)

Unterschuetz, Andreas
c/o IDRA (p. 330)

V K & C Partnership
2/2 248 Woodlands Road
Glasgow G3 6ND, UK
T/F +44 (0)141 332 2049

van Corven, René
c/o Design Academy
Eindhoven (p. 305)

van den Heuvel, Jessica
c/o Design Academy
Eindhoven (p. 305)

van der Kniff, Krista
c/o Design Academy
Eindhoven (p. 305)

van der Meulen, Jos
c/o Goods (p. 315)

van der Poll, Marijn
c/o Droog Design (p. 313)

van Dijke, Henk
c/o Design Academy
Eindhoven (p. 305)

van Severen, Maarten
Galgenberg 25
9000 Gent, Belgium
T +31 (0)32 92 33 89 99
F +31 (0)32 92 33 19 88
E maartenvanseveren@skynet.be

**VarioPac Disc Systems
GmbH**
Hangbaumstrasse 13
D 32257 Bünde, Germany
T +49 (0)5221 7684 17
F +49 (0)5221 7684 20
W www.variopac.com

Veloland Schweiz
Stiftung Veloland Schweiz
Postfach 8275
CH 3001 Bern, Switzerland
T +41 (0)31 307 47 40
E info@veloland.ch
W www.veloland.ch

Vuarnesson, Bernard
Sculptures-Jeux

18 Rue Domat
75005 Paris, France
T +33 (0)1 43 54 20 39
F +33 (0)1 43 54 83 32
W www.sculpturesjeux.fr

Wanders, Marcel
Marcel Wanders Studio
Jacob Catskade 35
1052 BT Amsterdam,
the Netherlands
T +31 (0)20 4221339
F +31 (0)20 4227519
E joy@marcelwanders.nl
W www.marcelwanders.com

**Weber, Jeff, Stumpf/Weber
+ Associates**
c/o Herman Miller, Inc.
(p. 316)

Wettstein, Robert A.
Josefstrasse 188
CH 8005 Zurich,
Switzerland
T +41 (0)44 272 97 25
F +41 (0)44 272 07 17
E robert.wettstein@gmx.ch
W www.wettstein.ws

Wiegand, Lorenz
Pool Products
Grosse Hamburger
Strasse 33
D 10115 Berlin, Germany
T +49 (0)30 440 555 16
W www.poolproducts-
design.com

Wiesendanger, Kobi
Avant de Dormir
Via Turato 3
20121 Milan, Italy
T +39 (0)2 659 9990
F +39 (0)2 657 1058
E marina@avantdedormir.com

Wiffels, Manuel
c/o Design Academy
Eindhoven (p. 305)

Willat, Boyd
c/o Willat Writing
Instruments (p. 323)

Wilson, Neil
180 Sackville Road
Heaton, Newcastle-
Tyne & Wear NE6 5TD, UK
T/F +44 (0)191 224 3850

Wolf, Michael
Fachhochschule Köln

Germany
E wolf@formlos.com

Wurz, Gerald
Teschniergasse 17
A 1170 Vienna, Austria
T +43 (0)407 21 25

Yoshioka, Tokujin
c/o Driade (p. 313)

Yeang, Ken
T R Hamzah & Yeang
Sdn Bhd
8 Jalan 1
Taman Sri Ukay
Off Jalan Ulu Keland
68000 Ampang
Selangor, Malaysia
T +60 (0)3 4257 1966
F +60 (0)3 4256 1005
E trhy@tm.net.my

Young, Michael
c/o Magis SpA (p. 318)

Zachau, Hans Philip
Nya Perspektiv Design AB
W www.nyaperspektiv.se

Zalotay, Elemer
CH 3054 Schüpfen
Ziegelried, Switzerland

21st Century Health Ltd
2 Fitzhardinge Street
London W1H 6EE, UK
T +44 (0)20 7935 5440
F +44 (0)20 7487 3710
E info@21stcenturyhealth.co.uk
W www.21centuryhealth.co.uk

2pm Limited
2 Shelford Place
London N16 9HS, UK
T +44 (0)20 7923 0222
F +44 (0)20 7923 2467
E info@2pm.co.uk
W www.2pm.co.uk

3M Deutschland GmbH
Carl-Schurz-Strasse 1
41453 Neuss, Germany
T +49 (0)2131 14 0
F +49 (0)2131 14 2649
W www.3m.com

AB Composites Pvt. Limited
1/1b/18 Ramkrishna
Naskar Lane
Calcutta 700 010, India
T +91 (0)2370 5982/6348
F +91 (0)2351 0305
E anukul@cal2.vsnl.net.in
W www.abcomposites.net

ABG Ltd
Unit E7,
Meltham Mills Road
Meltham,West Yorkshire
HD7 4DS, UK
T +44 (0)1484 852 096
W www.abg-geosynthetics.com

Acordis Fibres Ltd
PO BOX 111,
101 Lockhurst Lane
Coventry CV6 5RS, UK
T +44 (0)24 76582288
F +44 (0)24 76682737
W www.acordis.com
    www.acordisservices.com

Adept Polymers Ltd
Unit 7, Fairhills Industrial
Estate, Irlam
Manchester M44 6ZQ, UK
T +44 (0)161 777 4830
W www.adeptpolymers.co.uk

**Advanced Environmental
Recycling Technologies, Inc.**
PO Box 1237, Springdale
AR 72765, USA
T +1 476 756 7400
E sales@choicedek.com
W www.choicedek.com

Advanced Vehicle Design
The Barn,
Warrener Street
Sale Moor
Sale, Cheshire
M33 3GE, UK
T +44 (0)161 969 9692
E bob@windcheetah.co.uk
W www.windcheetah.co.uk

AEG Hausgeräte GmbH
Muggenhofer Strasse 135
D 90429 Nuremberg,
Germany
T +49 (0)911 323 0
F +49 (0)911 323 1770
E info@aeg.hausgeraete.de
W www.aeg hausgeraete.de

**aerodyn Energiesysteme
GmbH**
Provianthausstrasse 9
D 24768 Rendsburg,
Germany
T +49 (0)4331 12750
F +49 (0)4331 127555
E info@aerodyn.de
W www.aerodyn.de

AeroVironment, Inc.
Corporate HQ
825 S. Myrtle Drive
Monrovia
CA 91016, USA
T +1 626 357 9983
F +1 626 359 9628
W www.aerovironment.com

Aga-Rayburn
Station Road
Ketley
Telford
Shropshire TF1 5AQ, UK
T +44 (0)1952 642 000
F +44 (0)1952 641 961
W www.aga-rayburn.co.uk

Airnimal Designs Ltd UK
T/F +44 (0)1223 523973
E info@airnimal.co.uk
W www.airnimal.com

Alden & Ott
616 E. Brook Drive
Arlington Heights
IL 60005, USA
T +1 847 956 6509
W www.aldenottink.com

Alessi SpA
Via Privata Alessi 6
28882 Crusinallo (VB), Italy
T +39 (0)323 868 611
F +39 (0)323 866 132
E pub@alessi.it
W www.alessi.it

Allison Transmission
W www.allison
    transmission.com

Amasec Airfil Ltd
Unit 1, Colliery Lane
Exhall, Coventry
Warwickshire CV7 9NW, UK
T +44 (0)2476 367 994
F +44 (0)2476 644 325
W www.airfil.com

American Excelsior Company
PO Box 5067
850 Avenue H East
Arlington
TX 76011, USA
T +1 817 385 3500
F +1 817 649 7816
W www.amerexcel.com

Ampair Ltd
The Doughty Building
Crow Arch Land, Ringwood
Hampshire BH24 1NZ, UK
T +44 (0)1425 480 780
F +44 (0)1425 479 497
E sales@ampair.com
W www.ampair.com

The Amtico Company
Kingfield Road
Coventry
Warwickshire CV6 5AA, UK
T +44 (0)24 7686 1400
F +44 (0)24 7686 1552
E customer.services@
amtico.co.uk
W www.stratica.com
    www.amtico.com

Anstalten Thorberg
Postfach 1
CH 3326 Krauchtal,
Switzerland
T +41 (0)34 411 1417
F +41 (0)34 411 0019

Apple Computer, Inc.
1 Infinite Loop
Cupertino
CA 95014, USA
T +1 408 996 1010
W www.apple.com

**Armstrong World
Industries, Inc.**
2500 Columbia Avenue
PO Box 3001
Lancaster
PA 17603, USA
T +1 717 397 0611
F +1 717 396 2787
W www.armstrong.com

Artemide SpA
Via Bergamo 18
20010 Pregnana Milanese,
Italy
T +39 (0)2 935 181
F +39 (0)2 9359 0254
E info@artemide.com
W www.artemide.com

Atelier Alinea AG
Zähringerstrasse 14
CH 4007 Basel,
Switzerland
T +41 (0)61 690 97 97
F +41 (0)61 690 97 90

Auro Pflanzenchemie AG
Alte Frankfurter Strasse 211
D 38122 Braunschweig,
Germany
T +49 (0)531 281 41 0
F +49 (0)531 281 41 61
E info@auro.de
W www.auro.de

**Authentics artipresent
GmbH**
Max Eyth Strasse 30
D 71088 Holzerlingen,
Germany
T +49 (0)7031 6805 0
F +49 (0)7031 6805 99
W www.authentics.de

Avad
Persistence Works
21 Brown Street
Sheffield S1 2BS, UK
T +44 (0)114 2493695
E studio@avad.net
W www.avad.net

Baccarne
Gentbruggekouter 5
9050 Gent, Belgium
T +32 (0)9 232 44 21
F +32 (0)9 232 44 30
E baccarne@pi.be
W www.baccarne.be

Baleri Italia SpA
Via F. Cavallotti 8
20122 Milan, Italy
T +39 (0)2 76 0 23954
F +39 (0)2 76 023738
E info@baleri-italia.com
W www.baleri-italia.com

Bamboo Hardwoods, Inc.
510 S. Industrial Way
Seattle
WA 98108, USA
T +1 206 264 2414
W www.bamboohardwoods.com

Beacon Print Limited
Brambleside
Bellbrook Park
Uckfield
East Sussex
TN22 1PL, UK
T +44 (0)1825 768611
F +44 (0)1825 768042
E print@beaconpress.co.uk
W www.beaconpress.co.uk

Belkin Corporation
501 West Walnut Street
Compton
CA 90220, USA
T +1 877 523 5546
F +1 310 898 1111
W www.belkin.com

Benza, Inc.
413 W. 14 Street, #301
New York
NY 10014, USA
T +1 718 383 1334
E ingo@benzadesign.com
W www.benzadesign.com

BioChem Systems
3511 N. Ohio
Wichita
KS 67219, USA
T +1 316 838 4739
F +1 316 681 2168
W www.biochemsys.com

Biocorp
5155 West Rosecrans
Avenue, Suite 1116
Los Angeles
CA 90250, USA
T +1 310 491 3465
F +1 310 491 3338
E info@biocorpna.com
W www.biocorpna.com

Biofab LLC
PO Box 990556
Redding
CA 96099, USA
T +1 530 243 4032
F +1 530 246 0711
E info@ricestraw.com w
W www.strawboard.com

Biolan, Oy
PO Box 2
FIN 27501 Kauttua,
Finland
T +358 (0)2 549 1600
F +358 (0)2 549 1660
W www.naturum.fi

Biomega
Skoubogade 1,1
1158 Copenhagen,
Denmark
T +45 (0)70 20 49 18
F +45 (0)70 20 49 14
W www.biomegaphilosophy
.com

BioShield Paint Company
1330 Rufina Circle
Santa Fe
NM 87505, USA
T +1 505 438 3448
F +1 505 438 0199
E info@bioshieldpaint.com
W www.bioshieldpaint.com

Biotec
Biologische
Naturverpackungen GmbH
Werner-Heisenberg
Strasse 32
Emmerich
D 46422 Germany
T +49 (0)2822 92510
F +49 (0)2822 51840
E info@biotec.de
W www.biotec.de

Blackwall Ltd
Seacroft Estate, Coal Road
Leeds
W. Yorks LS11 5JP, UK
T +44 (0)113 201 8000
F +44 (0)113 201 8001
W www.blackwall-ltd.com

BMW AG
BMW-Haus Petuelring 130
D 80788 Munich,
Germany
T/F +49 (0)89 3820/
T +49 (0)89 382 24272
E bmwinfo@bmw.de
W www.bmw.com

Bodum AG
Kantonstrasse 100
CH 6234 Triengen,
Switzerland
T +41 (0)41 935 4500
F +41 (0)41 935 4580
W www.bodum.com

Body Shop International Plc
Watersmead
Littlehampton
West Sussex BN17 6LS, UK
T +44 (0)1903 731 500
F +44 (0)1903 726 250
W www.bodyshop.com

Bopp Leuchten GmbH
Postfach 1160
D 74835 Limbach,
Germany
T +49 (0)62 87 92 06 0

BP Amoco
Amoco plc
Britannic House
1 Finsbury Circus
London EC2M 7BA, UK
W www.bp.com

BREE Collection GmbH &
Co. KG
Gerberstrasse 3
D 30916 Isernhagen
Germany
T +49 (0)5136 8976 0
F +49 (0)5136 8976 229
E bree.collection@bree.de
W www.bree.de

British Polythene Industries
96 Port Glasgow Road
Greenock, PA15 TGL, UK
T +44 (0)1475 501 000
F +44 (0)1475 743 143
W www.bpipoly.com

Brompton Bicycle Ltd
Lionel Road South,
Brentford
Middlesex TW8 9QR, UK
T +44 (0)20 8232 8484
F +44 (0)20 8232 8181
W www.bromptonbicycle.co.uk

Brook Crompton
St Thomas Road
Huddersfield
West Yorkshire
HD1 3LJ, UK
T +44 (0)1484 557 200
W www.brookcrompton.com

Brüggli Produktion &
Dienstleistung
Hofstrasse 5
CH 8590 Romanshorn,
Switzerland
T +41 (0)71 466 94 94
F +41 (0)71 466 94 95
E info@brueggli.ch
W www.brueggli.ch

Buderus Heiztechnik GmbH
Sophienstrasse 30–32
D 35576 Wetzlar,
Germany
T +49 (0)6441 418 0
F +49 (0)6441 456 02
E info@buderus.de
W www.buderus.de

Bulo Office Furniture
Industriezone Noord B6
2800 Mechelem,
Belgium
T +32 (0)15 28 28 28
F +32 (0)15 28 28 29
E info@bulo.be
W www.bulo.be

Business Lines
The Old Motor House
Underley
Kearstwick
Kirby Lonsdale
Cumbria LA6 2DY, UK
T +44 (0)15242 71200
F +44 (0)15242 71209
W www.checkpoint-safety.com

Bute Fabrics Ltd
Barone Road
Rothesay
Isle of Bute
PA20 0DP, UK
T +44 (0)1700 50 37 34
F +44 (0)1700 50 44 45
W www.butefabrics.com

Canon, Inc.
30–2, Shimomaruko
3-Chome, Ohtaoku
Tokyo 146-8501
Japan
T +81(0)3 3758 2111
W www.canon.com

Cappellini SpA
Via Marconi 35
22060 Arosio, Italy
T +39 (0)31 759 111
F +39 (0)31 763 322
E cappellini@cappellini.it
W www.cappellini.it

Cargill Dow LLC
PO Box 5698, MS 114
Minneapolis
MN 55440, USA
T +1 989 633 1746
W www.cargilldow.com

Cecco Trading, Inc.
USA
T +1 414 445 8989
F +1 414 445 9155
E info@ironwoods.com
W www.ironwoods.com

Ceccotti Collezioni srl
PO Box 90
Viale Sicilia 4
Cascina
56021 Pisa, Italy
T +39 (0)50 701 955
F +39 (0)50 703 970
E info@ceccotti.it
W www.ceccotti.it

Celotex Ltd
Lady Lane Industrial Estate
Lady Lane
Hadleigh
Suffolk IP7 6BA, UK
T +44 (0)1473 822 093
F +44 (0)1473 820 880
W www.celotex.co.uk

Centriforce Products Ltd
14–16 Derby Road
Liverpool
L20 8EE, UK
T +44 (0)151 207 8109
F +44 (0)151 298 1319
E sales@centriforce.co.uk
W www.centriforce.com

Cerestar
Bedrijvenlaan 9
2800 Mechelen, Belgium
T +32 (0)15 400 532
F +32 (0)15 400 591
W www.cerestar.com

**Charles Lawrence
Recycling Ltd**
Jessop Way
Brunel House, Jesop Way
Newark, Nottinghamshire
NG24 2ER, UK
T +44 (0)1636 610 777
F +44 (0)1636 610 222
E recycling@clgplc.co.uk
W www.clgplc.co.uk

Ciel Ltd
T +44 (0)1273 720042
E ciel.ltd@btopenworld.com

ClassiCon
Sigmund-Riefler-Bogen 5
D 81829 Munich,
Germany
T +49 (0)89 74 81 330
F +49 (0)897 80 99 96
E info@classicon.com
W www.classicon.com

Clearvision Lighting Ltd
2 Elliot Park
Eastern Road
Aldershot
Hampshire
GU12 4TF, UK
T +44 (0)1252 344 011
F +44 (0)1252 344 066
E enquiries@virtualdaylight.com
W www.virtualdaylight.com

Clivus Multrum
15 Union Street
Lawrence
MA 01840, USA
T +1 978 725 5591
F +1 978 557 9658
W www.clivusmultrum.com

Color Trends, Inc.
5129 Ballard Avenue NW
Seattle
WA 98107, USA
T +1 206 789 1065

**Comatelec Schréder
Group GiE**
3 Rue du Cercle
Aéroport Charles de Gaulle
95700 Roissy en France,
France
T +33 (0)148 161788
F +33 (0)148 161789
W www.schreder.com

Concord Lighting
Avis Way
New Haven
E. Sussex BN9 0ED, UK
T +44 (0)870 606 2030
F +44 (0)1273 512 688
E info@concordmartin.com
W www.concord-lighting.com

Corbin Motors, Inc.
W www.sparrowelectriccars.com

Core Plastics
c/o Oxfam
274 Banbury Road
Oxford OX2 7DZ, UK
T +44 (0)1865 311 311
W www.oxfam.org.uk

Courtney Company
1935 S Plum Grove Rd
Suite 301
Palatine, IL 60067, USA
T +1 847 359 9401
F +1 847 359 9402
W www.soyclean.com

Crane & Company
30 South Street
Dalton
MA 01226, USA
T +1 413 684 6495
F +1 413 684 4278
E customerservice@crane.com
W www.crane.com

Curtis Fine Papers
Guardsbridge Mill
St Andrews
Fife KY16 0UU, UK
T +44 (0)1334 839 551
F +44 (0)1334 834 223
E curtis@curtisfinepapers.com
W www.curtisfinepapers.com

Daimler Chrysler
Epplestrasse 225
D 70546 Stuttgart,
Germany
T +49 (0)711 170
F +49 (0)711 17 222 44
E dialog@daimlerchrysler.com
W www.daimlerchrysler.com
www.daimlerchrysler.co.jp

Dalsouple Direct
PO Box 140
Bridgwater
Somerset TA5 1HT, UK
T +44 (0)1278 727733
F +44 (0)1278 727766
E info@dalsouple.com
W www.dalsouple.com

De Vecchi
Via Lombardini 20
20143 Milan, Italy
T +39 (0)28 32 33 65
F +39 (0)25 81 01 17 4
E info@devecchi.com
W www.devecchi.com

Designed to a 't' Ltd
11 Maxwell Gardens
Orpington
Kent BR6 9QR, UK
T +44 (0)1689 831 400
F +44 (0)1689 609 301

Deutsche Heraklith GmbH
Keraklith strasse 8
D 84353 Simbach am Inn,
Germany
T +49 (0)8571 40 440
E office@heraklith.de
W www.heraklith.com

Disc-O-Bed GmbH
Spitalstrasse 22
D-79539 Loerrach,
Germany
T +49 (0)7621 91523 0
W www.discobed.de

DMD
Grasweg 77
1031 HX Amsterdam,
the Netherlands
T +31 (0)20 43 50 250
F +31 (0)20 33 41 069

Domeau & Perés
21 rue Voltaire BP
92250 La Garenne
Colombes cedex, France
T +33 (0)1 47 60 93 86
F +33 (0)1 47 60 01 22
E info@domeauperes.com
W www.domeauperes.com

Dr Bronner's Magic Soaps
PO Box 28
Escondido
CA 92033, USA
T +1 760 743 2211
F +1 760 745 6675
E customers@drbronner.com
W www.drbronner.com

Driade SpA
Via Ancona 1/1
20121 Milan, Italy
W www.driade.com

Droog Design
Rusland 3
1012 CK Amsterdam,
the Netherlands
T +31 (0)20 626 9809
F +31 (0)20 638 8828
E info@droogdesign.nl
W www.droogdesign.nl

DuPont (UK) Ltd
Wedgewood Way
Stevenage
Hertfordshire
SG1 4QN, UK
T +44 (0)1438 734 000
F +44 (0)1438 374 836
E info@dupont.com
W www.dupont.com

Duralay
Interfloor Ltd
Broadway
Haslingden,
Rossendale
Lancashire BB4 4LS, UK
T +44 (0)1706 213 131
F +44 (0)1706 224 915
W www.interfloor.com/
    Duralay

Durawood
US Plastic Lumber, Ltd
2300 Glades Road
Suite 440W
Boca Raton,
FL 33431, USA
T +1 561 394 3511
F +1(561) 394 5335
W www.usplasticlumber.com

Durex Ltd
SSL International
Toft Hall
Toft Road
Knutsford
Cheshire
WA16 9PD, UK
T +44 (0)1565 624 000
F +44 (0)1565 624 099
W www.durex.com

Dyson Appliances
Dyson Ltd, Tetbury Hill
Malmesbury
SN16 0RP, UK
T +44 (0)8705 275104
E service@dyson.com
W www.dyson.co.uk

Earthshell Corporation
Heaver Plaza
1301 York Road,
Suite 200
Lutherville
MD 21093, USA
T +1 410 847 9420
F +1 410 847 9431
E inquiries@earthshell.com
W www.earthshell.com

Eastman Chemical Company
100 North Eastman Road
PO BOx 511
Kingsport
TN 37662, USA
T +1 423 229 2000
F +1 615 229 1193
W www.eastman.com

Eco Timber International
1611 Fourth Street
San Rafael
CA 94901, USA
T +1 415 258 8454
F +1 415 258 8455
W www.ecotimber.com

eco-ball
Unit 1, Tannery Close
Beckenham,
Kent BR3 4BY, UK
T +44 (0)20 8662 7200
F +44(0)20 8662 7222
E customerservice@
    ecozone.co.uk
W www.ecozone.co.uk

Ecos Organic Paints
Unit 34
Heysham Business Park
Middleton Road
Heysham
Lancashire LA3 3PP, UK
T +44 (0)1524 852 371
F +44 (0)1524 858 978
E mail@ecospaints.com
W www.ecospaints.com

Ecosynthetix
3900 Collins Road,
Lansing MI 48910, USA
T +1 517 336 4649
E info@ecosythetix.com
W www.ecosynthetix.com

Ecover NV
Industrieweg 3
2390 Malle
Belgium
E info.ecover@greenhq.co.uk
W www.ecover.com

edra SpA
PO Box 28
Perignano (PI), Italy
T +39 (0)587 616660
F +39 (0)587 617500
E edra@edra.com
W www.edra.com

Ejector GmbH
Hellerweg 180
32052 Herford, Germany
T +49 (0)5221 768415
F +49 (0)5221 768420
E info@ejector.de
W www.ejector.de

Eleksen Ltd
Pinewood Studios
Pinewood Road
Iver Heath,
Buckinghamshire
Sl0 0NH, UK
T +44 (0)870 0727 272
F +44 (0)870 0727 273
E incoming@eleksen.com
W www.electrotextiles.com

Emeco
805 Elm Avenue,
PO Box 223
Hannover
PA 17331, USA
T +1 717 637 5951
E info@emeco.net
W www.emeco.net

Ernst Schausberger
& Co. GmbH
Heidestrasse 19
A 4623 Gunskirchen,
Austria
T + 43 (0)7246 6493 0
E office@schausberger.com
W www.schausberger.com

Escofet 1886 SA
Ronda Universitat 20
E 08007 Barcelona,
Spain
T +34 (0)93 318 5050
F +34 (0)93 412 4465
E informacion@escofet.com
W www.escofet.com

Felicerossi
c/o Comunicabilita
Via Bertani 16
20154 Milan, Italy
E nocentini@tiscalinet.it

Festo & Co.
Corporate Design
KC-C1, Rechbergstrasse 3
D 73770 Denkendorf,
Germany
T +49 (0)711 347 3886
F +49 (0)711 347 3899
W www.festo.com

Fiam Italia SpA
Via Ancona 1/b
61010 Tavullia
Pesoro, Italy
T +39 (0)721 200 51
F +39 (0)721 202 432
E fiam@fiamitalia.it
W www.fiamitalia.it

Fiat Auto SpA
Corso G Agnelli
10010 Tonno, Italy
W www.fiat.com
    www.fiat.co.uk

Filsol Limited
Unit 15, Ponthenri
Industrial Estate
Ponthenri, Llanelli
Carmarthenshire
SA15 5RA, UK
T +44 (0)1269 860 229
F +44 (0)1269 860 979
E info@filsol.co.uk
W www.filsol.co.uk

Fingermax GmbH
Lindwurmstrasse 99
80337 Munich, Germany
T/F +49 (0)89 267417
W www.fingermax.de

Fisher Space Pen Co.
711 Yucca Street
Boulder City
NV 89005, USA
T +1 702 293 3011
F +1 702 293 6616
E fisher@spacepen.com
W www.spacepen.com

Fiskars Consumer Oy AB
10330 Billnäs,
Finland
T +358 (0)19 277721
F +358 (0)19 230986

Fiskeby Board AB
Box 1, Fiskeby,
SE-601 02 Norrköping,
Sweden
T +46 11 15 57 00
F +46 1115 59 95
E info@fiskeby.comT
W www.fiskeby.com

FlexForm Technologies
4955 Beck Drive
Elkhart
IN 46516, USA
T +1 586 598 7880
F +1 586 598 7859
W www.flexformtech.com

Flexipal Furniture
Flexipal Ltd
1 The Mount
76 Bedford Gardens
London W8 7EJ, UK
T +44 (0)20 7727 0486
E info@flexipal.com
W www.flexipal.com

Flint Ink
4600 Arrowhead Drive
Ann Arbor,
MI 48105 USA
T +1 734 622 6000
F +1 734 622 6060
E info@flintink.com
W www.flintink.com

**Flow Control Water
Conservation**
Conservation House
Brighton Street
Wallasey
Merseyside CH44 6QJ, UK
T +44 (0)151 638 8811
F +44 (0)151 638 4137
W www.waterconservation.co.uk

Forbo-Nairn Ltd
PO Box 1
Kirkcaldy
Fife KY1 2SB, UK
T +44 (0)1592 643 777
F +44(0)1592 643 999
W www.forbo-flooring.co.uk

Ford Motor Company
Customer Relationship
Center
PO Box 6248
Dearborn
MI 48126, USA
W www.ford.com

FP International
1090 Mills Way
Redwood City
CA 94063, USA
T +1 650 261 5300
F +1 650 361 1713
W www.fpintl.com

Franmar Chemical, Inc.
105 East Lincoln
PO Box 92
Normal
IL 61761, USA
T +1 309 452 7526
F +1 309 862 1005
E franmar@franmar.com
W www.franmar.com

Fredericia Furniture A/S
Treldevej 183
DK 7000 Fredericia,
Denmark
T +45 (0)75 92 33 44
F +45 (0)75 92 38 76
E sales@fredericia.com
W www.fredericia.com

Freeplay Energy Limited
Unit 12, Montague
Industrial Park
Montague Drive
Montague Gardens
7441, South Africa
T +27 (0)21 551 2002
F +27 (0)21 551 2096
W www.freeplay.net

**Freudenberg Building
Systems**
(Division of Freudenberg
Nonwovens LP)
Unit 6,
Wycliffe Industrial Park,
Leicester Road
Lutterworth, Leicestershire
LE17 4HG, UK
T +44 1455 204480
F +44 1455 556529
W www.freudenberg.com

Fritz Hansen A/S
Allerødvej 8
DK-3450 Allerød
Denmark
T +45 48 17 23 00
20–22 Rosebery Avenue
Clerkenwell
London EC1R 4SX, UK
T +44 (0)207 837 2030
W www.fritzhansen.com

Front Corporation
3-13-1 Takadanokata
Shinjuku-ku 169
Tokyo,
Japan
T +81 (0)3 3360 3391
F +81 (0)3 3362 6363

Furnature
86 Coolidge Avenue
Watertown
MA 02472, USA
T +1 617 926 0111
F +1 617 924 5432
E info@furnature.com
W www.furnature.com

Fusion
Unit 2.08 Oxo Tower Wharf
Bargehouse Street
London SE1 9PH, UK
T +44 (0)20 7928 4838
F +44 (0)20 7928 6969
E info@studiofusion.co.uk
W www.studiofusion.co.uk

Futureproof/ed
Hungaria Building, 1st floor
Vaartkom 35
B-3000 Leuven, Belgium
T +32 (0)16 50 37 50
F +32 (0)16 50 37 23
E info@futureproofed.com
W www.futureproofed.com

G T Design
Via del Barroccio 14/A
40138 Bologna, Italy
T +39 (0)51 535 951
F +39 (0)51 112
E customerservice@gtdesign.it
W www.gtdesign.it

Gebrüder Thonet GmbH
Michael Thonet Strasse 1
D-35066 Frankenberg,
Germany
T +49 (0)6451 508 119
F +49 (0)6451 508 108
E info@thonet.de
W www.thonet.de

General Motors (USA)
W www.gm.com

Gervasoni SpA
Zona Industriele Udinese
33050 Pavia di Udine, Italy
T +39 (0)432 656611
F +39 (0)432 656612
E ino@gervasoni1882.com
W www.gervasoni1882.com

Gibson Guitars
309 Plus Park Boulevard
Nashville, TN 37217, USA
T +1 615 871 4500
F +1 615 889 5509
W www.gibson.com

Gist
(formerly BOC Distribution
Services)
Rosewood, Crockford Lane
Chineham Business Park
Basingstoke,
Hants RG24 8UB, UK
T +44 (0)1256 891 111
W www.gistworld.com

Glas Platz
Auf den Pühlen 5
D 51674 Wiehl-Bomig,
Germany
T +49 (0)2261 7890 0
F +49 (0)2261 7890 10
E info@glas-platz.de
W www.glas-platz.de

Glindower Ziegelei GmbH
Alpenstrasse 47
D 14542 Glindow, Germany
T +49 (0)3327 66490
F +49 (0)3327 42662
E info@glindower-ziegelei.de
W www.glindower-ziegelei.de

Gloria-Werke
H.Schulte-Frankenfeld
GmbH & Co.
Diestedder Strasse 39
D 59329 Wadersloh,
Germany
T +49 (0)2523 77 0
F +49 (0)2523 77 120
E info@gloria.de
W www.gloria.de

Goods
218 Prinsengracht
1016 HD Amsterdam,
the Netherlands
T +31 (0)20 625 8405
F +31 (0)20 620 4457
E goods@goods.nl
W www.goods.nl

Govaerts Recycling NV
Kolmenstraat 1324
Industriepark Kolmen
Kolmenstraat 1324 B Alken,
Belgium
T +32 (0)11 59 01 60
F +32 (0)11 31 43 03
E info@govaertsrecycling.com
W www.govaertsrecycling.com

Green & Carter
Vulcan Works
Ashbrittle, Near Wellington
Somerset TA21 0LQ, UK
T +44 (0)1823 672365
E general@greenandcarter.com
W www.greenandcarter.com

GreenDisk
16398 NE 85th Street
Suite 100
Redmond
WA 98052, USA
T +1 425 883 9165
F +1 425 883 0425
W www.greendisk.com

GreenField Paper Company
1330 G Street
San Diego
CA 92101, USA
T +1 619 338 9432
F +1 619 338 0308
W www.greenfieldpaper.com

**Ground Support
Equipment (US)**
11 Broadway, space 905
New York
NY 10004, USA
T +1 212 809 4323
F +1 212 809 4324
W www.biomorphdesk.com

Grundig AG
Beuthener Strasse 43
D 90471 Nuremberg,
Germany
T +49(0)911 703 0
F +49 (0)911 703 9204
W www.grundig.de

Gusto Design Studio Ltd
6 Townmead Business
Centre
William Morris Way
London SW6 2SZ, UK
T +44 (0)20 7736 8828
F +44 (0)20 7736 7775
W www.gusto.co.uk

Hahn Kunststoffe GmbH
Gebäude 1027
D 55483 Hahn-Flughafen,
Germany
T +49 (0)6543 9886 0
F +49 (0)6543 9886 99
W www.hahnkunststoffe.de

Hans Grohe GmbH
& Co. KG
Auestrasse 5–9
D 77761 Schiltach,
Germany
T +49 (0)7836 51 0
F +49 (0)7836 51 1300
W www.hansgrohe.com

Happell
256 High Street
Glasgow G4 0QT, UK
T +44 (0)141 552 7723
F +44 (0)141 552 7813
E info@happell.co.uk
W www.happell.co.uk

Harwood Products
PO Box 224
Branscomb
CA 95417, USA
T +1 707 984 6181
F +1 707 984 6631
W www.harwoodproducts.com

Haworth, Inc.
One Haworth Center
Holland
MI 49423, USA
T +1 616 393 3000
F +1 616 393 1570
W www.haworth.com

**Hemp Textiles International
Corporation**
3200 30th Street
Bellingham
WA 98225, USA
T +1 360 650 1684
F +1 360 650 0523
E info@hemptex.com
W www.cantiva.com

Hemp Traders
2132 Colby Avenue, Suite#5
Los Angeles
CA 90025, USA
T +1 310 914 9557
F +1 310 478 2108
E contact@hemptraders.com
W www.hemptraders.com

Herman Miller, Inc.
855 East Main Avenue
PO Box 302
Zeeland
MI 49464, USA
T +1 616 654 6030
W www.hermanmiller.com

**Hock Vertriebs GmbH
& Co. KG**
Helmholtzstrasse 14
D 76297 Stutensee,
Germany
T +49 (0)7244 60 87 0
E info@thermo-hanf.de
W www.thermo-hanf.de

Homasote Company
PO Box 7240
West Trenton
NJ 08628, USA
T +1 609 883 3300
F +1 609 884 3497
W www.homasote.com

Honda
2-1-1 Minami Aoyama,
Minato-ku Tokyo 107-8556,
Japan
T +81 (0)3 3423 1111
W www.honda.com

Hoover Ltd
Pentrebach, Merthyr Tydfil
Mid-Glamorgan
CF48 4TU, UK
T +44 (0)1685 721 222
F +44 (0)1685 725 696
W www.hoover.co.uk

Hülsta GmbH & Co. KG
Karl-Hüls-Strasse 1
48703 Stadtlohn, Germany
T +49 (0)2563 86-0
F +49 (0)2563 86-1417
E huelsta@huelsta.de ›
W www.huelsta.de

Hunton Fiber (UK) Ltd
Rockleigh Court, Rock Road
Finedon
Northants NN9 5EL, UK
T +44 (0)1933 68 26 83
F +44 (0)1933 68 02 96
E admin@huntonfiber.co.uk
W www.hunton.no

**Husqvarna/The Electrolux
Group**
St. Göransgatan 143
SE-105 45 Stockholm,
Sweden
W www.husqvarna.com
    www.electrolux.com

IBM Corporation
1133 Westchester Avenue
White Plains
NY 10604, USA
W www.ibm.com

Ifö Sanitär AB
Box 140,
295 22 Bromölla, Sweden
T +46 (0)456 480 00
F +46 (0)456 480 48
E info@ifo.se
W www.ifo.se

Iform
Davidhallsgatan 20
Box 5055, SE-200 71 Malmö
Sweden
T +46 (0)40 303610
F +46 (0)40 302288
E info@iform.net
W www.iform.net

IKEA
IKEA of Sweden AB
Box 702 Älmhult
343 81 Småland,
Sweden
T +46 (0)47 68 10 00
F +46 (0)47 61 51 23
W www.ikea.com

Independence Technology
(a Johnston & Johnston
company)
W www.independencenow.com

Induced Energy Ltd
Westminster Road,
Brackley,
NN13 7EB, UK
T +44 (0)1280705 900
W www.inducedenergy.com

Ingeo/Cargill Dow
PO Box 5698,
MS 114
Minneapolis
MN 55440-5698, USA
T +1 989 633 1746
W www.cargilldow.com/ingeo

Inka Paletten GmbH
Bahnhofstrasse 21
85635 Siegertsbrunn /
München,
Germany
T +49 (0)81 02 77 42-0
F +49 (0)81 02 54 11
E info@inka-paletten.de
W www.inka-paletten.de

Interface, Inc.
2859 Paces Ferry Road
Suite 2000
Atlanta
GA 30399, USA
T +1 770 437 6800
W www.interfaceinc.com

Interlübke
Gebr. Lübke GmbH
& Co. KG
Ringstrasse 145
D 33373 Rheda-
Wiedenbrück,
Germany
T +49 (0)524 2121
F +49 (0)524 212 206
E info@interluebke.com
W www.interluebke.com

International Food Container Organization (IFCO)
Zugspitzstrasse 7
D 82049 Pullach, Germany
T +49 (0)89 744 91 0
F +49 (0)89 744 91 298
E info@ifcosystems.de
W www.ifcosystems.de

Interstuhl Büromöbel GmbH Co. KF
Brühlstrasse 21
D72469 Messstetten-Tieringen, Germany
T/F +49 (0) 7436 871
E info@interstuhl.de
W www.interstuhl.de

Inx International Ink Co.
651 Bonnie Lane
Elk Grove Village
IL 60007, USA
T +1 847 981 9399
F +1 847 981 9447
E inxinfo@inxintl.com
W www.inxink.com

Kayersberg Plastics
Madleaze Industrial Estate
Bristol Road
Gloucester
GL1 5SG, UK
T +44 (0)1452 316 500
F +44 (0)1452 300 436
W www.kayersberg-plastics.com

Keim Mineral Paints Ltd
Muckley Cross, Morville
Near Bridgnorth
Shropshire WV16 4RR, UK
T +44 (0)1746 714 543
F +44 (0)1746 714526
W www.keimpaints.co.uk

Kopf Solardesign GmbH + Co. KG
Stutzenstrasse 6
D 72172 Sulz-Bergfelden, Germany
T +49 (0)7454 75 0
F +49 (0)7454 75 302
E info@kopf-solardesign.com
W www.kopf-solardesign.com

Kronospan AG
Dekorative Holzwerkstoffe
Willisauerstrasse 37
CH 6122 Menznau, Switzerland
T +41 (0)41 494 94 94
F +41 (0)41 494 04 49
E info@kronospan.ch
W www.kronospan.ch

Kvadrat A/S
Lundbergsvej 10
DK-8400 Ebeltoft, Denmark
T +45 (0) 8953 1866
F +45 (0) 8953 1800
E kvadrat@kvadrat.dk
W www.kvadrat.de

K-X Industries
K-X Faswall Corporation
PO Box 180
Windsor
SC 29856, USA
T +1 803 642 9346
F +1 803 642 6361
W www.faswall.com

Kyocera Corporation
6 Takeda Tobadono-cho
Fushimi-ku
Kyoto 612-8501
Japan
T +81 (0)75 604 3500
F +81 (0)75 604 3501
W www.kyocera.com

Lampholder 2000 plc
Unit 8, Express Park
Garrard Way
Telford Way Industrial Estate (South)
Kettering
Northamptonshire
NN16 8TD, UK
T +44 (0)1536 520 101
F +44 (0)1536 523 014
E bc@lampholder.co.uk
W www.lampholder.co.uk

LCDA (La Chanvrière de l'Aube)
Rue du Général de Gaulle
B.P. 602
10208 Bar sur Aube, France
T +33 (0)3 25 92 31 92
F +33 (0)3 25 27 35 48
W www.chanvre.oxatis.com

Leclanché
48 Avenue de Grandson
CH 1401 Yverdon-les-Bains, Switzerland
T +41 (0)24 447 22 72
F +41 (0)24 445 24 42
E custservice@leclanche.ch
W www.leclanche.ch

LEDtronics
23105 Kashiwa Court
Torrance
CA 90505, USA
T +1 310 534 1505
F +1 310 534 1424
E info@ledtronics.com
W www.ledtronics.com

Levi Strauss & Co.
Global Headquarters
1155 Battery Street
San Francisco
CA 94111, USA
T +1 415 501 6000
F +1 415 501 7112
W www.levistrauss.com
www.levi.com

Lexon Design Concepts
98 ter. Boulevard Héloïse
BP 103
95103 Argenteuil Cedex, France
T +33 (0)1 39 47 04 00
F +33 (0)1 39 47 07 59
W www.lexon-design.com

Light Corporation
14800 172nd Avenue
Grand Haven
MI 49417, USA
T +1 616 842 5100
F +1 616 846 2144
E info@lightcorp.com
W www.lightcorp.com

Light Projects Ltd
23 Jacob Street
London SE1 2BG, UK
T +44 (0)20 7231 8282
F +44 (0)20 7237 4342
E info@lightprojects.co.uk
W www.lightprojects.co.uk

Ligne-Roset SA
Serrières de Briord
01471 Briord, France
T +33 (0)4 74 36 17 00
F +33 (0)4 74 36 16 95
W www.ligne-roset.com

LINPAC Environmental
Leafield Way
Leafield Industrial Estate
Corsham
Wiltshire SN13 9UD, UK
T +44 (0)1225 816 500
F +44 (0)1225 816 501
W www.linpac-environmental.com

Living Tree Paper Company
1430 Willamette Street, Suite 367
Eugene
OR 97401-4049, USA
T +1 541 342 2974
F +1 541 687 7744
E info@livingtreepaper.com
W www.livingtreepaper.com

Livos Pflanzenchemie GmbH und Co. KG
Auengrund 10
D 29568 Wieren, Germany
T +49 (0)5825 880
F +49 (0)5825 8860
E info@livos.de
W www.livos.de

Lloyd Loom of Spalding
Wardentree Lane
Pinchbeck
Spalding
Lincolnshire
PE11 3SY, UK
T +44 (0)1775 712 111
F +44 (0)1775 710 571
E info@lloydloom.com
W www.lloydloom.com

LSK Industries Pty Ltd
92 Woodfield Boulevarde
Caringbah
NSW 2229, Australia
T +61 (0)2 9525 8544
F +61 (0)2 9525 7601
W www.lsk.com.au

Luceplan SpA
Via E.T. Moneta 46
20161 Milan, Italy
T +39 (0)2 662 42 1
F +39 (0)2 662 03400
E luceplan@luceplan.it
W www.luceplan.it

Ludlow Coated Products
(formerly Simplex Products)
USA
W www.simplex-products.com

Lumatech Corporation
41636 Enterprise Circle
North Suite C
Temecula
CA 92590, USA
T +1 800 932 0637
F +1 800 345 5862
W www.carpenterlighting.com

LUMINO Licht Elektronik GmbH
Europark Fichtenhain A8
47807 Krefeld, Germany
T +49 (0)2151 8196 0
F +49 (0)2151 8196 359
E info@lumino.de
W www.lumino.de

Lusty's Lloyd Loom
W Lusty & Sons Ltd
Hoo Lane, Chipping
Campden, Gloucestershire
GL55 6AU, UK
T +44 (0)1386 841333
F +44 (0)1386 841322
E enquiries@lloyd-loom.co.uk
W www.lloyd-loom.co.uk

Magis SpA
Via Magnadola 15
31045 Motta di Livenza
(TV), Italy
T +39 (0)422 862600
F +39 (0)422 766 395
W www.magisdesign.com

Mallinson
Mangerton House
Mangerton, Bridport
Dorset DT6 3SG, UK
T +44 (0)1308 485 111
F +44 (0)1308 485 566
E info@bendywood.com
W www.mallinson.uk.com
www.bendywood.com

MAN Nutzfahrzeuge
Dachauer Straße 667
D-80995 Munich, Germany
T +49 (0)89 15 80 01
F +49 (0)89 150 39 72
E info@mn.man.de
W www.man-nutzfahrzeuge.de

Marlec Engineering Company
Rutland House
Trevithick Road , Corby
Northamptonshire
NN17 5XY, UK
T +44 (0)1536 201 588
F +44 (0)1536 400 211
E sales@marlec.co.uk
W www.marlec.co.uk

Masonite Corporation
Corporate Head Office
1600 Britannia Road East
Mississauga, Ontario
L4W 1J2, Canada
T +1 905 670 6500
W www.masonite.com

Mathmos
22–24 Old Street
London EC1V 9AP, UK
T +44 (0)20 7549 2700
E mathmos@mathmos.co.uk
W www.mathmos.com

Meadowood Industries, Inc.
33242 Red Bridge Road
OR 97321, USA
T +1 541 259 1303
F +1 541 259 1355
W www.meadowoodindustries
.com

Mercedes-Benz/Daimler Chrysler
See Daimler Chrysler

Metabo AG
Metabo-Allee 1
D 76622 Nürtingen,
Germany
T +49 (0)7022 72 0
F +49 (0)7022 72 2731
W www.metabo.de

Metabolix, Inc.
21 Erie Street
Cambridge
MA 02139, USA
T +1 617 492 0505
F +1 617 492 1996
E info@metabolix.com
W www.metabolix.com

MIO
1234 Hamilton Street,
Unit 2D
Philadelphia
PA 19123, USA
T +1 215 925 9359
E sales@mioculture.com
W www.mioculture.com

Monodraught Sunpipe Ltd
Halifax House
Cressex Business Park
High Wycombe
HP12 3SE, UK
T +44 (0)1494 897 700
F +44 (0)1494 532 465
E info@monodraught.com
W www.monodraught.com

Moonlight Aussenleuchten
Gewerbegebiet Hemmet
D 79664 Wehr, Germany
T +49 (0)7762 709 0
F +49 (0)7762 709 200
E info@moonlight.info
W www.moonlight.
outdoorlighting.de

Moooi
Minervum 7003, 4817 ZL
PO Box 5703
4801 EC, Breda,
the Netherlands
T +31 (0)76 578 4444
F +31 (0)76 571 0621
E info@moooi.com
W www.moooi.com

MOY Concept & Design
Klarendalseweg 530
6822 GZ Arnhem,
the Netherlands
T +31(0)26 443 9977
E hans@moy.nl

MUJI (Ryohin Keikatu Co. Ltd)
W www.mujionline.com
www.muji.co.uk

N Fornitore
Italy
c/o Purves and Purves
220–24 Tottenham
Court Road
London W1T 7QE, UK
T +44 (0)20 7580 8223

National Starch & Chemical Co.
10 Finderne Ave
Bridgewater
NJ 08807, USA
T +1 800 797 4992
F +1 609 409 5699
E nacinquiry@salessupport.com
W www.nationalstarch.com

The Natural Choice
See Bioshield

Natural Collection
PO Box 135,
Southampton,
Hampshire
SO14 0FQ, UK
T +44 (0)23 8024 8733
F +44 (0)870 331 3334
E info@naturalcollection.com
W www.naturalcollection.com

Nava, Italy
c/o Setsu Ito (p. 307)

NEC Deutschland GmbH
Reichenbachstrasse 1
D-85737 Ismaning
T +49 (0)8996 274 0
F +49 (0)89 96 274 500
W www.nec.com

New Leaf Paper
116 New Montgomery
Street
Suite 830
San Francisco
CA 94105, USA
T +1 415 291 9210
F +1 415 291 9353
E info@newleafpaper.com
W www.newleafpaper.com

News Design DfE AB
Stora Skuggans Väg 11
115 42 Stockholm,
Sweden
T +46 (0)8 15 39 29
F +46 (0)8 15 39 26

Nighteye GmbH/Nighteye Austria
Sporteye Austria
Lichteneckergasse 39
A-2511 Pfaffstätten,
Austria
T +43 (0)699 153 87 924
F +43 (0)2252 82 708
W www.nighteye.at

Nils Holger Moormann GmbH
An der Festhalle 2
D 83229 Aschau im
Chiemgau,
Germany
T +49 (0)80 52 9 04 50
F +49 (0)80 52 90 45 45
E info@moormann.de
W www.moormann.de

Nisso Engineering Co Ltd (NSE)
Nobuhiro Saito
Tokyo Takii Building
6-1, 1-chome,
Kanda , Kinbo-cho
Chiyoda-ku
Tokyo 101, Japan
T +81 (0)3 3296 9313/9204
F +81 (0)3 3296 9250

Nola Industrier AB
c/o Bernstrand & Co
Skånegatan 51
116 37 Stockholm,
Sweden
T +46 (0)8 641 91 10
F +46 (0)8 641 91 20
E mail@bernstrand.com
W www.bernstrand.com

Nordsjö
(Akzo Nobel Dekorativ)
205 17 Malmö,
Sweden
T +46 (0)40 35 50 00
F +46 (0)40 601 52 23
W www.nordsjo.no

Nova Cruz Products, Inc.
34 Sheep Road
Lee, NH 03824, USA
T +1 603 868 5985
F +1 603 868 5947
E info@novacruz.com
W www.novacruz.com

Nova Form/Kautzky
Mechanik
Schörgelgasse 21
A 8010 Graz, Austria
T +43 (0)316 822 263
F +43 (0)316 822 334
E novaform@novaform.com
W www.novaform.com

Novotex A/S
Ellehammervej 8
7430 Ikast, Denmark
T +45 (0)96 60 68 00
F +45 (0)96 60 68 10
E novotex@green-cotton.dk
W www.green-cotton.dk

Nuno Corporation
B1F Axis Building
5-17-1 Roppongi
Minato-ku
Tokyo 106-0032, Japan
T +81 (0)3 3582 7997
F +81 (0)3 3589 3439
W www.nuno.com

Nutshell Natural Paints
PO Box 72
South Brent
Devon TQ10 9YR, UK
T +44 (0)1364 73801
E info@nutshellpaints.com
W www.nutshellpaints.com

NYCEwheels
1603 York Avenue
New York
NY 10028, USA
T +1 212 737 3078
E info@NYCEwheels.com
W www.NYCEwheels.com

OFFECCT
Sweden
T +46 (0)504 415 00
E info@offect.se
W www.offecct.se

Ogallala Down Company
218 Prospector Drive
P.O. Box 830
Ogallala
NE 69153, USA
T +1 308 284 8403
E sales@ogallalacomfort
company.com
W www.ogallaladown.com

The Old Fashioned Milk
Paint Co. Inc
436 Main Street
Groton
MA 01450, USA
T +1 978 448 6336
F +1 978 448 2754
W www.milkpaint.com

Optare Group Ltd
Manston Lane
Leeds, West Yorkshire
LS15 8SU, UK
T +44 (0)113 264 5182
F +44 (0)113 260 6635
E info@optare.com
W www.optare.com

Ostermann & Scheiwe
Hafenweg 31
D 48155 Münster,
Germany

Oxo International
75 Ninth Avenue,
Fifth Floor
New York
NY 10001, USA
E info@oxo.com
W www.oxo.com

Pacific Northwest Fiber
PO Box 610
Plummer
ID 83851, USA
T +1 208 686 6800
F +1 208 686 6810

Pactiv Corporation
1900 West Field Court
Lake Forest,
IL 60045 USA
T +1 888 828 2850
W www.pactiv.com

Pallucco Italia SpA
Via Azzi 36
31040 Castagnole de Paese
(TV), Italy
T +39 (0)422 438800
E infopallucco@
palluccobellato.it
W www.palluccobellato.it

Panda Flooring Co.
1 Grange Park
Leicester LE7 9QQ, UK
T +44 (0)116 241 4816
F +44 (0)116 241 8889
W www.pandaflooring.co.uk

Paperback
Unit 2, Bow Triangle
Business Centre
Eleanor Street
London E3 4NP, UK
T +44 (0)20 8980 2233
F +44 (0)20 8980 2399
E sales@paperback.coop
W www.paperback.coop

Patagonia
8550 White Fir Street
P.O. Box 32050
Reno
NV 89523 USA
T +1-800 638 6464
W www.patagonia.com

PCD Maltron Ltd
15 Orchard Lane
East Molesey
Surrey KT8 0BN, UK
T/F +44 (0)20 8398 3265
E sales@maltron.com
W www.maltron.com

Phenix Biocomposites
PO Box 609
Mankato
MN 56002, USA
T +1 507 388 3434
F +1 507 344 522
E sales@phenixllc.com
W www.phenixbiocomposites
.com

Philips Design & Philips
Electronics
Building hwd
PO Box 218
5600 MD Eindhoven,
the Netherlands
T +31 (0)40 275 9066
F +31 (0)40 275 9091
W www.philips.com

Philips Lighting
Haarmanweg 25
4538 AN Temeuzen,
the Netherlands
T +31 (0)115 684 318
F +31 (0)115 684 448
W www.lighting.philips.com

Pinturas Proa
San Salvador de Budiño
Gánderas de Prado
36475 Porriño,
Spain
T +34 (0)986 34 6525
F +34 (0)986 34 6589
E proa@pinturasproa.com
W www.pinturasproa.com

Planet
c/o Creative Energy
Technologies
POB 149/ 2872 State Rt 10
Summit
NY 12175, USA
W www.cetsolar.com

Polti SpA
Via Ferloni 83
22070 Bulgarograsso (CO),
Italy
T +39 (0)31 939 111
F +39 (0)31 890 513
E contabilita@polti.it
W www.polti.it

Porous Pipe Ltd
Calder Mill,
Green Road, Colne
Lancashire BB8 7BW, UK
T +44 (0)1282 871 778
F +44 (0)1282 871 785
W www.porouspipe.co.uk

Potatopak Ltd
'Seafire Hanger'
Henstridge Airfield
Henstridge
Somerset BA8 0TN, UK
T +44 (0)1963 362744
F +44 (0)1963 364648
W www.potatoplatescom

Powabyke Ltd
3 Wood Street
Queen Square
Bath BA1 2JQ, UK
T +44 (0)1225 443 737
F +44 (0)1225 446 878
E sales@powabyke.com
W www.powabyke.com

Prairie Forest Products
P.O. Box 279, Neepawa
Manitoba, Canada
T +1 204 476 7700
E info@prairieforest.com
W www.prairieforest.com

Primeboard, Inc.
2441 North 15th Street
Wahpeton, ND 58075, USA
T +1 701 642 3286
F +1 701 642 3287
E sales@primeboard.com
W www.primeboard.com

PSA Peugeot Citroën
France
W www.peugeot.com

Prismo Travel Products
No. 1 Cavendish Court
South Parade, Doncaster
DN1 2DJ, UK
T +44 (0)1904 712 050
F +44 (0)1904 712 601
W prismo.co.uk

Publique Living
Westeinde 135B
2512 GW Den Haag,
the Netherlands
T +31 (0)70 361 6623
F +31 (0)84 221 9351
E euro@publiqueliving.com
W www.publiqueliving.com

Quinton & Kaines Ltd
Creeting Road, Stowmarket
Suffolk IP14 5AS, UK
T +44 (0)1449 612 145
F + 44 (0)1449 677 604
E sales@quintonkaines.co.uk
W www.quintonkaines.co.uk

Radius GmbH
Hamburger Strasse 8a
50321 Brühl, Germany
T +49(0)2232 7636 0
F +49(0)2232 7636 20
E info@radius-design.de
W www.radius-design.de

Recycline, Inc.
681 Main Street
Waltham
MA 02451, USA
T +1 781 893 1032
F +1 781 893 1036
E info@recycline.com
W www.recycline.com

**Red Bank Manufacturing
Company Ltd**
Atherstone Road
Measham, Swadlincote
Derbyshire DE12 7EL, UK
T +44 (0)1530 270 333
F + 44 (0)1530 273667
E sales@redbankmfg.co.uk
W www.redbankmfg.co.uk

Remarkable Pencils Ltd
56 Glentham Road
London SW13 9JJ, UK
T +44 (0)20 8741 1234
F +44 (0)20 8741 7615
E info@remarkable.co.uk
W www.remarkable.co.uk

Re-New Wood
104 N. W. 8th,
PO Box 1093,
Wagoner, OK 74467, USA
T +1 800 420 7576
F +1 918 485 5803
W www.renewwood.com

Renfe
Avda. Pio XXI
E 110-28036 Madrid, Spain
T+34 (0)91 300 66 00
W www.renfe.es

Rexite SpA
Via Edison 7
20090 Cusago (Milan), Italy
T +39 (0)29 039 0013
F +39 (0)29 039 0018
W www.rexite.it

riese und müller GmbH
Haasstrasse 6
64293 Darmstadt, Germany
T +49 (0)6151 366 86 0
F +49 (0)6151 366 86 20
E team@r-m.de
W www.rieseundmueller.de

**Ritter Energie-und
Umwelttechnik GmbH KG**
Ettlinger Strasse 30
D 76307 Karlsbad,
Germany
T +49 (0)7202 922 0
F +49 (0)7202 922 100
E ritter@paradigma.de
W www.paradigma.de

Robert Cullen & Sons
10 Dawsholm Avenue
Glasgow G20 0TS, UK
T +44 (0)141 945 2222
F +44 (0)141 945 3567
E sales@cullen.co.uk
W www.cullen.co.uk

Rodman Industries
P.O. Box 88
Oconomowoc,
WI 53066, USA
T +1 262 569 5820
E info@rodmanindustries.com
W www.rodmanindustries.com

Rohner Textil AG
CH 9435 Heerbrugg,
Switzerland
T +41 (0)61 722 2218
F +41 (0)61 722 7152
E info@climatex.com
W www.climatex.com

Roland Plastics
The High Street
Wickham Market
Woodbridge
Suffolk
IP13 0QZ, UK
T +44 (0)1728 747 777
F +44 (0)1728 748 222
W www.grouproland.com

**Rolls-Royce International
Limited**
65 Buckingham Gate
London SW1E 6AT, UK
T +44 (0)20 7222 9020
F +44 (0)20 7227 9178
W www.rolls-royce.com

Ron Ink Company
200 Trade Court
Rochester
NY 14624, USA
T +1 800 833 7383
F +1 716 529 3519

Röthlisberger
Sägeweg 11
CH-3073 Gümligen,
Switzerland
T +41 (0)31 950 21 40
F +41 (0)31 950 21 49
E kollektion@roethlisberger.ch
W www.roethlisberger.ch

Sandhill Industries
6898 S. Supply Way,
Suite 100
Boise
ID 83716, USA
T +1 208 345-6508
F +1 208 345-4424
E sales@sandhillind.com
W www.sandhillind.com

Sanford UK Ltd
Berol House
Oldmeadow Road
King's Lynn
Norfolk
PE30 4JR, UK
T +44 (0)1553 761 221
F +44 (0)1553 766 534
E mail@sanford.co.uk
W www.sanford.co.uk

**Save A Cup Recycling
Company**
Suite 2, Bridge House
Bridge Street,
High Wycombe
HP11 2EL, UK
T +44 (0)1494 510167
F +44 (0) 1494 510168
E info@save-a-cup.co.uk
W www.save-a-cup.co.uk

S C Bourgeois with Halton
c/o 3bornes Architectes
70 rue Jean Pierre Timbaud
75011 Paris, France
T + 33 (0)1 47 00 78 27
F + 33 (0)1 47 00 78 57
E 3bornes@magic.fr

Schäfer Werke GmbH
Postfach 1120
D 57290 Neunkirchen,
Germany
T +49 (0)2735 787 01
F +49 (0)2735 787 249
W www.schaefer-werke.de

**Schiebel Elektronische
Geräte AG**
Margaretenstrasse 112
A 1050 Vienna, Austria
T +43 (0)1 546260
F +43 (0)1 5452339
W www.schiebel.net

Schmitz Recycling BV
PO Box 1277
(Prodktieweg 6)
6040 KG Roermond
the Netherlands
T +31 (0)475 370 270
F +31 (0)475 340 212
E sales@schmitzfoam.com
W www.schmitzfoam.com

sdb Industries BV
De Beverspijken 20
PO Box 2197
5202 CD 's-Hertogenbosch
the Netherlands
T +31 (0) 633 91 33
F +31 (0) 633 91 45
E info@sdb-industries.nl
W www.sdb-industries.nl

Second Nature UK Ltd
Soulands Gate
Dacre, Penrith
Cumbria CA11 0JF, UK
T +44 (0)1768 486285
F +44 (0)1768 486285
E info@secondnatureuk.com
W www.secondnatureuk.com

Segway LLC
14 Technology Drive
Bedford
NH 03110, USA
T +1 603 222 6000
F +1603 222 6001
W www.segway.com

Serralunga
Via Serralunga,9
13900 Biella, Italy
T +39 (0)15 2435711
F +39 (0)15 31081
E serralunga1825@bmm.it
W www.serralunga.it

Shimano, Inc.
3-77 Oimatsu-cho,
Sakai Osaka 590-8577,
Japan
T +81 (0)72 223 3210
F +81(0)72 223 3258
W www.shimano.com

Shortstraw
Unit 1,
Avon Valley Country Park
Stidham Farm
Stidham Lane
Keynsham
BS31 1TS, UK
T +44 (0)117 972 0109
E nigel@short-straw.com
W www.short-straw.com

Simplicitas
Grävlingsbacken 2
SE 131 50 Saltsjö-Duvnäs,
Sweden
T +46 (0)8 661 00 91
F +46 (0)8 667 37 45
E info@simplicitas.se
W www.simplicitas.se

SLI Lighting Ltd
Otley Road
Charlestown
Shipley BD17 7SN, UK
T +44 (0)1274 532 552
F +44 (0)1274 531 672
E sylvana.uk.info@sylvana-lighting.com
W www.sylvana-lighting.com

Smart Deck Systems/Plastic Lumber
2600 W. Roosevelt Road
Chicago
IL 60608, USA
E info@smartdeck.com
W www.smartdeck.com

Smile Plastics
Mansion House
Ford, Shrewsbury
Shropshire SY5 9LZ, UK
T +44 (0)1743 850267
F +44 (0)1743 851067
E SmilePlas@aol.com
W www.smile-plastics.co.uk

Smith & Fong Company
Plyboo Bamboo Products
375 Oyster Point Blvd, #3
San Francisco
CA 94080, USA
T +1 650 872 1184
F +1 650 872 1185
E dino@plyboo.com
W www.plyboo.com

Smith Anderson & Co. Ltd
Fettykil Mills
Leslie
Fife KY6 3AQ, UK
T +44 (0)1592 746 000
F +44 (0)1592 743 888
W www.smithanderson.com

Solar Cookers International
1919 21st Street, Suite 101
Sacramento
CA 95814, USA
T +1 916 455 4499
F +1 916 455 4498
E info@solarcookers.org
W solarcooking.org

Sony Corporation Design Center
6-7-35 Kitashinagawa
Shinagawa-ku
Tokyo 141, Japan
T +81 (0)3 5448 7758
F +81 (0)3 5448 7822
W www.sony.net
www.sony.co.uk

Sony Ericsson Mobile Communications AB
Nya Vattentornet
22188 Lund, Sweden
T +46 (0)46 193000
F +46 (0)46 194390
W www.sonyericsson.com

Sound Service (Oxford)
55 West End
Witney
Oxfordshire OX8 1NJ, UK
T +44 (0)1993 704981
F +44 (0)1993 779569
E stephen@soundservice.co.uk
W www.acousticinsulation.co.uk

SRAM Corporation
Global Headquarters
1333 N. Kingsbury,
4th Floor
Chicago IL 60622, USA
T +1 312 664 8800
F +1 312 664 8826
W www.sram.com

Staber Industries, Inc.
4800 Homer Ohio Lane
Groveport OH 43125, USA
T +1 614 836 5995
F +1 614 836 9524
E info@staber.com
W www.staber.com

Starch Tech, Inc.
720 Florida Avenue
Mineapolis
MN 55426, USA
T +1 800 597 7225
E sti@startech.com
W www.starchtech.com

Steelcase, Inc.
901 44th Street SE
Grand Rapids
MI 49508, USA
T +1 247 2710
W www.steelcase.com

Stiletto DESIGN Vertrieb
Auguststrasse 2
D 10117 Berlin, Germany
T +49 (0)30 280 94 614
F +49 (0)30 280 94 615
E hallo@siltetto .de
W www.stiletto.de

Stokke Gruppen AS
Håhjem
N 6260 Skodje, Norway
T +47 (0)70 24 49 00
F +47 (0)70 24 49 90
E info@stokke.com
W www.stokke.com

Stomatex
Cornwall, UK
T +44 (0)1579 362 566
F +44 (0)1759 362 958
E info@stomatex.com
W www.stomatex.com

Strandwood Molding, Inc.
22705 Highway M-26
PO Box 360
Hancock
MI 49930, USA
T +1 906 487 9768
F +1 906 487 9770
W www.strandwood.com

Sun-Mar Corporation
600 Main Street
Tonawanda
NY 14150, USA
T +1 905 332 1314
F +1 905 332 1315
E compost@sun-mar.com
W www.sun-mar.com

Sundeala
Middle Mill
Cam, Dursley
Gloucestershire
GL11 5LQ, UK
T +44 (0)1453 542 286
F +44 (0)1453 549 085
W www.sundeala.co.u

Sunways AG
Macairestrasse 3–5
D 78467 Konstanz,
Germany
T +49 (0)7531 99677 0
F +49 (0)7531 99677 10
E info@sunways.de
W www.sunways.de

Supercool AB
Box 27, S-401 20 Göteborg.
Sweden
T +46 (0)31 42 05 30
F +46 (0)31 24 79 09
W www.supercool.se

Syndesis, Inc.
2908 Colorado Avenue
Santa Monica
CA 90403, USA
T +1 310 829 9932
F +1 310 829 5641
E inquiries@syndesisinc.com
W www.syndesisinc.com

Teisen Products Limited
Bradley Green
Redditch
Worcestershire
B96 6RP, UK
T +44 (0)1527 821 621
F +44 (0)1527 821 665

Teknos Group Oy
Takkatie 3
PO Box 107
FIN 00371 Helsinki,
Finland
T +358 (0)9 506 091
F +358 (0)9 5060
W www.teknos-group.com

Tendo Co. Ltd
1-3-10 Midarekawa
Tendo
Yamagata 994-8601,
Japan
T +81 (0)23 653 3121
F +81 (0)23 653 3454
E suga@tendo-mokko.co.jp
W www.tendo-mokko.co.jp

Tensar International
New Wellington Street
Blackburn
Lancashire BB2 4PJ, UK
T +44 (0)1254 262 431
F +44 (0)1254 266 868
E info@tensar-
international.com
W www.tensar.com

Tetrapak Ltd
Bedwell Road
Cross Lanes
Wrexham LL13 0UT, UK
T +44 (0)870 442 6000
F +44 (0)870 442 6001
W www.tetrapak.co.uk
www.tetrapak.com

Thermo Technologies
5360 Sterett Place,
Suite 115
Columbia
MD 21044, USA
T +1 410 997 0778
F +1 997 997 0779
W www.thermomax.com

Tonester
Huhdantie 4
FIN-21140 Rymättylä,
Finland
T +358 (0)2 252 1000
F +358 (0)2 252 1022
E tonester@durat.com
W www.durat.com

Tonwerk Lausen AG
Hauptstrasse 74
CH 4415 Lausen,
Switzerland
T +41 (0)61 927 95 55
F +41 (0)61 927 95 58
E twlag@twlag.ch
W www.twlag.ch

Toshiba Europe GmbH
Hammfelddamm 8
D 41460 Neuss
Germany
T +49 (0)2131 158 01
F +49 2131 158 341
W www.toshiba-teg.com

Toyota Motor Corporation
1 Toyota-cho
Toyota City
Aichi Prefecture 471-8571,
Japan
W www.toyota.com

Trevira GmbH & Co. KG
Philipp-Reis-Strasse 2
D 65795 Hattersheim,
Germany
T +49 (0)69 305 18 108
F +49 (0)69 305 16 341
E info@fra.trevira.com
W www.trevira.de

TrusJoist
USA
W www.trusjoist.com

TSA Inox
c/o 3bornes Architectes
70 rue Jean Pierre Timbaud
75011 Paris, France
T + 33 (0)1 47 00 78 27
F + 33 (0)1 47 00 78 57
E 3bornes@magic.fr

Ultra Plastics Ltd
Units 1–3 Heys Street
Bacup
Lancashire
OL13 9QL, UK
T +44 (0)1706 878984
F +44 (0) 1706 879463
E info@ultraplus.co.uk
W www.ultraplas.co.uk

Uni-Solar
3800 Lapeer Road
Auburn Hills
MI 48326, USA
T +1 248 475 0100
F +1 248 364 0510
E info@uni-solar.com
W www.uni-solar.com

Universal Pulp & Packaging
Milton of Campsie
Glasgow G65 8EE, UK
T +44 (0)1360 310 322
F +44 (0)1360 311 975

**Urethane Soy Systems
Company (USSC)**
100 Caspian Avenue
PO Box 500
Volga, SD 57071
T +1 888 514 9096
F +1 605 627 5869
W www.soyoyl.com

Vaccari Ltd
52 Greenway
Crediton
Devon EX17 3LP, UK
T +44 (0)1363 777746
E info@vaccari.co.uk
W www.vaccari.co.uk

Vauxhall
Griffin House
Osborne Road, Luton
Bedfordshire,
LU1 3YT, UK
T +44 (0)1582 721 122
F +44 (0)1582 427 400
W www.vauxhall.co.uk

Velcro
406 Brown Avenue
Manchester
NH 03103, USA
T +1 800 225 0180
F +1 603 669 9271
W www.velcro.com

Vestfrost A/S
Spangsbjerg Møllevej 100
DK 6705 Esbjerg Ø,
Denmark
T +45 (0)79 14 22 22
F +45 (0)79 14 23 55
E info@vestfrost.dk
W www.vestfrost.com

Vision Paper
KP Products Inc.
4930 Jefferson
Albuquerque
NM 87109, USA
T +1 505 294 0293
F +1 505 294 7040
E info@nmia.com
W www.visonpaper.com

Vitra (International) AG
Charles-Eames-Strasse 2
79576 Weil am Rhein,
Germany
T +49 (0)7621 702 0
F +49 (0)7621 702 17 20
E info@vitra.com
W www.vitra.com

Volkswagen (UK)
Yeomans Drive
Blakelands
Milton Keynes
Buckinghamshire
MK14 5AN, UK
T +44 (0)1908 601 777
F +44 (0)1908 663 936
W www.volkswagen.co.uk

Volvo Articulated Haulers
SE 405 08 Göteborg,
Sweden
T +46 (0)31 660 000
W www.volvo.com

WaterFilm Energy Inc
PO Box 128
Medford
NY 11763, USA
T +1 631 758 6271
F +1 631 758 0438
E info@gfxtechnology.com
W www.gfxtechnology.com

Wellman, Inc.
212 7th Street
Jersey City
NJ 07302, USA
E home@wellmaninc.com
W www.wellmaninc.com

Werth Forsttechnik
Seelbach 5
D 66687 Wadern,
Germany
T +49 (0)68 71 90 90 0
F +49 (0)68 71 90 90 50
W www.weihnachtsbaum.de

Weyerhaeuser Europe Ltd
Redmonstown,
Clonmel,
Co. Tipperary,
Ireland
T +353 (0)52 21166
F +353 (0)52 21815
W www.weyerhaeuser-
europe.com

Wiggly Wigglers (UK)
Lower Blakemere Farm
Blakemere
Herefordshire
HR2 9PX, UK
T +44 (0)1981 500 391
F +44 (0)1981 500 108
E wiggly@wigglywigglers.co.uk
W www.wigglywigglers.co.uk

Wilde & Spieth GmbH & Co.
Zeppelinstrasse 126
D 73730 Esslingen,
Germany
T +49 (0)711 31971 0
F +49 (0)711 31971 55
E info@wilde-spieth.com
W www.wilde-spieth.com

Wilkhahn
Wilkening + Hahne
GmbH& Co.
Fritz-Hahn-Strasse 8
D 31844 Bad Münder,
Germany
T +49 (0)5042 9990
F +49 (0)5042 999 226
E info@wilkhahn.de
W www.wilkhahn.com

Willat Writing Instruments
8548 Washington
Boulevard Culver City
CA 90232, USA
T +1 310 202 6000
F +1 310 202 0405
W www.sensa.com

WISA Wood Products
P.O. Box 203
(Niemenkatu 16)
FIN-15141 Lahti,
Finland
T +35(8 0)204 15 113
F +358 (0)204 15 112
E wood@upm-kymmene.com
W www.wisa.fi

**WKR Altkunst-
Stoffproduktions-
u.Vertriebsgellschaft mbH**
Entenpfuhl 10
D 67547 Worms,
Germany
T +49 (0)6241 43451
F +49 (0)6241 49579
E kontakt.wkr@t-online.de

Woodnotes Oy
Tallberginkatu 8
00180 Helsinki,
Finland
T +358 (0)9 694 2200
F +358 (0)9 694 2221
E woodnotes@woodnotes.fi
W www.woodnotes.fi

X-City Marketing
Hannover GmbH
Reuterstrasse 9
30159 Hannover,
Germany
E info@xcima.de
W www.xcima.de

Xerox Corporation
800 Long Ridge Road
Stamford
CT 06904, USA
W www.xerox.com

XO
RN 19
77170 Servon, France
T +33 (0)1 60 62 60 60
F +33 (0)1 60 62 60 62
W www.xo-design.com

Yemm & Hart
1417 Madison 308
Marquand
MO 63655, USA
T +1 573 783 5454
F +1 573 783 7544
E info@yemmhart.com
W www.yemmhart.com

**YKK Architectural Products,
Inc.**
1, Kanda Izumi-cho
Chiyoda-ku
Tokyo 101-8642, Japan
W www.ykk.com

Zanotta SpA
Via Vittorio Veneto 57
20054 Nova Milanese
(MI), Italy
T +39 (0)362 4981
F +39 (0)362 451 038
E zanottaspa@zanotta.it
W www.zanotta.it

ZEM Europe GmbH
Wohllebgasse 11
Postfach 1252
CH 8034 Zurich,
Switzerland
T +41 (0)1 210 4774
F +41 (0)1 210 4770
E zem@zem.ch
W www.zem.ch

The design strategies described with each product in Objects for Living, Objects for Working and Materials are listed below. They are grouped according to one of five lifecycle phases: Pre-production, including materials selection; Manufacturing/Making/ Fabrication; Distribution/ Transportation; Functionality and use; and Disposal/ End-of-life. Other strategies that do not easily fit into this product lifecycle are described under the heading Miscellaneous. Extended descriptions of each design strategy are given where appropriate. Reference should also be made to the Glossary of eco-design terms (p. 337).

PRE-PRODUCTION PHASE

Anti-fashion – a design that avoids temporary, fashionable styles.

Anti-obsolescence – a design that is easily repaired, maintained and upgraded so it is not made obsolete with changes in technology or taste.

Appropriate/intermediate technology – designs that apply appropriate levels of technology for the local economic, political, environmental and socio-cultural conditions.

Classic design – creating a design that will have socio-cultural durability.

Dematerialization – the process of converting products into services. A good example of dematerialization through timeshare of a product is a local community sharing a car 'pool' in which all individuals have the opportunity to use or hire a car when needed rather than own a car that stands idle for a large part of its life. Other examples include digital cameras where silver halide film is replaced by CCD chips, dematerializing part of the consumables cycle. Designing products used in the context of a dematerialized service may place unusual constraints on the design such as concentration on maintenance and longevity of parts.

Lifecycle Analysis (LCA) – the calculation on paper or with the use of software of the key areas of environmental impact of a product or building and subsequent effort to minimize such impacts by design.

Local economy/employment focused – creating products or buildings that generate local employment and/or nurture the local economy using local materials and resources.

Product-service-system (PSS) – creating a service (based upon infrastructure, network and ICT provision) whose products have less environmental impact than individually owned and consumed products.

Product take-back – a system under which manufacturers agree to take back a product when it has reached the end of its useful life so that components and/or materials can be reused or recycled (see also Producer responsibility). This can fundamentally change the essence of the design and engage the designer in examining design for assembly (DfA), disassembly (DfD) and remanufacture.

Retention of craft skills/hand-making – designs that encourage a diverse skill base.

Reusable product – a product that can be reused at the end of its initial lifespan for an identical, similar or new use.

Universal design – the application of widely accepted practices, components, fixtures, materials and technologies suitable for a wide range of end-uses.

Use of existing manufacturing capacity – using existing factories and plant with a proven low environmental impact to produce other goods.

PRE-PRODUCTION: MATERIALS SELECTION

Abundant materials from the lithosphere/geosphere – inorganic materials, such as stone, clay, minerals and metals from the earth's crust.

Biodegradable – decomposed by the action of microbes such as bacteria and fungi.

Biopolymers – plastics made from plants. Biopolymers can be composted and returned to nature.

Certified sources – materials that are independently certified as originating from sustainably managed resources, from recycled materials or conforming to a national or international eco-label.

Compostable – can be decomposed by microbes such as bacteria and fungi to release nutrients and organic matter.

Durable/extremely durable – tough, strong materials that do not break or wear and survive the life of the product or well beyond.

Lightweight – materials with a high strength-to-weight ratio.

Locally sourced materials – materials originating from close proximity to the point of manufacturing or production.

Low-embodied-energy materials – materials that require relatively little energy to extract and manufacture.

Non-toxic/Non-hazardous – not likely to cause loss of life or ill health to people or degradation of living ecosystems.

Precious materials – precious materials can serve to ensure the socio-cultural longevity of a product or building.

Reclaimed – materials saved for reuse on demolition of the built environment.

Recyclable materials – components of products that can be used in a new product.

Recyclate – material that has been made into a new material comprising wholly or partially recycled materials. An alternative term is 'recycled feedstock'.

Recycled – materials that have been processed (such as cleaned, graded, shredded, blended), then remanufactured.

Recycled content – materials that include some recycled and some virgin content. If a material has 100%-recycled content, it is a recycled material.

Reduction in materials used
– reducing the materials required to deliver the required design functions.

Renewable – a material that can be extracted from resources which absorb energy from the sun to synthesize or create matter. These resources include primary producers, such as plants and bacteria, and secondary producers, such as fish and mammals.

Single or mono-materials – consist of pure materials rather than mixtures. This facilitates recycling.

Stewardship sourcing – materials from certified sources and supply chain management.

Supply-chain management (green procurement) is the process of specifying that the goods or materials of suppliers meet minimum environmental standards. The specification may be that the goods will come from certified sources (e.g., the Forest Stewardship Council, national or international eco-labels), carry recognized accreditation (e.g. ISO 14001, EMAS) or meet trade association standards (e.g. National Association of Paper Manufacturers' recycled-paper logo in the UK).

Sustainable/from sustainable sources – materials that originate from managed resources which are forecast to last for a very long time and/ or are renewable resources (see above).

Waste materials – materials fabricated from production (factory), agricultural or consumer waste.

Production processes

Avoidance of toxic/ hazardous substances – avoiding substances liable to damage human health and living ecosystems.

Bio-manufacturing – using nature to help fabricate products in situ. For example, 'manufacturing' natural gourds by training them in special shapes for later use as packaging; growing plants to produce biopolymers (natural plastics).

Clean production – systems are put in place to reduce the impact of manufacturing goods by minimizing the production of waste and emissions to land, air and water. Closed-loop recycling (see below) technologies are often incorporated into clean production.

Closed-loop recycling/ production – the process of introducing waste streams back into the manufacturing process in a continuous cycle without loss of waste from that cycle. The textile and chemical industries often recycle chemical compounds used in processing their end-products, resulting in cleaner production.

Cold construction/ manufacturing – methods that require no heat or pressure and hence reduce energy consumption and facilitate disassembly.

Design for assembly (DfA) is a method of rationalizing and standardizing parts to facilitate the fixing together of components during production or manufacture.

Design for disassembly (DfD) is a method of designing products to facilitate cost-effective, non-destructive breakdown of the component parts of a product at the end of its life so that they can be recycled or reused.

Efficient use of raw and manufactured materials – reducing materials used and minimizing waste production.

Innovation of traditional (low impact) technologies – using inherently low environmental impact, traditional or craft, technologies in an innovative way, e.g. weaving plant fibres for furniture or boat construction.

Lightweight construction – reducing materials used but maintaining strength.

Low-energy manufacturing/ production/construction techniques/assembly – reducing the energy required to make components or products.

Reduced resource consumption – reducing materials used, especially raw materials extracted from the environment.

Reduction in embodied energy of materials and construction – considering the production process as an energy flow and trying to reduce the total energy used.

Reduction in use of consumables – reducing consumables used during the manufacturing process.

Reduction of production waste – achieved by more efficient designs and/or manufacturing processes.

Reusable buildings – demountable, modular buildings, which can be transported and reassembled in new locations.

Self-assembly – the final assembly is done by the consumer, thereby saving energy in the fabrication process.

Simple, low-cost construction – manufacturing with simple, inexpensive tooling and low-energy processes.

Zero waste production – the elimination of waste from the production process.

Recycling and reuse

Design for recyclability (DfR) – is a design philosophy that tries to maximize positive environmental attributes of a product, such as ease of disassembly, recyclability, maintenance, reuse or refurbishment, without compromising the product's functionality and performance.

Design for recycling (DfR) – considers the best methods to improve recycling of raw materials or components by facilitating assembly and disassembly, ensuring that materials are not mixed and appropriately labelling materials and components.

Materials labelling – assists with improved identification of materials for recycling.

Materials recycled at source – use of office, factory or domestic waste to make new products in situ.

Reuse of end-of-life components (remanufacturing) – taking back worn-out or old components/ products and refurbishing them to an 'as-new' standard for resale.

Reusing materials – reusing materials with out changing their original state. By comparison, recycling involves some reorganization or partial destruction of the material followed by reconstitution.

Reuse of redundant components – components formally manufactured for another use are re-employed in a new product.

Reused objects – any complete object reused in a new product.

Single material components – components made of one material (a mono-material component).

Use of ready-mades/ready-made components – components made for one product reapplied to a new or different type of product.

DISTRIBUTION/ TRANSPORTATION PHASE

Flat-pack products –products that can be stored flat to maximize use of transport and storage space.

Lightweight products – products that have been designed to be lightweight, yet retain full functionality, and as a result require less energy to transport.

Reduced energy use during transport/reduction in transport energy – this can be achieved by careful design of products to maximize packing per unit area and minimize weight per product.

Reusable packaging – packaging that can provide protection on more than one trip.

Self-assembly – designs that are assembled by the consumer, therefore saving valuable space in transport and storage.

FUNCTIONALITY AND USE PHASE

Socially beneficial designs

Affordable and key worker housing – high-quality living spaces for sectors of society that need support.

Alternative modes of transport for improved choice of mobility – reduces dependency on high-environmental-impact products such as the car and affords improved mobility options for minority groups, such as the disabled.

An aid to reduce population growth – helps keep the balance between population and resource availability and so slows environmental degradation, social exclusion and other problems.

Community ownership – encourages group rather than individual ownership and so improves the efficiency of product usage.

Design for need – A concept that emerged in the 1970s and was promoted by exponents such as the design academic Victor Papanek and by a landmark exhibition at the Royal College of Art, London, in 1976. Design for need concentrates on design for social needs rather than for creating 'lifestyle' products.

Emergency provision/ distribution of clean, safe water – products designed to reduce human mortality and disease.

Encourages recycling – products designed to facilitate recycling.

Equal access to information resources – products to enable minority groups, such as the disabled, to gain access to information resources.

Equal access for public services – products to enable minority groups, such as the disabled, full access to public services, such as transport.

Hire rather than ownership – products designed for hire rather than for personal ownership, receiving more efficient and economical use.

Reduced noise/noise pollution – products designed to minimize distress and disturbance caused by excessive noise.

Self-help design, design democracy – designs that encourage people to examine and implement their own design skills.

Tools for education, communication – designs that are, in themselves, tools to expand educational and communication possibilities.

Designs to reduce emissions/pollution/toxins

Free of CFCs and HCFCs – products, generally associated with the use of refrigerants, that do not use either chlorofluorocarbons (CFCs), which are greenhouse gases, or hydrochlorofluorocarbons (HCFCs), which are greenhouse gases and ozone-depleting gases.

Reduction in /avoidance of emissions (to water) – products whose production and use avoids or minimizes emissions of hazardous and toxic substances to water.

Reduction in /avoidance of emissions/pollution (to air) – products whose production and use avoids or minimizes emissions of hazardous and toxic substances to air, including greenhouse gases, hydrocarbons, particulate matter and cancer-causing substances (carcinogens).

Reduction in /avoidance of hazardous/toxic substances – products that are safe for human use because they contain little or no hazardous or toxic substances. There are international and national lists of banned substances. including chemicals and pesticides. Some companies produce their own lists, in addition to those that they legally have to comply with. Safe for human use does not necessarily mean safe for plant and other wildlife.

Zero emissions – refers to vehicles powered with electric motors or with hydrogen fuel-cell power systems that do not produce exhaust emissions of greenhouse gases (such as carbon dioxide, carbon monoxide, methane or oxides of nitrogen) or particulate matter (such as PM10s). A true zero-emission electric vehicle (EV) is one that uses electricity generated from renewable power rather than fossil fuel or nuclear sources.

Designs for improved functionality

Customizable – describes a product that the consumer can alter to his or her own specification or configuration.

Dual function – one product with two functions.

Improved ergonomics – products that are easier and more comfortable to use.

Improved health and safety – products that don't endanger health or safety or that promote better health.

Improved social well-being – products that encourage social interaction or deeper, more meaningful experience or where users contribute to making the experience.

Improved user-friendliness – products that are easier to understand and more fun to use.

Improved user functionality – products that serve their purpose better than previous designs.

Interactivity/ user involvement – engaging the user's abilities and skills in the product or building to improve the experience.

Modular design/modularity – products that can be configured in many ways to suit the user by changing the arrangement of individual modules. Modular design also offers the user the possibility of adding modules as needs require.

Multifunctional – a product capable of more than two functions.

Multi-use space – a space capable of being used for different types of functions.

Portable – a product that is easily transported for use in different locations.

Safe, i.e., non-toxic and non hazardous – a product without adverse effects on human health.

Universal/inclusive design – design that encourages use by a wide range of people with varying abilities; design that enables rather than disables.

Upgradable/upgradability – a product that is easy to upgrade by replacing old components/elements with new. This is especially important for technological products.

Designs to increase product lifespan/longevity

Design for ease of maintenance/ maintainability – products with good instructions and easy access to maintain or service parts that wear.

Durable – products that are tough, owing to strong materials and high-quality manufacturing, and so resistant to use and wear.

Ease of repair/repairability – products easy to assemble/ disassemble to repair worn or broken parts.

Designs to reduce energy consumption

Energy conservation – products designed to prevent loss of energy.

Energy efficient – products/ buildings designed to use energy efficiently.

Energy label standards – Products/buildings certified to independent energy consumption and conservation standards e.g. EU Energy Label, Energy Star rating, Blue Angel Eco-label, BREAM/ Eco-home rating by the Building Research Establishment (UK).

Energy neutral – products/ buildings that generate as much energy as they consume.

Fuel economy – products that use less fossil fuel energy than an earlier generation of products and so cause reduced emissions to air over their lifetime.

Human-powered products – products that use energy supplied by humans.

Hybrid power – products that combine two or more power sources, for example, hybrid electric/petrol or fuel cell/electric cars.

Improved energy efficiency – products with improved usage or output per unit of energy expended.

Integrated energy control systems – conservation, generation, reclamation of energy in a product/ building using an integrated ICT-based system.

Integrated or intelligent transport systems – transport systems that permit a range of mobility products to be used to offer a choice of mobility paths for the user.

Low voltage – products capable of operating on 12-volt or 24-volt electricity supply rather than higher voltages.

Natural lighting – products that encourage the use of natural lighting (rather than consuming electricity).

Rechargeable (batteries) – products that encourage repeat battery use by recharging from a mains or renewable power supply, and so reduce waste production.

Renewable power – electricity generated from products that convert the energy of the sun, wind, water or geothermal heat from the earth's crust.

Solar power (passive) – products that produce light or heat by absorbing the energy of the sun.

Solar power (generation) – products that generate electricity by absorbing the energy of the sun. These typically include products equipped with photovoltaic panels.

Recycling and reduction of waste production

Recyclable packaging/ containers – packaging and containers made of materials that can be recycled.

Reduction in use of consumables – products that reduce the use of consumables such as paper, inks, batteries, oils and detergents.

Reusable packaging/ containers – packaging and containers that can be reused for repeat trips.

Designs to improve water usage

Water conservation – products that reduce water usage, and/or facilitate water collection.

Water generation (freshwater) – products that generate fresh water from contaminated surface or ground water, seawater or water-saturated air.

DISPOSAL/END-OF-LIFE PHASE

Conservation of landfill space – products that decompose to release landfill space or products that can be recycled, reused or remanufactured to avoid being sent to landfill.

Encouraging local composting/local biodegradation of waste – products that can be locally decomposed by the owner, so saving on the transport energy of waste collection and landfill space.

Producer responsibility/ product take-back – a system under which manufacturers agree to take back a product when it has reached the end of its useful life so that it can be disassembled and components or materials can be reused or recycled. (See also Pre-production phase).

Recycling – products that are designed to be easily recyclable by being made of single materials or by being easily disassembled into materials or components that can be recycled.

Remanufacture – products that are easily disassembled for refurbishment or to remanufacture new products.

Reuse – products that are easily reused for the same or a new purpose or are easily disassembled for the components and/or materials to be reused.

MISCELLANEOUS STRATEGIES

Certification of products (see also Green Organizations, p. 329)

Eco-labels – labels attached to products which confirm that the manufacturers conform to independently certified standards in terms of reduced environmental impacts.

Independently certified labels – a variety of labels applied to products which signify that the products meet specific criteria for reduced environmental impacts, inclusion of recycled materials, and/or materials/ products from sustainable sources.

Environmental management and business systems

Corporate environmental policy – a written statement defining a company's position on the environment with an on-going audit of progress over time. Existence of a corporate environmental policy *usually* indicates inclusion of environmental management systems and/or the use of basic eco-design strategies in everyday business.

Eco Management and Audit Scheme (EMAS) – an independently certified environmental management system, which operates in the European Union. Certification is awarded by national bodies in individual EU countries verified by the EMAS organization.

ISO 14001 – an international standard for environmental management schemes maintained by the International Standards Organization (ISO) in Geneva, Switzerland. New standards are emerging for lifecycle assessment (ISO 14040) and eco-labelling and environmental labels (draft ISO 14021).

ISO 9001 – an international standard for quality assurance maintained by the International Standards Organization (ISO) in Geneva, Switzerland. Certification is granted by independent national organizations accredited by the ISO.

Biodiversity

Animal-friendly products – products that are manufactured without harm to animals.

Conservation of land resources – using 'brownfield' or under-used urban sites/locations rather than building on virgin, agricultural or rural land.

Encouragement of conservation and biodiversity – products that assist in promoting conservation and diversity as a result of a corporate environmental or supply-chain management policy or by sourcing materials from habitats managed to maintain diversity.

Low ecological footprint – minimizing inputs and outputs of materials and energy per capita for a product/building

Protection against soil erosion – products used to avoid or reduce soil erosion by water or wind.

Spatial economy – minimizing spatial requirements per capita.

ACADEMIC AND RESEARCH

Centre for Design at Royal Melbourne Institute of Technology (RMIT)
GPO Box 2476V
Melbourne
Victoria 3001, Australia
W www.cfd.rmit.edu.au

Centre for Environmental Assessment of Product and Material Systems (CPM)
Chalmers University
of Technology
Gothenburg, Sweden
T +46 (0)31 772 56 40
F +46 (0)31 772 56 49
E info@cpm.chalmers.se
W www.cpm.chalmers.se

Centre for Sustainable Design
Surrey Institute of Art
& Design
Farnham
Surrey GU9 7DS, UK
T +44 (0)1252 892 772
F +44 (0)1252 892 747
E cfsd@surrart.ac.uk
W www.cfsd.org.uk

Consortium on Green Design and Manufacturing (CGDM)
134 Hesse Hall
University of California
Berkeley
CA 94720, USA
T +1 510 643 5114
W greenmfg.me.berkeley.edu/

DEMI – Design for the Environment Multimedia Implementation Project
E info@demi.org.uk
W www.demi.org.uk
DEMI is a consortium of institutions comprising CTI Art & Design, the Design Council, Falmouth College of Arts, Forum for the Future, Goldsmiths College (lead institution), the Open University, the Royal Society for the Encouragement of Arts, Manufacturing and Business, Surrey Institute of Art and Design and the University of Brighton. The aim of the DEMI project is to create a multimedia

design and environmental teaching and learning resource for higher education. It is funded by HEFCE, the Higher Education Funding Councils for England, Wales and Northern Ireland.

Design Academy Eindhoven
Emmasingel 14
PO Box 2125
5600 CC Eindhoven,
the Netherlands
T +31 (0)40 239 3939
F +31 (0)40 239 3940
E info@designacademy.nl
W www.designacademy.nl

Design for Sustainability Program
TU Delft Subfaculty
of Industrial Design
Engineering
Jaffalaan 2628 BX Delft,
the Netherlands
T +31 (0)15 278 2738
F +31 (0)15 278 2956
E dfs@io.tudelft.nl
W www.io.tudelft.nl/research

The Designition Forum
Frank Wuggenig (Director)
Surbiton
Surrey KT5 9HE, UK
T +44 (0)20 8390 7682
F +44 (0)20 8390 7682
E info@designition.org.uk
W www.designition.org.uk

EcoDesign C@mpus
Luigi Bistanino
(Coordinator)
Politecnico di Torino
Turin, Italy
W www.ecodesigncampus.com

Environment Conscious Design and Manufacturing Lab (ECDM)
Department of Industrial
& Manufacturing Systems
Engineering
University of Windsor
Windsor
Ontario N9B 3P4, Canada
W athena.uwindsor.ca/units/
eng-industrial/IMSE.nsf

European Design Centre
PO Box 6279
5600 HG Eindhoven,
the Netherlands

T +31 (0)40 848 4848
F +31 (0)40 848 4844
E info@edc.nl
W www.edc.nl
Promotes high-quality integrated product development by provision of educational facilities, a database and support for engineers, product designers and industrialists.

Institute for Engineering Design – Austrian Ecodesign Information Point
Austrian Ministry of
Transport, Innovation and
Technology with Vienna
University of Technology
Austria
W www.ecodesign.at

Interduct/Clean Technology Institute (CTI)
Delft University of
Technology
Rotterdamseweg 145
2628 AL Delft,
the Netherlands
T +31 (0)15 27 83341
F +31 (0)15 27 86682
E Mail@Interduct.TUDelft.nl
W www.interduct.tudelft.nl

IVAM Environmental Research
University of Amsterdam
Postbus 18180
1001 ZB Amsterdam,
the Netherlands
T +31 (0)20 525 5080
F +31 (0)20 525 5850
E office@ivambv.uva.nl
W www.ivambv.uva.nl

ARCHITECTURE

American Institute of Architects
1735 New York Avenue NW
Washington
DC 20006, USA
T +1 202 626 7300
F +1 202 626 7547
E infocentral@aia.org
W www.aia.org
www.e-architect.com
The AIA publishes the Environmental Resource Guide, in cooperation with

the US Environmental Protection Agency (US EPA), which includes articles and case studies of the environmental impacts and resource issues for different building methods and materials.

Association for Environment Conscious Building (AECB)
PO Box 32
Llandysul, Carmarthenshire
SA44 5ZA, UK
E info@aecb.net
W www.aecb.net

Energy Efficiency and Renewable Energy
Mail Stop EE-1
Department of Energy
Washington
DC 20585, USA
T +1 202 586-9220
W www.eere.energy.gov

Centre for Alternative Technology
Machynlleth
Powys SY20 9AZ, UK
T +44 (0)1654 705 950
F +44 (0)1654 702 782
E info@cat.org.uk
W www.cat.org.uk
Established in 1975 as a resource centre to encourage a more ecological way of living, CAT now physically demonstrates ways in which buildings, renewable-energy technology and waste-water treatment can reduce environmental impact. CAT has also published extensive DIY and professional guides on all aspects of low-impact technology.

Centre for Sustainable Construction (CSC)
BRE – Building Research
Establishment
Garston, Watford
Hertfordshire WD2 7JR, UK
T +44 (0)1923 664 000
E enquiries@bre.co.uk
W www.bre.co.uk/sustainable

The Ecological Design Group
Fionn Stevenson
Scott Sutherland School
of Architecture

Faculty of Design
The Robert Gordon
University
Garthdee Road
Aberdeen AB9 2QB, UK
T +44 (0)1224 263 713
F +44 (0)1224 263 535
E f.stevenson@rgu.ac.uk
W www.rgu.ac.uk/subj/ecoldes/
edg1.htm

**Environmental Design
Research Association
(EDRA)**
PO Box 7146
Edmond
OK 73083-7146, USA
T +1 405 330 4863
F +1 405 330 4150
E edra@edra.org
W www.edra.org
*Founded in 1968 for the
advancement of the art and
science of environmental
design and research to
improve the understanding
of the interrelationships
between people and their
built and natural
surroundings.*

**Royal Institute of British
Architects**
66 Portland Place
London W1N 4AD, UK
T +44 (0)20 7580 5533
F +44 (0)20 7580 1541
E info@inst.riba.org
W www.architecture.com

US Green Building Council
1015 18th Street, NW,
Suite 805
Washington
DC 20036, USA
T +1 202 828 7422
F +1 202 828 5110
E info@usgbc.org
W www.usgbc.org

**ASSOCIATIONS – ECO-DESIGN,
GREEN DESIGN**

**Alternative Technology
Association**
PO Box 2919
Fitzroy
Victoria 3065, Australia
E ata@ata.org.au
W www.ata.org.au

EcoDesign Association
The British School
Slad Road
Stroud, Gloucestershire
GL5 1QW, UK
T +44 (0)1453 765 575
F +44 (0)1453 759 211
E ecological@designassociation.
freeserve.co.uk
W www.edaweb.org

Ecodesign Foundation
PO Box 369
Rozelle, Sydney
NSW 2039, Australia
T +61 (0)2 9555 7028
E edf@edf.edu.au
W www.edf.edu.au

O2 Network
W www.o2.org
*O2 Network coordinates
participating O2 groups
in sixteen countries
sharing information and
promoting discussion about
eco-design and sustainable
design in order to integrate
sustainability into the
design process.*

**Scottish EcoDesign
Association (SEDA)**
c/o Richard Atkins,
Secretary, SEDA
PO Box 14167
Tranent
East Lothian EH33 2YG, UK
T +44 (0)1875 614 105
E info@seda2.org
W www.seda2.org

**Society for Responsible
Design (SRD)**
PO Box 288
Leichhardt, Sydney
NSW 2040, Australia
T +61 (0)500 589 500
F +61 (0)2 9358 4151
E srd@green.net.au
W www.green.net.au/srd

AWARDS

Design Preis Schweiz
c/o D'S Design Center AG
Postfach 852
CH 4901 Langenthal,
Switzerland
T +41 (0)62 923 03 33
F +41 (0)62 923 16 22

E designcenter@designnet.ch
W www.designpreis.ch

**Good Design Award –
Ecology Design Prize**
Japan Industrial Design
Promotion Organization
(JIDPO)
G-Mark Division
4th Floor Annex
World Trade Center
Building
2-4-1 Hamamatsu-cho
Minato-ku
Tokyo 105, Japan
T +81 (0)3 3435 5633
F +81 (0)3 3432 7346
E p-div@jidpo.or.jp
W www.jidpo.or.jp/
*The Good Design Selection
System, with its G-Mark
logo for winning products,
was launched in 1957 and
became the Good Design
Award from April 1998.
Product categories are
wide-ranging and attract
thousands of entrants
from Japan and the
international design
community. A special
category, the Ecology Design
Prize, is awarded to
products with reduced
impact on the environment.*

**Industrie Forum Design
Hannover (iF)**
Messegelände
30521 Hannover, Germany
T +49 (0)511 89 32 402
F +49 (0)511 89 32 401
E info@ifdesign.de
W www.ifdesign.de
*This is one of the most
prestigious annual design
awards in Germany.
Categories include a
special Ecology Design
Award and Interaction
Design Award, as well as
more traditional themes
such as Product Design
Awards for office, business,
communications, home,
household, lighting,
consumer electronics,
lifestyle, public design,
packaging design, textile
design, building technology,
industry, transport, medical
and leisure. Winners of
the Ecology Design Award*

*are selected from any of
the subcategories in the
competition.*

**International Design
Resource Awards (IDRA)**
Design Resource Institute
347 NW 105th Street
Seattle
WA 98177, USA
T +1 206 289 0949
F +1 206 7893 144
E Designwithmemory@cs.com
W www.designresource.org
*The Design Resource
Institute, founded in 1993
by Tom and Barbara
Johnson in Seattle, USA,
operates the International
Design Resource Awards
competition that is
devoted to furthering
the development of
commercially viable
sustainable design in
products and architecture.
Now in its eighth year, the
IDRA competition requires
that entries have a high
degree of post-consumer
recycled or sustainably
harvested materials and
demonstrate added value
as a result, be designed
for future reuse or recycling,
and be suitable for
commercial production.*

**BUSINESS AND THE
ENVIRONMENT**

Enviroene
US Environmental
Protection Agency (EPA)
Ariel Rios Building
1200 Pennsylvania
Avenue NW
Washington
DC 20460, USA
T +1 202 272-0167
W http://es.epa.gov
*A government web resource
for pollution prevention,
compliance assurance,
enforcement information
and databases.*

The Future 500
415 Jackson Street,
2nd Floor
San Francisco

CA 94111, USA
T +1 415 364 3803
F +1 415 693 9163
E infor@globalff.org
W www.globalff.org/

International Institute of Sustainable Development (IISD)
Head Office
161 Portage Avenue East, 6th floor
Winnipeg
Manitoba R3B 0Y4, Canada
T +1 204 958 7700
F +1 204 958 7710
W www.iisd.org
www.iisd.ca

National Centre for Business and Sustainability
The Peel Building
University of Salford
Greater Manchester
M5 4WT, UK
T +44 (0)161 295 5276
F +44 (0)161 295 5041
W www.thencbs.co.uk

World Business Council for Sustainable Development (WBCSD)
4, chemin de Conches
Conches
CH 1231 Geneva, Switzerland
T +41 (0)22 839 3100
F +41 (0)22 839 3131
E info@wbcsd.ch
W www.wbcsd.ch

CERTIFICATION, ECO-LABELS AND ENERGY LABELS

British Standards Institute
BSI Quality Assurance
389 Chiswick High Road
London W4 4AL, UK
T +44 (0)20 8996 9000
F +44 (0)20 8996 7001
E cservices@bsi.org.uk
W www.bsi-global.com

Certified Forest Products Council
14780 SW Osprey Drive,
Suite 285
Beaverton
OR 97007, USA
T +1 503 590 6600
F +1 503 590 6655

E info@certifiedwood.org
W www.certifiedwood.org
Established in 1997, the Certified Forest Products Council is an independent, not-for-profit, voluntary business initiative committed to encouraging responsible forest management and the manufacture of environmentally responsible forest products in North America. The CFPC endorses the 'well-managed' standards defined by the Forest Stewardship Council. Members include suppliers, manufacturers, specifiers and individuals.

Duales System Deutschland AG – The Green Dot (Der Grüne Punkt)
Frankfurter Strasse 720-726
D 51145 Cologne (Porz Eil), Germany
T +49 (0)2203 93 70
F +49 (0)2203 93 7190
E pressestelle@gruener-punkt.de
W www.gruener-punkt.de
Founded in 1990, this not-for-profit organization administers Der Grüne Punkt (the Green Dot) packaging recycling scheme to comply with the 1991 German Packaging Ordinance. Any packaging marked with the Green Dot is acceptable for recycling. All types of packaging are accepted including glass, wood, ceramics, ferrous and non-ferrous metals, plastics and paper. This scheme is now licensed to a number of organizations in other EU countries – the ARA System (Austria), Ecoembalajes España (Spain), FOST Plus (Belgium), Repak Ltd (UK), Sociedade Ponto Verde SA (Portugal) and VALORLUX asbl (Luxembourg).

EMAS (Eco-Management and Audit Scheme)
c/o Bradley Dunbar Associates
Scotland House
Rond-Point Schuman 6
1040 Brussels, Belgium
T +32 (0)2 282 84 54
F +32 (0)2 282 84 54
E emas@cec.eu.int

EU Energy Label Scheme/European Commission
Environment DG
Information Centre Office:
BU-9 01/11
1049 Brussels
Belgium
F +32 (0)2 299 61 98
W europa.eu.int/comm/environment

FSC

FSC Forest Stewardship Council
UK Working Group:
Unit D, Station Building
Llanidloes
Powys SY18 6EB, UK
T +44 (0)1686 413 916
F +44 (0)1686 412 176
E hannah@fsc-uk.demon.co.uk
W www.fsc-uk.demon.co.uk
USA:
1155 30th Street NW,
Suite 300
Washington
DC 20007, USA
T +1 202 342 04113
F +1 202 342 6589
E info@fscus.org
W www.fscus.org
International HQ:
Avenida Hidalgo 502
Oaxaca 68000
Oaxaca, Mexico
F 52 951 62110.
E FSC@laneta.apc.org
Founded in 1993, the Forest Stewardship Council (FSC) is an independent, not-for-profit, non-governmental organization, which is responsible for administering, monitoring and tracking a programme to certify timber produced from well-managed

woodlands and labelling for products originating from such timber. It is an international programme implemented by independent organizations that are evaluated, accredited and monitored by the FSC. In the UK the Soil Association Woodmark Scheme and the SGS Forestry QUALIFOR Programme are both accredited. Other accredited organizations include the Rainforest Alliance SmartWood Program and Scientific Certification Systems Forest Conservation Program (USA), Silva Forest Foundation (Canada), Skal (Netherlands) and the Institut für Marktökologie (Switzerland). In the Directory of FSC Endorsed Forests Worldwide, Spring 2000, a total of 234 forests covering millions of hectares were certified in temperate, subtropical and tropical regions spanning thirty-four countries. A companion Directory of Manufacturers of FSC Endorsed Products Worldwide, Spring 2000, includes 825 manufacturers from forty-three countries. Products include sawn timber, veneers and finished products. The chain of custody is also inspected by the FSC, ensuring that endorsement with the FSC logo is not abused by agents, distributors, wholesalers or retailers. The extent of manufacturers' certification is most advanced in the Netherlands (seventy-five manufacturers), the UK (158) and the USA (185).

Global Ecolabelling Network (GEN)
GEN Secretariat
Terra Choice Environmental Services, Inc.
Mr John C. Polak
Suite 801,
1280 Old Innes Road
Ottawa

Ontario K1B 5M7, Canada
T +1 613 247 1900
F +1 613 247 2228
E jpolak@terrachoice.ca
www.terrachoice.ca
GEN is not accredited to
issue eco-labels but keeps
the most up-to-date list
of all eco-labelling
organizations worldwide
on its website and details
of the type of products and
materials currently covered.
GEN links directly with most
eco-labelling organizations'
websites.

**Group for Efficient Appliances
(GEA)**
c/o RAMBØLL
Teknikerbyen 31
DK 2830 Virum, Denmark
T +45 (0)45 98 87 92
F +45 (0)45 98 85 15
E kimj@ramboll.dk
W www.gealabel.org/
An association of energy-
labelling authorities in
European countries (includes
Austria, Denmark, Finland,
France, Germany, Sweden,
the Netherlands and
Switzerland), the European
Energy Network and the
European Association of
Consumer Electronics
Manufacturers (EACEM).
Labels are available for a
range of electronic
equipment from PCs to TVs.

**International Organization for
Standardization (ISO)**
Central Secretariat
1 Rue de Varembé
Case postale 56
CH 1211 Geneva 20
Switzerland
T +41 (0)22 749 01 11
F +41 (0)22 733 34 30
E central@iso.ch
W www.iso.ch

**National Association of Paper
Merchants (NAPM)**
PO Box 2850
Nottingham

NG5 2WW, UK
T +44 (0)115 841 2129
F +44 (0)115 841 0831
E info@napm.org.uk
W www.napm.org.uk
NAPM-approved recycled
paper and boards are
guaranteed to contain a
minimum of 75%-recycled
fibre content from genuine
paper and board waste, not
mill waste.

**Pan European Forest
Certification (PEFC)**
W www.pefc.org
The PEFC is a new scheme
initiated by the private
forestry sector. Forests
(and their timber and wood
product output) are certified
by independent auditors to
be managed in accordance
with the Pan European
Criteria on the Protection
of Forests in Europe,
which were resolved at
the Helsinki and Lisbon
Ministerial Conferences
in 1993 and 1998. It is
a scheme that offers a
common European
framework, in contrast to
the FSC scheme, which is
applied to forests worldwide.

ReSy GmbH
Postfach 101541
64215 Darmstadt, Germany
T +49 (0)6151 92 94 22
F +49 (0)6151 92 94 522
W www.resy.de/ind-eng.htm
This company certifies that
the content of paper and
corrugated board packaging
is suitable for recycling in
the German paper industry.
The ReSy logo is used with
the international recycling
logo of the Mobius loop.

SmartWood® Program
Rainforest Alliance
65 Bleecker Street
New York
NY 10012, USA
T +1 212 677 1900

E info@smartwood.org
W www.smartwood.org
www.rainforest-alliance.org
SmartWood is a program
of the Rainforest Alliance®,
which encourages
environmentally and
socially responsible forestry
management. SmartWood
has certified up to one
hundred operations
worldwide, which produce
a wide range of certified
lumber and products. The
Forest Stewardship Council
has accredited SmartWood
for its certification of forestry
operations. The SmartWood
Rediscovered Program
certifies salvaged or recycled
wood from demolished
buildings or waste sources.

Soil Association
Bristol House
40-56 Victoria Street
Bristol BS1 6BY, UK
T +44 (0)117 929 0661
F +44 (0)117 925 2504
E info@soilassociation.org
W www.soilassociation.org
WoodMark is the name
of the Soil Association's
international forestry and
chain of custody scheme.
The Soil Association is an
accredited organization to
the Forest Stewardship
Council and is permitted
to inspect and certify forests
and their products as
sustainably managed under
the FSC scheme. It is also
the leading organization in
the UK that independently
certifies farm produce as
being organically grown
and certified to bear the
Soil Association logo.

**US EPA Climate Protection
Partnerships Division**
ENERGY STAR Programs
Hotline & Distribution
(MS-6202J),
1200 Pennsylvania Ave NW
Washington

DC 20460, USA
E energystar@
optimuscorp.com
W www.energystar.gov
Energy Star labels for office
equipment, buildings and
more.

ECO-MATERIALS

**ATHENA™ Sustainable
Materials Institute Canada**
E wbtrusty@fox.nstn.ca and
E jkmeil@fox.nstn.ca
W www.athenasmi.ca

The BioComposites Centre
University of Wales
Bangor
Gwynedd LL57 2UW, UK
T +44 (0)1248 370 588
F +44 (0)1248 370 594
E biocomposites@bangor.ac.uk
W www.bc.bangor.ac.uk
Specializes in industrial
contract research on the
processing of wood and
plant materials to facilitate
the production of new
materials.

**Building Research
Establishment BRE**
Garston
Watford
Hertfordshire WD25 9XX,
UK
T +44 (0)1923 664000
E enquiries@bre.co.uk
W www.bre.org.uk
The BRE holds the National
Database of Environmental
Profiles for a wide range
of common building and
construction materials.
These Environmental
Profiles document the
materials' inputs, outputs
and lifecycle assessment,
enabling architects and
their clients, specifiers and
manufacturers to assess
the impacts of different
materials. Full access to the
database is subject to a fee.

**The Carbohydrate Economy
Clearinghouse**
c/o The Institute for Local
Self-Reliance
1313 5th Street SE

Minneapolis
MN 55414-1546, USA
T +1 612 379 3815
F +1 612 379 3920
E info@ilsr.org
W www.carbohydrate
economy.org

**Center for Environmentally
Appropriate Materials**
Department of Work
Environment
University of
Massachusetts, Lowell
1 University Avenue,
Lowell
MA 01854, USA
T +1 978 934 3275
F +1 978 934 3250
E kgeiser@turi.org
W www.uml.edu/Dept/WE/
centers.htm

**Certified Forest Products
Council**
14780 SW Osprey Drive
Suite 285
Beaverton
OR 97007, USA
T +1 503 590 6600
F +1 503 590 6655
E info@certifiedwood.org
W www.certifiedwood.org

EcoDesign Resource Society
Vancouver, Canada
E edrs@4sustainability.com
W www.vcn.bc.ca/edrs/

**Institute for Local Self
Reliance (USA)**
927 15th St NW, 4th floor
Washington
DC 20005, USA
T +1 202 898 1610
E info@ilsr.org
W www.ilsr.org
*The ILSR maintains
an online database of
materials called The
Carbohydrate Economy,
which lists state by state
the companies in the USA
that are manufacturing
materials from biological
sources. This includes
biofuels, biocomposites,
biopolymers, paints,
finishes and cleaners,
with examples of the use
of waste or recycled raw
materials.*

Material ConneXion
127 West 25th Street,
2nd floor
New York
NY 10001, USA
T +1 212 842 2050
F +1 212 842 1090
E info@materialconnexion.com
W www.materialconnexion.com
*Material ConneXion
maintains a database
of over three thousand
materials, including
materials derived from
or containing recycled
content. This privately
operated database is
available online and can
be visited in New York.*

**National Non-Food Crops
Centre (NNFCC)**
Innovation Centre
York Science Park
Innovation Way
Heslington
York YO10 5DG, UK
T +44 (0)1904 435182
F +44 (0)1904 435345
E enquiries@nnfcc.co.uk
W www.nnfcc.co.uk

New Uses Council
c/o Doane Agricultural
Services
11701 Borman Drive,
Suite 300
St Louis
MO 63146-4193, USA
T +1 314 372 3519
F +1 314 569 1083
E info@newuses.org
W www.newuses.org
*The New Uses Council
is dedicated to developing
and commercialising
new industrial, energy
and non-food consumer
uses of renewable
agricultural, forestry,
livestock and marine
products. It publishes
The BioProducts
Directory, an extensive
online listing of bio-
products, such as biofuels,
biocomposites
and biopolymers.*

**Pan European Forest
Certification (PEFC)**
PEFC Council asbl
2ème Etage

17 rue des Girondins
L 1626 Merl-Hollerich,
Luxembourg
T +352 (0)26 25 90 59
F +352 (0)26 25 92 58
E pefc@pt.lu
W www.pefc.org/internet/html

Proterra BV
Nude 54 A
6702 DN Wageningen
the Netherlands
T +31 (0)317 467 661
F +31 (0)317 467 660
E info@proterra.nl
W www.proterra.nl

Salvo (UK)
W www.salvo.co.uk
*Established in 1992, Salvo
is Europe's only association
coordinating the activities
of architectural salvage
companies and reclaimed
building materials suppliers.
Although members are
predominantly from the
UK, listings include
companies in Australia,
Belgium, Canada, France,
Ireland and the USA.*

Waste Watch
Lornamead House
1-5 Newington Causeway
London SE1 6ED, UK
E info@wastewatch.org
W www.wastewatch.org.uk/
*The UK Recycled Products
Guide was jointly published
by the National Recycling
Forum and Waste Watch
in 1998. It is available as
a bound copy or online
at www.nrf.org.uk and lists
over a thousand products
and materials. Data
includes type of material,
percentage of post-
consumer waste or
recovered material, brand
names, accreditation and
contact details of suppliers.*

ECO SHOPS

**Centre for Alternative
Technology**
Machynlleth
Powys SY20 9AZ, UK
T +44 (0)1654 705940

F +44 (0)1654 702 782
E info@cat.org.uk
W www.cat.org.uk

EcoMall
New York, USA
E ecomall@ecomall.com
W www.ecomall.com

**The Green Stationery
Company**
Studio One
114 Walcot Street
Bath BA1 5BG, UK
T +44 (0)1225 480 556
F +44 (0)1225 481 211
E jay@greenstat.demon.co.uk
W www.greenstat.co.uk

**Millennium Whole Earth
Catalog**
Whole Earth
PO Box 3000
Denville,
NJ 07834-9879, USA
T +1 888 732 6739
E info@wholeearthmag.com
W www.wholeearthmag.com

Natural Collection
PO Box 135
Southampton
Hampshire SO14 0FQ, UK
T +44 (0)8703 313 333
F +44 (0)1225 331 3334
E info@naturalcollection.com
W www.naturalcollection.com

**Real Goods and Jade
Mountain**
Real Goods
360 Interlocken Boulevard,
Suite 300, Broomfield
CO 80021-3440, USA
E techs@realgoods.com
W www.jademountain.com
www.realgoods.com

**Sustainability Souce™,
Inc., USA**
E info@sustainability
source.com
W www.sustainability
source.com

ENERGY

**Amazing Environmental
Organization Web
Directory – Alternative
Energy**

California, USA
W www.webdirectory.com/
Science/Energy/Alternative_
Energy

The British Wind Energy Association
Renewable Energy House
1 Aztec Row
Berners Road
London N1 0PW, UK
T +44 (0)20 7689 1960
F +44 (0)20 7689 1969
E info@bwea.com
W www.bwea.com
*Promotes the use of
renewable wind power
and has an extensive list
of publications for
commercial and domestic
generation, plus a list of
members and suppliers.*

Centre for Sustainable Energy, CSE
Create Centre
Smeaton Road
Bristol BS1 6XN, UK
T +44 (0)117 929 9950
F +44 (0)117 929 9114
E info@cse.org.uk
W www.cse.org.uk
*CSE provides research,
consultancy, education
and training in sustainable
energy technology and
systems. It also has
experience of delivering
local and regional initiatives
and lobbying to assist
development of appropriate
energy policies.*

Energy Efficiency and Renewable Energy Network (EREN) (USA)
W www.eere.energy.gov

GENERAL

Center for Renewable Energy and Sustainable Technology (CREST) and Renewable Energy Policy Program (REPP)
1612 K Street NW,
Suite 202
Washington
DC 20006, USA
T +1 202 293 2898
F +1 202 293 5857

E info2@repp
W www.CREST.org
*CREST operates Solstice,
an Internet information
service about renewable
energy.*

International Network for Environment Management (INEM)
Dr George Winter,
Chairman
Osterstrasse 58
D 20259 Hamburg,
Germany
T +49 (0)40 4907 1600
F +49 (0)40 4907 1601
E office@inem.org
W www.inem.org/

INTERNATIONAL, NATIONAL AND FEDERAL AGENCIES

Department for Environment, Food and Rural Affairs
Nobel House
17 Smith Square
London SW1P 3JR, UK
T +44 (0)20 7238 6000
W www.defra.gov.uk

Envirowise
UK
W www.envirowise.gov.uk
*A government
programme run by the
Department of the
Environment offering
practical environmental
advice for business,*

European Environment Agency
Kongens Nytorv 6
DK 1050 Copenhagen K
Denmark
T +45 (0)3336 7100
F +45 (0)3336 7199
E eea@eea.eu.int
W www.eea.eu.int

US Environmental Protection Agency
Ariel Rios Building
1200 Pennsylvania
Avenue NW
Washington
DC 20460, USA
T +1 202 272 0167
W www.epa.gov

LIFECYCLE ANALYSIS, LINKS AND ORGANIZATIONS

Life-Cycle Links by Thomas Gloria
E tgloria@icfconsulting.com
W www.life-
cycle.org/Academia.htm
*A comprehensive list of the
links for LCA from
academia, research
institutes, government and
international organizations,
together with companies
applying LCA.*

SETAC (Society of Environmental Toxicology and Chemistry)
Avenue de la
Toison d'Or 67
1060 Brussels, Belgium
T +32 2 772 72 81
F +32 2 770 53 86
E setac@setaceu.org
W www.setac.org

SETAC North America
1010 North 12th Avenue
Pensacola
FL 32501-3367, USA
T +1 850 469 1500
F +1 850 469 9778
W www.setac.org

SETAC Asia/Pacific
Christopher Hickey
National Institute of Water
& Atmospheric Resources -
NIWA Ecosystems
PO Box 11-115
Hamilton, New Zealand
F +64 (0)7 856 0151
E c.hickey@niwa.cri.nz
W www.setac.org

Society of the Promotion of Life-Cycle Assessment Development (SPOLD)
W lca-net.com/spold/

LCA SOFTWARE REVIEWS AND SUPPLIERS

Boustead Consulting
Black Cottage
East Grinstead
Horsham
West Sussex RH13 7BD, UK
T +44 (0)1403 864 561
F +44 (0)1403 865 284

E sales@boustead-
consulting.co.uk
W www.boustead-
consulting.co.uk
*Boustead Model version 4
is the most extensive, up-
to-date, lifecycle inventory
tool on the market today,
drawing on over twenty-
seven years' experience to
define inputs and outputs
for thousands of raw and
manufactured materials
and processes.*

Cambridge Engineering Selector (CES3)
Granta Design Ltd
Unit 300, Rustat House
62 Clifton Road
Cambridge CB1 7EG, UK
T +44 (0)1223 518 895
F +44 (0)1223 506 432
E info@grantadesign.com
W www.grantadesign.com
*Cambridge Engineering
Selector permits
simultaneous selection of
material, manufacturing
process and shape from
three interlinked
comprehensive databases.
An accompanying CD-ROM
provides access to online
documentation and
web links. Recently a new
Eco-data module has
been developed by
researchers at Cambridge
University, together with
connector software, which
allows interconnectivity
with Boustead Consulting's
Version 4 Life Cycle
Analysis programme.*

PRé Consultants BV
E info@pre.nl, support@pre.nl
W www.pre.nl
*Suppliers of ECO-it and
entry level LCA software
and SimaPro, a professional
package based upon the
Eco-indicator 99
methodology.*

TNO Institute of Industrial Technology
PO Box 6235
5600 HE Eindhoven
the Netherlands
T +31 (0)40 265 00 00
F +31 (0)40 265 03 01

E info@ind.tno.nl
W www.ind.tno.nl/en/
Suppliers of EcoScan3.0
entry level LCA software
based upon the Eco-
indicator 95 and 99
methodologies.

RECYCLING

Aluminium Packaging Recycling Organisation (Alupro)
1 Brockhill Court
Brockhill Lane
Redditch
Worcestershire
B97 6RB, UK
T +44 (0)1527 597757
F +44 (0)1527 594140
E info@alupro.org.uk
W www.alupro.org.uk
Alupro is a national
organization dedicated to
the collection and recycling
of aluminium drinks cans.
It claims a recycling rate of
36% (1998 data) of all the
aluminium cans sold, which
means that this material is
the most recycled type of
packaging in the UK.

Amazing Environmental Organization Web Directory – Recycling
California, USA
W www.webdirectory.com
/Recycling

American Plastics Council
1300 Wilson Boulevard,
Suite 800
Arlington
VA 22209, USA
T +1 703 741 5000
W www.plasticresource.com
The Council maintains an
online database of sources
of recycled plastics and
plastics feedstock in the
USA and Canada in
cooperation with the
Environment and Plastics
Industry Council (EPIC)
of Canada.

British Glass
9 Churchill Way
Chapeltown
Sheffield S35 2PY, UK

T +44 (0)114 290 1850
F +44 (0)114 290 1851
E info@britglass.co.uk
W www.britglass.co.uk
In 1998 glass recycling in
the EU exceeded 8 million
metric tonnes, of which
British glass constituted 6%,
or 476,000 tonnes. The
organization encourages
post-consumer collection
and recycling of glass.

British Metals Recycling Association (BMRA)
E info@recyclemetals.org
W www.recyclemetals.org
The British Metals Recycling
Association encourages
recycling of ferrous and
non-ferrous metals in the
UK. It publishes a directory
of members and provides
links to other associations
and organizations in the
metals recycling industry
worldwide.

Bureau of International Recycling
24 Avenue Franklin
Roosevelt
1050 Brussels, Belgium
T +32 (0)2 627 5770
F +32 (0)2 627 5773
E bir@bir.org
W www.bir.org
BIR is an international
trade association of the
recycling industries.

Deutsche Gesellschaft für Kunststoff-Recycling mbH
Frankfurter Strasse 720–726
51145 Cologne,
Germany
T +49 (0)22 0393 17745
F +49 (0)22 0393 17774
E info@dkr.de
W www.dkr.de
DKR recycled around
600,000 tonnes of plastics
in 1998 collected from
plastic packaging under
the Green Dot system (see
Duales System Deutschland
AG) and encourages
recycling and reuse of this
waste. DKR maintains an
online database of mainly
German companies that
manufacture materials
and products from recycled

plastics. The organization
works closely with the
design agency Bär + Knell
and organizes touring
exhibitions of their diverse
range of furniture, lighting
and fittings using recycled
plastics.

Industry Council for Electronic Equipment Recycling (ICER)
6 Bath Place
Rivington Street
London EC2A 3JE, UK
T +44 (0)20 7729 4766
F +44 (0)20 7729 9121
W www.icer.org.uk
The ICER is a cross-industry
group examining the best
way to improve recycling
and reuse of end-of-life
electronic equipment.

RECOUP
1 Metro Centre
Welbeck Way
Woodston, Peterborough
Cambridgeshire
PE2 7UH, UK
T +44 (0)1733 390 021
F +44 (0)1733 390 031
E enquiry@recoup.org
W www.recoup.org
RECOUP is the UK's
national plastic-bottle
recycling organization
with seventy-five members
including plastics
manufacturers, beverage
companies, retailers and
local authorities.

Waste Watch
Lornamead House
1-5 Newington Causeway
London SE1 6ED, UK
T +44 (0)20 7089 2100
and +44 (0)20 7939 0780
F +44 (0)20 7403 4802
E info@wastewatch.org.uk
W www.wastewatch.org.uk
Waste Watch publishes
an online directory of
products and materials in
the UK made from recycled
materials. Waste Watch
also manages the
independent National
Recycling Forum, which
promotes recycling.

SUSTAINABLE DEVELOPMENT

Centre for Environmental Strategy (CES)
University of Surrey
Guildford
Surrey GU2 5XH, UK
T +44 (0)1483 686670
F +44 (0)1483 686671
E cesinfo@surrey.ac.uk
W www.surrey.ac.uk/eng/ces
/contact.htm

European Foundation for the Improvement of Living and Working Conditions
Wyattville Road
Loughlinstown
Dublin 18, Ireland
T + 353 (0)1 2043100
F + 353 (0)1 2826456,
+ 353 (0)1 2824209
E postmaster@eurofound
.eu.int
information@eurofound.eu.int
W www.eurofound.eu.int

International Institute of Sustainable Development (IISD)
61 Portage Avenue East,
6th Floor
Winnipeg, Manitoba
R3B 0Y4 Canada
T +1 204 958-7700
F +1 204 958-7710
E info@iisd.ca
W www.iisd.org

National Councils for Sustainable Development
NCSD Program
Earth Council Secretariat
PO Box 319-6100
San Jose, Costa Rica
T +506 205 1600
F +506 249 3500
E info@ncsdnetwork.org
W www.ncsdnetwork.org

The Product-Life Institute (Institut de la Durée)
Geneva, Switzerland
W www.product-life.org
An independent contract
research institute developing
innovative strategies
and policies to encourage a
sustainable society. The
Institute provides
consultancy services to
government, industry and
universities.

Rocky Mountain Institute
1739 Snowmass Creek
Road
Snowmass
CO 81654-9199, USA
T +1 970 927 3851
F +1 970 927 3420
E info@rmi.org
W www.rmi.org

Sustainability Web Ring
Sustainability Development
Communications Network
(SDCN), c/o IISD
161 Portage Avenue East,
6th floor
Winnipeg
Manitoba R3B 0Y4,
Canada
T +1 204 958 7700
F +1 204 958 7710
W sdgateway.net

Tellus Institute
11 Arlington Street
Boston
MA 02116-3411, USA
T +1 617 266 5400
F +1 617 266 8303
E info@tellus.org
W www.tellus.org

United Nations Sustainable Development
United Nations Plaza
Room DC2-2220
New York
NY 10017, USA
T +1 212 963 3170
F +1 212 963 4260
E dsd@un.org
W www.un.org/esa/sustdev/

World Resources Institute
10 G Street NE, Suite 800
Washington
DC 20002, USA
T +1 202 729 7600
F +1 202 729 7610
E lauralee@wri.org
W www.wri.org

Wuppertal Institute
(Wuppertal Institut für
Klima, Umwelt, Energie)
PO Box 10 04 80
D 42204 Wuppertal,
Germany
T +49 (0)202 2492 0
F +49 (0)202 2492 108
E info@wupperinst.org
W www.wupperinst.org

TRADE AND BUSINESS
ASSOCIATIONS

Alliance for Beverage Cartons and the Environment (ACE)
250 Avenue Louise, Box
106
1050 Brussels, Belgium
T + 32 (0)2 504 07 10
F + 32 (0)2 504 07 19
E info@ace.be
W www.ace.be
ACE is an association of leading producers of beverage cartons and paperboard, which provides information on the impact of these products on the environment.

American Plastics Council
1300 Wilson Boulevard,
Suite 800
Arlington
VA 22209, USA
T +1 703 741 5000
W www.americanplastics council.org

Association of Plastics Manufacturers Europe (APME)
Avenue E van
Nieuwenhuyse 4, Box 3
1160 Brussels, Belgium
T +32 (0)2 676 17 32
F +32 (0)2 675 39 35
E info.apme@apme.org
W www.apme.org
An extensive resource on plastics including information about plastics and the environment together with detailed eco-profiles of common plastics.

British Plastics Federation
6 Bath Place
Rivington Street
London EC2A 3JE, UK
T +44 (0)20 7457 5000
F +44 (0)20 7457 5045
W www.bpf.co.uk

Composite Panel Association (CPA) and Composite Wood Council (CWC)
18928 Premiere Court
Gaithersburg
MD 20879-1574, USA

T +1 301 670 0604
F +1 301 840 1252
E info@pdmdf.com
W www.pbmdf.com
The CPA is a US and Canadian organization devoted to promoting the use and acceptance of particleboard, MDF and other similar products. Operating from the same headquarters, the CWC is an international organization that provides a forum for members of the particleboard and fibreboard industries and promotes their products.

The Institute of Packaging
Sysonby Lodge
Nottingham Road
Melton Mowbray
Leicestershire
LE13 0NU, UK
W www.iop.co.uk

National Association of Paper Merchants (NAPM)
PO Box 2850
Nottingham
NG5 2WW, UK
T +44(0)115 841 2129
F +44 (0)115 841 0831
E info@napm.org.uk
W www.napm.org.uk
This is the trade association representing UK paper merchants. It operates the NAPM Approved Recycled scheme, in which use of its logo guarantees a minimum content of 75% paper and board waste.

TRANSPORT AND HUMAN-
POWERED VEHICLES

Calstart
2181 E. Foothill Boulevard
Pasadena
CA 91107, USA
T +1 626 744-5600
E calstart@calstart.org
W www.calstart.org
Commercial company with an extensive catalogue of electric and hybrid bicycles, cars and commercial vehicles.

International Human Powered Vehicle Association (IHPVA)
PO Box 1307
San Luis Obispo
CA 93406-1307, USA
E bwilson@ihpva.org
W www.ihpva.org

Also refer to Eco-Design Strategies (p. 324)

5Rs is a concept with five cornerstones aimed at reducing the impact of design, manufacturing and products on the environment - to reduce, remanufacture, reuse, recycle and recover (energy by incineration). 'Reduce' implies designing to use fewer raw materials and less energy.

Agenda 21 is a comprehensive blueprint for global action drafted by the 172 governments present at the 1992 Earth Summit organized by the United Nations in Rio de Janiero, Brazil. It is often interpreted and implemented at a local level in 'Local Agenda 21' plans.

Atmosphere refers to the gaseous components at and above the world's surface, including the important gases oxygen, hydrogen, nitrogen, carbon dioxide, methane and ozone.

Biosphere is the term for the living components of the world that meet the seven characteristics of life – movement, feeding, respiration, excretion, growth, reproduction and sensitivity.

Carcinogens are chemicals that are definite or potential agents in causing cancer in humans. They are classified by the World Health Organization according to their perceived risk. Group 1 chemicals carry clear evidence of risk, Group 3 chemicals may have some associated risk.

Carrying capacity is a finite quantity (K) that equates to the ecosystem resources of a defined area such as a locality, habitat, region, country or planet. A given carrying capacity can support a finite population of organisms. Stable populations in harmony with the carrying capacity are sustainable, but excessive population growth can lead to sudden decline and/or permanent reduction in the carrying capacity.

Corporate social responsibility (CSR) is the integration of social and environmental policies into day-to-day corporate business and the involvement of internal and external stakeholders to deliver these policies.

Design for environment (DfE) is the analysis and optimization of the environmental, health and safety issues considered over the entire life of the product. DfE permits resource depletion, waste production and energy usage to be reduced or even eliminated during the manufacture, use and disposal or reuse of the product.

Design for manufacturing (DfM) examines the relationship between resource usage and product design using computer-aided design (CAD) and computer-aided manufacturing (CAM) tools for cost-effectiveness and reduced environmental impacts.

Design for X (DfX) is a generic term where X denotes the specific focus of a design strategy, such as DfD (Design for disassembly) or DfE (Design for environment).

Downcycling refers to the recycling of a waste stream to create a new material that has properties inferior to those of the original virgin materials. A good example is recycled plastic (HDPE) panels made of multicoloured waste sources.

Eco-efficiency embodies the concept of more efficient use of resources with reduced environmental impacts resulting in improved resource productivity, i.e., doing more with less.

Eco-label refers to labels applied to products and materials that conform to standards set by independent organizations to reduce environmental impacts. There are national and international eco-labels – see p. 331 for a detailed listing.

Eco-tools A generic name for software or non-software tools that help with the analysis of the environmental impact of products, manufacturing processes, activities and construction. Tools generally fall into several main categories: lifecycle analysis, design, environmental management or eco-audits and energy flow management.

Eco-wheel, or eco-design strategy wheel, is a means of identifying strategies that will assist in making environmental improvements to existing products. It embraces eight strategies: 1) selection of low-impact materials; 2) reduction of materials usage; 3) optimization of production techniques; 4) optimization of distribution system; 5) reduction of impact during use; 6) optimization of initial lifetime; 7) optimization of end-of-life system; and 8) new concept development.

Eco-design is a design process that considers the environmental impacts associated with a product throughout its entire life from acquisition of raw materials through production, manufacturing and use to end of life. At the same time as reducing environmental impacts, eco-design seeks to improve the aesthetic and functional aspects of the product with due consideration to social and ethical needs. Eco-design is synonymous with the terms design for environment (DfE), often used by the engineering design profession, and lifecycle design (LCD) in North America.

Eco-materials are materials that have minimal impact on the environment at the same time as providing maximum performance for the required design task. Eco-materials originating from components from the biosphere are biodegradable and cyclic, whereas eco-materials originating from the technosphere are easily recyclable and can be contained within 'closed-loop' systems.

Ecological footprint is a measure of the resource use by a population within a defined area of land, including imported resources. Assessment of the ecological footprints of nation states or other defined geographic areas reveals the true environmental impact of those states and their ability to survive on their own resources in the long term. The term ecological footprint can also be applied to products but is more commonly referred to as the environmental 'rucksack' associated with product manufacturing.

EcoReDesign (ERD) was first coined by the Royal Melbourne Institute of Technology, Australia, to denote the redesigning of existing products to reduce the environmental impact of one or more components of the product.

Embodied energy is the total energy stored in a product or material and includes the energy in the raw materials, transport to the place of production, energy in manufacturing and (sometimes) transport energy used in the distribution and retail chain. It is measured in MJ per kg or GJ per tonne.

End of life (EoL) describes both the end of the life of the actual product and the cessation of the environmental impacts associated with the product. Disassembly and recycling of components and materials at a product's EoL are preferable to disposal via landfill or incineration.

Environment-conscious manufacturing (ECM) is the application of green engineering techniques to manufacturing to encourage greater efficiency and reduction of emissions and waste.

Environmental management systems (EMS) are aimed at improving the environmental performance of organizations in a systematic way integrated with legislative and compliance requirements. The international bench-mark for EMS is the

International Standard ISO 14001, which more and more organizations each year are meeting, but national EMS standards also play a significant role, such as the British Standard for Environmental Management, BS 7750. Other independently certified systems exist, such as EMAS operated in the European Union.

EU Energy label is a classification applied to domestic appliances such as washing machines and refrigerators according to their energy use, expressed as kWh per year. Group A are the most energy-efficient and Group G are the least efficient. This scheme is due to be expanded to other types of appliances.

Geosphere consists of the inorganic, geological components of the world such as minerals, rocks and stone, sea and fresh water.

Green design is a design process in which the focus is on assessing and dealing with individual environmental impacts of a product rather than on the product's entire life.

Greenhouse gases are any man-made gaseous emission that contributes to a rise in the average temperature of the earth, a phenomenon known as global warming, by trapping the heat of the sun in the earth's atmosphere. The key greenhouse gases include carbon dioxide, mainly from fossil-fuel burning activities; methane from landfill sites, agriculture and coal production; chlorofluorocarbons (CFCs), hydro-chlorofluorocarbons,

(HCFCs) and hydrofluoro-carbons (HFCs), used in refrigerants and aerosols; nitrous oxide from nylon and nitric acid production, fossil-fuel burning and agriculture; and sulphur hexafluoride from the chemical industry.

Grey water is the waste water from personal or general domestic washing activities.

Industrial ecology is a holistic approach that considers the interaction between natural, economic and industrial systems. It is also termed industrial metabolism.

Information and Communications Technology (ICT) is the deployment of computers, telecommunications, networks and skills to create systems that deliver more than the sum of their parts.

Intelligent transport system (ITS) is a series of integrated transport networks in which individual networks use specific transport modes but allow easy interconnection to facilitate efficient movement of people.

Lifecycle analysis or Lifecycle assessment (LCA) is the process of analyzing the environmental impact of a product from the cradle to the grave in four major phases: production; transport/distribution/ packaging; usage; disposal or end of life/design for disassembly or recycling.

Lifecycle inventory (LCI) is the practice of analyzing the environmental consequences of inputs required and outputs generated during a product's life.

Lifecycle matrix is a tool or checklist to analyze potential environmental impacts at each phase in the product's lifecycle. Different types of industry create specific lifecycle matrices related to the peculiarities of the manufacturing process of their products.

Lithosphere is the geological strata that make up the earth's crust.

Mobility path describes a route an individual can take travelling between two points using one or more forms of transport which are, preferably, integrated into a flexible system (see ITS).

Non-renewable resources are those in finite supply that cannot be regenerated or renewed by synthesizing the energy of the sun. Such resources include fossil fuels, metals and plastics. Improving the rate of recycling will extend the longevity of these resources.

Off-gasing is the term for emissions of volatile compounds to the air from synthetic or natural polymers. Emissions usually derive from the additives, elastomers, fillers and residual chemicals from the manufacturing process rather than from the long, molecular-chain polymers.

Post-consumer waste is waste that is collected and sorted after the product has been used by the consumer. It includes glass, newspaper and cans from special roadside 'banks' or disposal facilities. It is generally much more variable in composition than pre-consumer waste (see below).

Pre-consumer waste is waste generated at the manufacturing plant or production facility.

Producer responsibility (PR) prescribes the legal responsibilities of producers/manufacturers for their products from the cradle to the grave. Recent European legislation for certain product sectors, such as electronic and electrical goods, packaging and vehicles, sets specific requirements regarding 'take-back' of products and targets for recycling components and materials.

Product lifecycle (PLC) is the result of a lifecycle assessment of an individual product, which analyzes its environmental impact.

Product-service-systems (PSS) are designed service systems comprising products, along with supporting infrastructure and networks, that deliver less environmental impact than individually consumed products to meet similar needs.

Product stewardship is the concept of manufacturing responsibility extending beyond the retail or business purchase to include the entire life of the product including its disposal or take-back.

Rebound effect refers to the undesirable environmental impacts that may be generated directly or indirectly by eco-efficient products, such as causing increased demand or new behavioural tendencies for other linked products, services or opportunities.

Renewable resources refer to those resources that originate from storage of energy from the sun by living organisms including plants, animals and humans. Providing that sufficient water, nutrients and sunshine are available, renewable resources can be grown in continuous cycles.

Smart products are those with in-built sensors to control the function of the product automatically or to make the user aware of the condition of the product.

Sustainable is an adjective applied to diverse subjects including populations, cities, development, businesses, communities and habitats; it means that the subject can persist a long time into the future.

Sustainable development: According to the most widely quoted definition, published in the 1987 report 'Our Common Future' by the World Commission on Environment and Development chaired by Gro Harlem Brundtland, the Norwegian Prime Minister, sustainable development is development that meets the needs of the present without compromising the ability of future generations to meet their own needs. The term contains within it two key concepts: the concept of 'needs', in particular the essential needs of the world's poor, to which overriding priority should be given; and the idea of limitations imposed by the state of technology and social organization on the environment's ability to meet present and future needs.

Sustainable product design (SPD) is a design philosophy and practice in which products contribute to social and economic well-being, have negligible impacts on the environment and can be produced from a sustainable resource base. It embodies the practice of eco-design, with due attention to environmental, ethical and social factors, but also includes economic considerations and assessments of resource availability in relation to sustainable production.

Sustainable products serve human needs without depleting natural and man-made resources, without damage to the carrying capacity of ecosystems and without restricting the options available to present and future generations.

Technosphere consists of the synthetic and composite components and materials formed by human intervention in re-ordering and combining components and materials of the biosphere, geosphere and atmosphere. True technosphere materials cannot re-enter the biosphere through the process of biodegradation alone. Synthetic polymers such as plastics are examples of such materials.

Transport energy is the energy expended to transport or distribute a product from the manufacturer to the wholesaler or retailer. Locally manufactured and locally purchased products tend to have much lower transport energies than imported products. The unit of measure is MJ per kilogram.

Use-impact products are consumer products that create (major) environmental impacts, such as cars and electrical appliances.

Volatile organic compounds (VOCs) are natural and synthetic organic chemicals that can easily move between the solid, liquid and gaseous phase.

Materials, chemicals
ABS acrylonitrile-butadiene-styrene
CFCs chlorinated fluorocarbons – compounds containing chlorine, fluorine and carbon
CO carbon monoxide
CO_2 carbon dioxide
GRP glass-reinforced plastic (polymer)
HC hydrocarbon
HCFCs hydrochlorofluorocarbons – compounds containing hydrogen, chlorine, fluorine and carbon
HDPE high-density polyethylene
HFCs hydrofluorocarbons – compounds containing hydrogen, fluorine and carbon
LDPE low-density polyethylene
NiCd nickel cadmium
NiMH nickel metal hydride
NO nitrous oxide
NO_x oxides of nitrogen
O_3 ozone
PE polyethylene (polythene)
PET polyethylene terephthalate
PP polypropylene
PS polystyrene
PU polyurethane
PVC polyvinyl chloride
VOC volatile organic compound

Miscellaneous
AC alternating current
CFL compact fluorescent lamp
DC direct current
EV electric vehicle
LED light emitting diode
PM10s particulate matter (dust, acids and other types) suspended in the air and measuring less than 0.00001 mm diameter
PRN packaging recovery note
PV photovoltaic
UV ultraviolet light

Further Reading

Books

EARLY VISIONARIES

Carson, Rachel,
Silent Spring, Hamish
Hamilton, UK, 1962

Ecologist, The (eds),
A Blueprint for Survival,
Penguin Books, UK and
Australia, 1972

Fuller, Richard Buckminster,
*Operating Manual for
Spaceship Earth,* Feffer
& Simons, London and
Amsterdam, 1969

**Meadows, Donella, Dennis
Meadows, Jørgen Randers
and William Behrens III,**
*The Limits to Growth,
A Report for the Club of
Rome's Project on
the Predicament of
Mankind,* Earth Island,
London, 1972

Meller, James (ed.),
*The Buckminster Fuller
Reader,* Jonathan Cape,
London, 1970

Packard, Vance,
The Hidden Persuaders,
Penguin Books, UK, 1957

Packard, Vance,
The Waste Makers,
Penguin Books, UK and
Australia, 1960

Papanek, Victor,
*Design for the Real World,
Human Ecology and Social
Change,* Thames &
Hudson, London, 1972

Wright, Frank Lloyd,
The Natural House,
Horizon Press, New York,
1963

ARCHITECTURE

Baggs, Sydney and Joan,
The Healthy House,
Thames & Hudson,
London, 1996

Behling, Sophia and Stefan,
*Solar Power: The Evolution
of Sustainable Architecture,*
Prestel Verlag, Munich,
2000

**Howard, Nigel and David
Sheirs,**
*The Green Guide to
Specification, An
Environmental Profiling
System for Building
Materials and Components,*
BRE Report 351, Building
Research Establishment,
UK, 1998

Jones, David Lloyd,
*Architecture and the
Environment: Bioclimatic
Building Design,* Lawrence
King Publishing, London,
1998

Slessor, Catherine,
*Eco-Tech: Sustainable
Architecture and High
Technology,* Thames
& Hudson, London, 1997

Vale, Robert and Brenda,
*Green Architecture: Design
for a Sustainable Future,*
Thames & Hudson,
London, 1991

**Wines, James and Philip
Jodidio** (ed.),
Green Architecture,
Taschen, Cologne, 2000

**Woolley, Tom, Sam
Kimmins, Paul Harrison
and Rob Harrison,**
*Green Building
Handbook, A guide to
building products and
their impact on the
environment,* E & F N
Spon, London, 1997

BUSINESS AND
SUSTAINABILITY

Allenby, B. and D. Richards,
(eds),
*The Greening of Industrial
Ecosystems,* National
Academy Press,
Washington DC, 1994

**Charter, Martin and Ursula
Tischner** (eds),
*Sustainable Solutions:
Developing Products and
Services for the Future,*
Greenleaf Publishing, UK,
2001

Datschefski, Edwin,
*Sustainable Products: The
Trillion Dollar Opportunity,*
J L Publishing, Hitchin, UK,
1999

Davis, John,
*Greening Business:
Managing for Sustainable
Development,* Blackwell,
Oxford, 1991

**Fussler, Claude with Peter
James,**
Driving Eco-innovation,
Pitman Publishing,
London, 1996

**Hawken, P., A. B. Lovins
and L. H. Lovins,**
*Natural Capitalism:
Creating the Next Industrial
Revolution,* Little & Brown,
Boston, and Earthscan,
London, 1999

**Institute of Materials and
Glasgow Caledonian
University,**
*Manufacturing and the
Environment,* Institute of
Materials, London, 1997

**Kirkwood, R. C. and A. J.
Longley,**
*Clean Technology and
the Environment,* Blackie
Academic & Professional,
London, 1995

**McDonough, William and
Michael Braungart,**
*Cradle to Cradle: Remaking
the Way We Make Things,*
North Point Press, New
York, 2002

ECO-DESIGN AND
SUSTAINABLE PRODUCT
DESIGN

Balcioglu, Tevfik (ed.),
The Role of Product Design

in Post Industrial Society,
Middle East Technical
University Faculty of
Architecture Press, Ankara,
and Kent Institute of Art &
Design, Rochester, UK,
1998

**Beukers, Adriaan and Ed van
Hinte,**
*Lightness: The Inevitable
Renaissance of Minimum
Energy Structures,* 010
Publishers, Rotterdam,
1999

**Billastos, Samir
and Nadia A Basaly,**
*Green Technology and
Design for the Environment,*
Taylor and Francis,
Washington DC, 1997

**Brezet, Han and Carolein
van Hemel,**
*Ecodesign: A Promising
Approach to Sustainable
Production and
Consumption,* United
Nations Environment
Programme, Paris, France,
1997

Burrell, P.,
*Product Development
and the Environment,*
Design Council & Gower
Publications, London, 1996

**Commission of the European
Communities,**
*Green Paper on Integrated
Product Policy,* COM,
Brussels, 2001

Datschefski, Edwin,
*The Total Beauty of
Sustainable Products,*
Rotovision, Brighton, UK,
2001

Henstock, M.,
Design for Recyclability,
Institute of Metals,
London, 1988

**Krause, F. and Helmut
Jansen** (eds),
*Life Cycle Modelling for
Innovative Products &
Processes,* Chapman & Hall,
London, 1996

Lyle, John,
*Regenerative Design for
Sustainable Design,* Wiley,
New York, 1994

MacKenzie, Dorothy,
*Green Design: Design for
the Environment,* Rizzoli,
New York, 1991

**Manzini, Ezio and Francois
Jegou,**
*Sustainable Everyday:
Scenarios of Urban Life,*
Edizione Ambiente, Milan,
Italy, 2003

Papanek, Victor,
*The Green Imperative:
Ecology & Ethics in Design
and Architecture,* Thames
& Hudson, London, 1995

**Van der Ryn, Sim & Stuart
Cowan,**
Ecological Design, Island
Press, Washington DC,
1996

**Van Hinte, Ed and Conny
Bakker,**
*Trespassers: Inspirations for
Eco-efficient Design,*
010 Publishers, Rotterdam,
1999

Whiteley, Nigel,
Design For Society,
Reaktion Books, London,
1993

**ENVIRONMENTAL ISSUES
AND DATA**

**Brundtland, Gro Harlem et al,
World Commission on
Environment and
Development,**
Our Common Future,
Oxford University Press,
UK and USA, 1987

Curran, Susan,
Environment Handbook,
The Stationery Office, UK,
1998

**Lees, Nigel and Helen
Woolston,**
*Environmental Information:
A Guide to Sources,* The

British Library, London,
1997

**McLaren, Duncan, Simon
Bullock and Nusrat
Yousuf,**
*Tomorrow's World, Britain's
Share in a Sustainable
Future,* Earthscan, London,
1998

GENERAL

**Benson, John F. and Maggie
H. Roe** (eds),
*Landscape and
Sustainability,* Spon Press,
London, 2000

Birkland, Janis,
*Design for Sustainability:
A Sourcebook of Integrated
Ecological Solutions,*
Earthscan Publications,
London, 2002

**Magazines, journals,
e-zines and
newsletters**

**BUSINESS AND
SUSTAINABILITY**

*Greener Management
International – The Journal
of Corporate Environmental
Strategy and Practice*
(Greenleaf Publishing, UK)
A quarterly journal,
which discusses the
developments around key
strategic environmental
and sustainability issues
and their effects on public-
and private-sector
organizations.
W www.greenleaf-publishing.com

Sustain
(World Business Council for
Sustainable Development,
Switzerland)
Quarterly magazine
providing examples of how
members are tackling the
issue of sustainable
development and
discussing current issues.
W www.wbcsd.ch

Sustainable Business
(USA)
A monthly online
magazine, which collates
news, features and regular
columns from the growing
arena of sustainable
business.
W www.sustainable
business.com

**ECO-DESIGN, GREEN DESIGN,
DfX, SUSTAINABLE
PRODUCT DESIGN**

The BioThinker
(Biothinking International,
UK)
A newsletter that promotes
the philosophy of cyclic,
solar and safe practices
in relation to the design
of products and services.
W www.biothinking.com

Ecocycle
An online newsletter
dedicated to product lifecycle management (LCM)
and the dissemination of
information on policy
and technical issues.
W www.ec.gc.ca/ecocycle

*International Journal of
Environmentally Conscious
Design and Manufacturing*
(ECDM Lab, University of
Windsor, Canada)
Examines the short- and
long-term effects of design
and manufacturing on
the environment and
reports recent trends,
advances and research
results.
W www.ijecdm.com

*The International Journal of
Life Cycle Assessment*
A journal devoted
entirely to lifecycle
assessment for
practitioners, product
managers and all those
interested in reducing
the ecological burdens of
products and systems.
W www.environmental-expert.
com/magazine/ecomed/lca

Journal of Industrial Ecology
(Massachusetts Institute of
Technology, USA)
A quarterly hard-copy and
online journal published
by MIT, which
encompasses material and
energy-flow studies,
lifecycle analysis, design
for the environment,
product stewardship and
much more. Although
aimed at academia, it is a
good source of technical
information, statistics and
contacts.
W www.mitpress.mit.edu/JIE

*Journal of Sustainable Product
Design*
(The Centre for Sustainable
Design, UK)
A quarterly publication
that includes contributions
from academia and
industry to encourage
business towards
sustainable practices,
products and services.
W www.cfsd.org.uk/journal

**ENVIRONMENTAL NEWS,
POLICY AND INFORMATION**

ENDS Report, The
(Environmental Data
Services Ltd, UK)
In continuous publication
since 1978, *The ENDS
Report* is a comprehensive
monthly print and web
media journal offering
news, analysis and
features on environmental
policy and business,
with a UK focus informed
by developments in
the EU.
W www.endsreport.com

ENDS Environment Daily
(Environmental Data
Services Ltd, UK)
A daily electronic news
service focusing on
environmental policy
developments in Europe.
W www.environmentdaily.com

Further Reading

GENERAL

Green Futures
(Forum for the Future, UK)
Magazine that focuses on
issues of sustainable
development illustrated by
case studies and initiatives
in business, industry and
local government.
W www.forumforthefuture.org.
ukItch (magazine)

Recycler's World
(RecycleNet Corporation,
Canada)
This is a worldwide trading
site for information about
reusable and recyclable
products, by-products and
materials.
W www.recycle.net

Warmer Bulletin
Focuses on resource
recovery and waste
management, including
reports on legislative,
technical and policy
developments around the
world.
W www.residua.com/WB.html

Sustainable Design Networks

Alternative Technology Association
W www.ata.org.au

Association for Environment Conscious Building
W www.aecb.net

Doors of Perception
W www.doorsofperception.com

Ecodesign Foundation
W www.edf.edu.au

EcoDesign Resource Society
W www.vcn.bc.ca/edrs

Eternally Yours
W www.eternally-yours.org

International Design Resource Awards
W www.designresource.org

O2 Network
W www.o2.org

Scottish Ecodesign Association
W www.seda2.org

Society for Responsible Design
W www.green.net.au/srd

Tempo
W www.tempodesign.net

Thinkcycle
W www.thinkcycle.com

Index

Illustration Credits

Illustration Credits
t–top; b–bottom; l–left;
r–right; c–centre; c1, c2, c3,
c4–column 1, column 2,
column 3, column 4

1, Keith Parry; 2 Iform; 6
(from top to bottom) Ceccotti
Collezioni, Hans van der
Mars, Dyson Appliances,
Carpet-Burns, Daimler
Chrysler, Freeplay Energy
Group, Studio Brown; 7 (from
top to bottom) Ground
Support Equipment (US) Ltd,
Mak D. Guynn, Magis SpA,
Steffan Jänicke, Bartsch
Industrial Design, Alastair
Fuad-Luke, Alastair Fuad-
Luke; 9 Wharington
International Pty Ltd; 16–17
Julienne Dolphin-Wilding;
19bl, Viaduct; 19br, Ceccotti
Collezioni; 20 tl, Fernando
and Humberto Campana;
20 br, Viaduct; 21 Fernando
and Humberto Campana;
22 tl, Studio Ilkka Suppanen;
22 bl, Robert A. Wettstein; 22
cr, IKEA; 23 tl, Marre Moerel;
23 bl, David Trubridge; 23 cr,
Gianni Antoniali, Ikon Udine;
24 David Trubridge; 25 Marcel
Loermans; 26 Michael
Gerlach; 27 tl, Magis Design;
27 r, MIO; 28 Stokke
Gruppen; 29 tl, Tsujimura
Hisanoubu Design Office;
29 b, Stokke Gruppen; 30
tr, c, Deutsche Gesellschaft
für Kunststoff-Recycling; 30 b,
Mads Flummer; 31 t, Tendo
Co. Ltd; 31 b, Marino
Ramazzotti; 32 t, Driade,
photo by Nacasa & Partners;
32 b, David Colwell Design;
33 tl, Fernando and
Humberto Campana; 33 br,
Guy Martin; 34 t, Tokujin
Yoshioka; 34 bl, br, Johan
Kalén; 35 tl, b, Stokke
Gruppen; 35 tr, Vitra; 36 t,
Iform; 36 c, b, Lloyd Loom of
Spalding; 37 t, Design
Academy Eindhoven; 37 b,
IDRA; 38 Lino Codato
Collection; 39 tl, cr, Lorenz-
Kaz; 39 bl, Lusty's Lloyd
Loom; 40, Iform; 41 t,
Maarten van Houten/Henk
Jan Kamerbeek; 41 b, Robert
A. Wettstein; 42 l, Robert A.

Wettstein; 42 tr, br, Frans
Feijn; 43 tl, Cappellini SpA;
43 tr, b, Wilde & Spieth; 44 t,
Mads Flummer; 44 bl, IDRA;
44 r, Henk Jan Kamerbeek;
45 bl, Fiam Italia SpA; 45 tr,
Jane Atfield; 46 t, Dominique
Uldry; 46 b, Lloyd Loom of
Spalding; 47 a, IKEA; 47 b,
Tonester; 48 t, Natanel
Gluska; 48 b, Iform; 49 cl, Erik
Krogh Design; 49 cr, Startup
Design; 49 bl, Droog Design;
50 tl, Isabelle Moore; 50 bl,
Serralunga; 50 cr, Wilde &
Spieth GmbH; 51 tl, tr,
Viaduct; 51 bc, Cappellini SpA;
52 Pawel Grunert; 53 tl,
Michael Marriott; 53 bl, Miro
Zagnoli; 53 tr, Julienne
Dolphin-Wilding; 54 Corné
Bastiaansen; 55 Emeco; 56 t
Pallucco Italia; 56 bl, br, Front
Corporation; 57 tl, Verne/
Bulo; 57 tr, Serralunga; 58 tl,
tr, David Hertz/Syndesis Inc.;
58 b, Design Academy
Eindhoven; 59 t, Nils Holger
Moormann GmbH; 59 b,
John Houshmand; 60 t, Beat
Karrer; 60 b, David Colwell
Design; 61 Malcom Baker;
62 t, Pawel Grunert; 62 b,
Gusto Design; 63 Shortstraw;
64 l, Kim Ahm; 64 tr,
DMD/Droog Design; 64 br,
Haberli/ Huwiler/ Marchand;
65 tr, Avant de Dormir; 65 bl,
Inga Sempé; 66 t, Magis SpA;
66 br, Atelier Alinea; 67 t,
Atelier Alinea; 67 b, Fernando
and Humberto Campana;
68 tc, tr, Iform; 68 b,
Dominique Uldry; 69 l, Susi &
Ueli Berger/Röthlisberger;
69 tc, br, Dominique Uldry;
70 tc, bl, Jörg Boner, N2;
70 br, Avad; 71 tl, Startup
Design; 71 b, Lorenz
Wiegand/POOL-products;
72 tl, tr, Lorenz-Kaz; 72 c, b,
Deutsche Gesellschaft für
Kunststoff-Recycling; 73 tl, tr,
Design Academy Eindhoven;
73 b, Design Academy
Eindhoven; 74 Pallucco Italia;
75 Interlübke; 76 ClassiCon;
77 t, Retail Place Ltd; 77 b,
Design Academy Eindhoven;
78 tc, l, Droog Design; 78r,
Domeau & Perés; 79 Chiara
Cantono; 80 tl, IDRA;

80 b, N2/sdb industries; 81
Felicerossi; 82 tl, Hülsta;
82 tr, Dominique Uldry; 83 tl,
Natural Collection; 83 cr,
Lauren Moriarty; 84 Droog
Design; 85 t, Robert A.
Wettstein; 85 bl, br, Cappellini
SpA, Italy; 86 Andrés Otero;
87 tl, Gianni Antoniali, Ikon
Udine; 87 tr, Tom Vack; 87 br,
Atelier Alinea; 88 cl, Setsu Ito;
88 br, Peter Gabriel; 89
V K & C Partnership;
90 Luceplan; 91 t, Sebastian
Bergne/Radius GmbH; 91 b,
Tiffany Tomato Designs; 92 t,
Polly Farquharson; 92 b,
Hans van der Mars; 93 t,
Deborah Thomas; 93 b, Leo
Scarff; 94 t, Stiletto DESIGN
VERTReiB; 94 b, IDRA; 94 r,
Uwe Walter; 95 t, Deutsche
Gesellschaft für Kunststoff-
Recycling; 95 b, Lauren
Moriarty; 96 tl, Sculptures-
Jeux; 96 r, Alastair Fuad-
Luke; 96 b, Sophie Krier and
Hans van der Mars; 97 bl,
Robert A. Wettstein; 97 tr,
Pallucco Italia; 97 br, Fire &
Water; 98 Gianni Antoniali,
Ikon Udine; 99 tl, br, Design
Museum, London; 100 tl,
Designed to a 't'; 100 bc, N2;
101 t, Cappellini SpA; 101 b,
Katherine Fawsett; 102 tl, bl,
Cappellini SpA; 102 br, Steven
Krause; 103 t, 2pm Limited;
103 bl, Lampholder 2000 plc;
104 t, Deutsche Gesellschaft
für Kunststoff-Recycling;
104 b, Juanico; 105 t, Aki
Kotkas; 105 b, Deutsche
Gesellschaft für Kunststoff-
Recycling; 106 Stiletto
DESIGN VERTReiB; 107 t,
Glas Platz; 107 b, Stiletto
DESIGN VERTReiB; 108 bl,
Studio Jacob de Baan and
Frank de Ruwe; 108 tr, br,
Arnold Photography; 109 t,
Viaduct; 109 b, Moonlight
Aussenleuchten; 110 tl, bl,
Mark Sutton Vane; 110 tr,
Viaduct; 111 tl, Freeplay
Energy Group; 111 br,
Luceplan; 112 cl, cr, Chiara
Cantono; 112 bl, Nisso
Engineering; 113 t, K.C.Lo;
113 bl, Dominique Uldry;
113 br, Light Corporation;
114 tl, Philips Lighting;

114 bl, SLI Lighting; 114 cr,
Lampholder 2000 plc; 115 t,
LEDtronics; 115 br, Toine van
den Nieuwendijk; 116 t,
Induced Energy; 116 bl, Aga-
Rayburn; 116 cr, Philips
Design; 117 tl, Philips Design;
117 tr, br, Solar Cookers
International; 118 Polti;
119 Dyson Appliances; 120
GDDS; 121 Staber Industries;
122 tr, c, Supercool AB; 122
br, Nicholas Albertus; 123 tl,
Supercool AB; 123 bl, Gianni
Antoniali/Gervasoni SpA;
123 br, Planet; 124 Gabriella
Dahlman; 125 tl, br, Acordis
Fibres (Holdings) Ltd; 126 t,
Ben Gold; 126 b, Acordis
Fibres (Holdings) Ltd; 127 tr,
Royal Philips Electronics;
127 bl, Fusion; 128 t, Emiliana
Design Studio; 128 b, Design
Academy Eindhoven; 129 tr,
Vikram Mitra; 129 c, IDRA;
129 bl, IDRA; 130 tc, tr,
Antonia Roth; 130 bl, Alastair
Fuad-Luke; 131 t, BREE
Collection; 131 b, Carpet-
Burns; 132 BMW (GB) Ltd;
133 Fiat Auto UK; 134
Peugeot; 135 t, Peugeot; 135 b,
General Motors/Vauxhall
Motors; 136 t, General
Motors; 136 b, Honda; 137 t,
Mercedes-Benz/Daimler
Chrysler; 137 b, Volkswagen
AG; 138 Honda; 139 t, Fiat
Auto UK; 139 b, Ford Motor
Company; 140 Toyota; 141 t,
Corbin Motors; 141 b, Daimler
Chrysler AG; 142 Segway;
143 t, NYCEwheel; 143 b,
Behind-the-Wheel Product-
Design; 144 NASA/Dryden
Flight Research Station; 145
Independence Technology;
146 t, Biomega; 146 b,
Daimler Chrysler AG; 147 t,
riese und müller GmbH;
147 b, Airnimal; 148 t, Roland
Plastics; 148 b, Advanced
Vehicle Design; 149 t, ZEM;
149 bl, br, Brompton Bicycle
Ltd; 150 tr, riese und müller
GmbH; 150 c, b, Nova Cruz
Products; 151 Powabyke Ltd;
152 tr, Design Academy
Eindhoven; 152 b, SRAM
Corporation; 153 cl, Design
Academy Eindhoven; 153 br,
Dominique Uldry;